MASS-ENERGY CONVERSION FACTORS

	SYMBOL		kg	J	eV	u	erg	cal	Btu	MWd	lb
1 kilogram	kg		×	8.988E 16	5.610E 35	6.022E 26	8.988E 23	2.148E 16	8.519E 13	1.040E 06	2.205E 00
1 joule	J		1.113E−17	×	6.242E 18	6.701E 09	1.000E 07	2.390E 01	9.478E−04	1.157E−11	2.453E−17
1 electron volt	eV		1.783E−36	1.602E−19	×	1.074E−09	1.602E−12	3.829E−20	1.518E−22	1.854E−30	3.929E−36
1 atomic mass unit	u		1.660E−27	1.492E−10	9.315E 08	×	1.492E−03	3.567E−11	1.414E−13	1.727E−21	3.660E−27
1 erg	erg	EQUALS	1.113E−24	1.000E−07	6.242E 11	6.701E 02	×	2.390E−08	9.478E−11	1.157E−18	2.453E−24
1 gram calorie	cal		4.655E−17	4.184E 00	2.612E 19	2.804E 10	4.184E 07	×	3.694E−03	4.843E−11	1.026E−16
1 British thermal unit	Btu		1.174E−14	1.055E 03	6.586E 21	7.070E 12	1.055E 10	2.522E 02	×	1.221E−08	2.588E−14
1 megawatt day	MWd		9.616E−07	8.640E 10	5.393E 29	5.790E 20	8.640E 17	2.065E 10	8.189E 07	×	2.120E−06
1 pound	lb		4.536E−01	4.077E 16	2.545E 35	2.732E 26	4.077E 23	9.743E 15	3.864E 13	4.717E 05	×

6/48

FOUNDATIONS OF NUCLEAR ENGINEERING

Thomas J. Connolly
Mechanical Engineering Department
Stanford University

John Wiley & Sons
New York, Santa Barbara, Chichester,
Brisbane, Toronto

Library of Congress Cataloging in Publication Data

Connolly, Thomas J 1923–
Foundations of nuclear engineering.
Includes indexes.
1. Nuclear engineering. I. Title.
TK9145.C59 621.48 77-26916
ISBN 0-471-16858-0

Printed in the United States of America

10 9 8 7 6 5 4 3 2 1

Preface

The national efforts in the United States and other countries to develop nuclear energy sources for electric energy generation have resulted in remarkable achievement and widespread controversy. One measure of the achievement is the forecast that by 1980 there will be 100 million kilowatts of fission electric capacity in the United States and by 1990 perhaps 200 million kilowatts; these values represent 20 percent and 30 percent, respectively, of the projected total U.S. electrical capacity for those years. Moreover, there is a widely held belief that nuclear fusion, which has proved more difficult to tame than fission, offers the best possibility for a long-term energy source. The controversy centers on the question of whether the radioactive and fissionable materials produced in fission reactors can be kept out of the environment to the extent necessary to avoid excessive harmful effects to human health over the centuries they will exist. A derivative concern is that the national and international regulations required to achieve adequate environmental control of nuclear industrial activities may lead to highly centralized or even totalitarian forms of government.

This book, however, is not directly concerned with the controversy. It is, instead, an effort to explain in quantitative terms the nuclear reactions and nuclear reactor systems that convert, or give promise of converting, nuclear energy to more useful energy forms. Given the very large investments in research and development in operating nuclear energy installations, it is desirable that a large proportion of students of engineering and the sciences have a basic quantitative understanding of the field.

This book is an introduction to the field of nuclear energy; it provides the student with the foundation for additional study without creating the necessity for "unlearning" overly simplified fundamentals. Its explanation of the principal features of nuclear energy systems also meets the needs of individuals whose primary professional interest is in other engineering fields. This book, therefore, is intended to be used by engineering or science students who may range anywhere from their sophomore year in college to graduate school and by engineers in practice who are working on nuclear energy problems but who have not had any formal introduction to their concepts. The book's presentation is designed to be both understandable and interesting to this wide audience.

The formal scientific knowledge assumed is that generally covered in an advanced high school chemistry course or a first-year college chemistry course, a first physics course covering mechanics and heat, and a course on differential and integral calculus. It is assumed, for instance, that the reader understands the Bohr model of the atom and has a concept of elements, atoms, and molecules. First-order linear differential equations are used and are developed in the chapter discussions. In terms of sophistication of analysis a course based on this text should be comparable to a first course in circuits, heat transfer, fluid mechanics, or thermodynamics.

The subject material is purposefully limited to that which can be covered in a three-semester-hour lecture course. A conscious decision was made to rely, to some extent on data sources beyond the appendix of the book. Thus, occasionally, students and instructors will find an outside reference unavailable. On the other hand, students should become acquainted with and, it is hoped, comfortable with standard references in the nuclear field.

The purpose throughout the book is to present a model of the nuclear reactions involved in nuclear energy systems that is sufficient for all design calculations. Although the model does not have the detail presented in nuclear physics courses, neither does it do any violence to basic nuclear theory. The three classes of energy-producing reactions—radioactive decay, fission, and fusion—are treated in a consistent manner. The other neutron and gamma ray reactions, so important to reactor systems, are also discussed. The very important topic of radiation effects on humans and the interrelation between nuclear industry activities and radiation in the environment is considered in some detail.

The plan of this book can be summarized briefly as follows. The first two chapters are devoted primarily to energy concepts and analysis. With this background, the subjects of radioactivity, fission, and fusion are introduced in a single chapter. Emphasis is on the energetics of these reactions and the similarities and contrasts between them. The next two chapters are devoted to the interactions of nuclear particles in matter and the effects of radiation on humans. With the analytical techniques developed in Chapter 4, each of the subjects of radioactivity, fission, and fusion is examined again in more detail in the final three chapters. The reader can proceed from Chapter 4 to the remaining chapters in any order that may be of interest. There is a minimum of interdependency between the last four chapters.

Thanks are due to Ruth Korb and Nancy Silvis for their careful typing and other preparatory work in connection with the manuscript. I am grateful to the many students who provided helpful criticisms during the preparation of the manuscript.

Thomas J. Connolly

Contents

Chapter 1

Energy: its forms and transformations

In spite of our being what we regard as a sophisticated, technology-oriented society, it is interesting that a quarter century after the first man-caused release of nuclear energy in a self-sustained reaction, the word "nuclear" appended to "energy" still has very mysterious implications for the majority of educated people. Certainly there is a good reason for much of this mystery. Most people personally deal with chemical energy every day: they light a match, start a car, turn the gas up under the coffee, and so on. This familiarity creates the notion—partly illusory—that they understand the combustion reaction. Nuclear reactions, on the other hand, are not part of the average person's conscious daily experience. There is, of course, the sun, a very fine nuclear reactor; but at 93 million miles distance, its impact on the average person's consciousness is not different from that of a combustion reactor. Here on earth there are few naturally occurring nuclear reactions. Even those that do occur do not produce visible effects to compare with chemical reactions, such as, for example, a forest fire. The discovery of nuclear reactions came, in fact, from an accidental exposure of photographic film by radiation. In the intervening three fourths of a century, artificially induced nuclear reactions have become commonplace in laboratories and nuclear power plants. It has not, however, been found practicable to develop nuclear devices that are as widely used as matches or automobiles. As a result, the average person lacks a familiarity with or a feeling of understanding of nuclear reactions and energy. One object of this work is to present a quantitative treatment of nuclear energy that includes an operational understanding of the atomic nucleus and nuclear reactions and their by-products. There are many similarities to other, more familiar energy forms. This chapter presents a review of energy principles that the nuclear form shares in common with other forms.

UNITS

A conscious decision was made to use the metric system or, more formally, the SI system of units (Système International d'Unités) throughout this book. There

Table 1.1 Some Common SI Units

QUANTITY	UNIT	SYMBOL
Length	Meter	m
Mass	Kilogram	kg
Time	Second	s
Temperature	Kelvin	°K
Energy	Joule	J
Force	Newton	N
Power	Watt	W
Pressure	Pascal	Pa

will be an occasional use of a non-SI unit in cases where it is widely used in a specialized field. The nuclear energy field, in common with other areas of technology, contains many units ad hoc or peculiar to the field. The use and significance of these will be illustrated in the examples throughout the text. In the initial examples, considerable care will be taken in manipulating and converting these units in order to demonstrate one technique. The student should also adopt some system of handling unit conversions. Table 1.1 gives some of the common SI units. More extensive tables are in Appendix A.

THERMODYNAMICS

Thermodynamics is the science of energy transformations. Thermodynamic analysis requires a set of very carefully defined terms such as energy, heat, work, power, and friction. A brief treatment of the subject is complicated by the fact that these are words in everyone's daily vocabulary, and that in this context they are not well defined but used with a variety of meanings. For a rigorous treatment, the student must devote more time to the subject of thermodynamics than we can give here. The following treatment is intended merely to summarize some information about energy that the student has learned in previous chemistry and physics courses.

Energy

Perhaps the most difficult of the terms to define is energy. We do not so much define it as build up a concept. Our use of the word in the vernacular tells us that energy is something that enables us to *do* something. Do what? Well, boil water, lift a weight, pull a plow, heat a room, operate a television set, and so on. The list could go on almost without end.

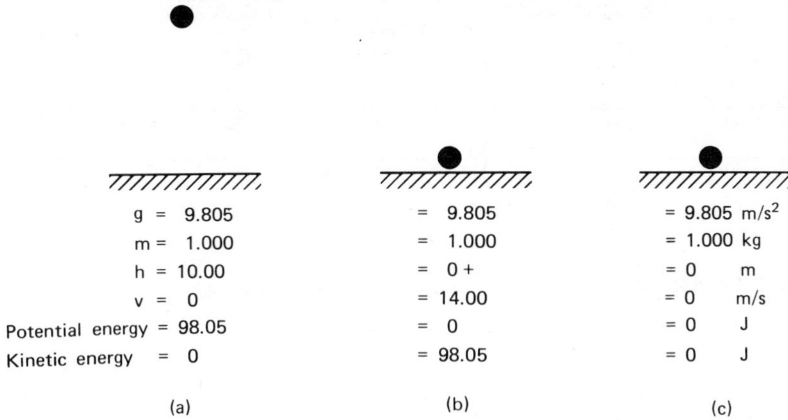

Figure 1.1 Representation of gravitational potential energy and kinetic energy.

Probably the easiest forms of energy to recognize because we constantly observe them are the kinetic energy and the gravitational potential energy of macroscopic objects. Macroscopic, as used here, means simply that we can see the objects. Figure 1.1 is a representation of these forms. In Figure 1.1*a* we have a stationary ball with a mass of one kilogram located 10 meters above ground level, or datum. We recognize that the ball possesses energy because it can do something. For example, with the aid of a simple machine, it can raise another object to its own height. If the machine is frictionless, the second object can even have the same mass (one kilogram). If the object is motionless on the ground (Figure 1.1*c*), we also recognize that in order to give it the energy it has in Figure 1.1*a*, we would have to act on it over the distance h with a force at least equal to its weight (Mg). Thermodynamically, this lifting or acting on an object is called work. We set the gravitational potential energy of Figure 1.1*a* equal to the work and claim we have added that much energy to the object.

$$E = Fh = Mgh$$

The SI unit of energy is the joule (J), which is the result of a force of one newton (N) acting through one meter (m). A newton is the force that will give a one-kilogram mass acceleration of one m/s^2. In the case of Figure 1.1*a* we can evaluate the energy

$$E = 1 \times 9.805 \times 10 = 98.05 \text{ J}$$

Kinetic energy (that is, energy possessed by virtue of motion) is depicted in Figure 1.1*b*, which shows the same object about to strike the ground after falling from the height of 10 meters. We know (from any elementary physics text) that

Figure 1.2 Representation of kinetic and potential energy exchange in a spring-mass system.

its velocity is

$$v = \sqrt{2gh}$$

and that its kinetic energy is

$$\frac{Mv^2}{2} = Mgh$$

The gravitational potential energy has been changed quantitatively to kinetic energy. We have assumed, of course, that there is no air resistance or other form of friction in the fall.

Figures 1.2 and 1.3 depict other forms of energy possessed by objects by virtue of their position in other kinds of force fields—that is, systems in which the rules relating force to position are different. In Figure 1.2 we recognize that the ball will undergo repetitive exchanges between kinetic energy and potential energy. It will also exchange energy with the spring. The case of the bar magnet system in Figure 1.3 is quite analogous to Figure 1.1 except for the force field

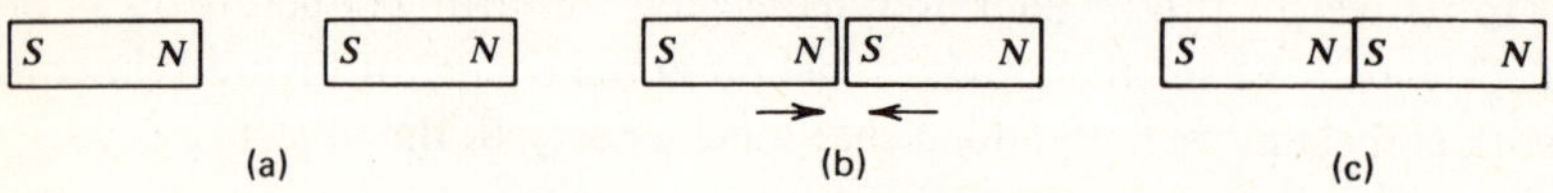

Figure 1.3 Representation of kinetic and potential energy exchange in a magnetic field.

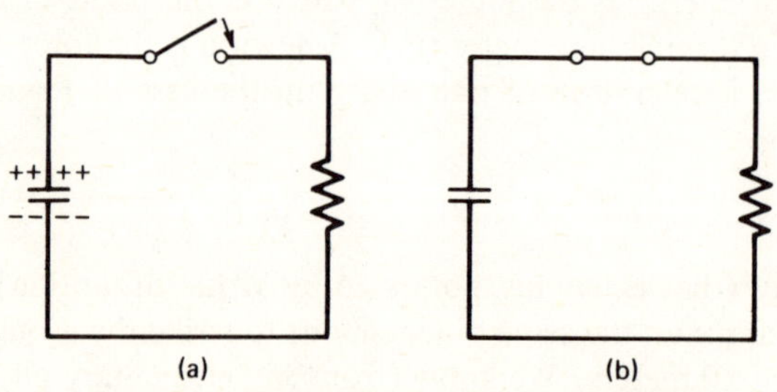

Figure 1.4 Potential energy in an electric field.

rules. In Figure 1.4 we depict a charged and discharged capacitor in order to include a case in which we would have to rely on a voltage detector instead of our eyes to be aware of the potential.

The Conservation of Energy

With reference to Figure 1.1, we know that when the ball strikes the ground it will come to rest, perhaps after bouncing a bit. When it does so, there will be neither kinetic nor potential energy. That would be the end of the story, were it not for the fact that all of our experience indicates that the energy in question is not gone; it is simply in a different form. This empirical finding was articulated as a law of nature after many painstaking investigations in the nineteenth century. It is called the First Law of Thermodynamics. It may be stated in various ways: energy can neither be created nor destroyed; energy is conserved; the energy of an isolated system is constant. There is no formal proof of this law. It simply is in agreement with a vast amount of experimental evidence; no credible evidence to the contrary has been found.

Thermal Energy

If energy is conserved, then the question remains: where is the potential energy in Figure 1.1*a* or the kinetic energy in Figure 1.1*b* to be found in Figure 1.1*c*? One might say, in the vernacular, that it went to heat up the ball. Our experience in more perceptible cases, such as a hot brake drum on a car, tends to support such a conclusion. In more careful language we say that the kinetic energy in Figure 1.1*b* has been converted to the thermal energy of the ball (or the ball and the ground, if the energy is shared between the two).

Example 1.1
Assume that the one-kilogram ball of Figure 1 has a specific heat capacity identical to water and that all of its kinetic energy in Figure 1.1*b* is converted to thermal energy. What will be its temperature increase?

Solution: The heat capacity of water is

$$C = 4187\ \text{J/kg} \cdot {}^\circ\text{K}$$

The energy of the object in Figure 1.1*a* can be computed

$$E = Mgh$$

$$E = \underbrace{1}_{\text{kg}} \times \underbrace{9.805}_{\frac{m}{s^2}} \times \underbrace{10}_{m} = \underbrace{98.05}_{\text{J}}$$

The rise in temperature is calculated from the equation

$$E = MC(T_2 - T_1)$$

$$T_2 - T_1 = \frac{E}{MC} = \frac{98.05}{1 \times 4187} = 0.0234\ ^\circ\mathrm{K}$$

We see that the temperature rise is so small that it is hardly perceptible. In such cases we rely on the validity of the First Law rather than our senses when we say that the kinetic energy is converted to thermal energy.

We define thermal energy as that form of energy which is transferred from one object to another or one region of space to another by virtue of a temperature difference. That is a macroscopic perspective. It will be helpful to an understanding of energy conversion if we also have a microscopic picture of thermal energy. The picture is that the microscopic constituents of matter—electrons, nuclei, atoms, molecules—possess energy in the same way that macroscopic objects do. Some of these ways are shown in Figure 1.5. The similarity to Figures 1.1–1.4 is obvious. As the thermal energy of a system is increased or decreased, the energies in these various modes will increase or decrease accordingly. When liquid water is heated under some constant pressure, for example, the energy will initially go into increased molecular speeds, intramolecular vibra-

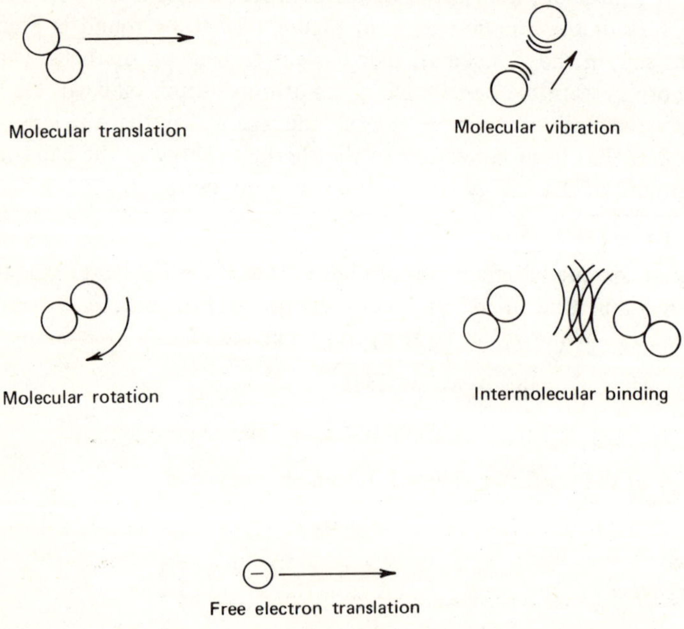

Figure 1.5 Some forms of microscopic components of thermal energy.

tion, and so on. At the boiling point, with continued heating, the water vaporizes at constant temperature. This means that intermolecular bonds or attractions are being broken; the energy is going there rather than into increased molecular speeds. The vaporization process is analogous to the magnet system of Figure 1.3, with Figure 1.3*c* representing the liquid and 1.3*a*, the vapor.

Thermal energy is unique among energy forms. An understanding of its distinction is very important. We said previously that energy enables us to do something. The ball of Figure 1.1*a* and the magnet pair of Figure 1.3*a* can exert a force through a distance by virtue of their potential energy. The energy they possess can be transferred quantitatively as work. An object that possesses the same value of thermal energy cannot transfer that energy quantitatively as work. If, in the microscopic view, thermal energy is simply the sum of mechanical forms of energy possessed by the microscopic constituents, why is it that it cannot be transferred completely as work? The answer is that the mechanical energy of the microscopic constituents is not organized. It is not uniformly directed. The atoms, for example, are moving every which way. We cannot get them organized to transfer their energy quantitatively as a force through a distance, that is, to do work. Note the contrast to the ball in Figure 1.1*b*. Every electron, every molecule in the ball has a velocity component of 14 m/s downward. That's organization. True, each constituent, due to thermal energy, has another velocity component in a random direction. The 14 m/s downward component, however, is shared by all constituents. It is this velocity only that is used in computing the macroscopic kinetic energy. It is this energy that can, with the aid of a suitable mechanism, be converted quantitatively to work. Thermal energy cannot. We can, however, get some work out of thermal energy. We will get to the question of how much shortly.

Work

For many thousands of years, human energy requirements were primarily the thermal energy needed to keep warm. This energy came from food, solar energy, and the combustion of wood. The energy transfer process was heat, which we shall discuss in the next section. A relatively small amount of energy was needed for such tasks as raising water from a well or hurling a projectile. The energy required came from people themselves or from domesticated animals. Note that, unlike a transfer of thermal energy, the energy transfer involved in a task such as raising water calls for a measurable force acting through a distance. This kind of energy transfer we call work. For better or worse, the historical development of civilization has been accompanied by humanity's increasing knowledge of how to transfer various energy forms as work. Early civilizations, for example, were able to develop mechanisms to convert the energy in wind and flowing water. The development of the steam engine, which permitted the energy in

England's coal deposits to be harnessed to do work, is regarded as the trigger to the Industrial Revolution. The subject of thermodynamics also developed, of course, and the concept of work became more sophisticated. Work is still defined as the product of a force and a displacement, the direction of these two quantities being the same. The concept of a force has been extended beyond the simple push-pull force, which we physically experience, to include electrical and magnetic field forces and so on. We shall illustrate the evaluation of work with a few examples.

Example 1.2
If the magnet pair in Figure 1.3*c* has an attractive force of one newton when the effective separation is 0.001 m, how much work will be required to separate the two magnets an indefinitely large distance?

Solution: We must know the attractive force as a condition of the separation distance s. The rule that applies is

$$F = \frac{k}{s^2}$$

We can evaluate k from the information given.

$$k = Fs^2 = 1 \times (0.001)^2 = 10^{-6}\ N \cdot m^2$$

The differential work dW for a movement ds at any value of s is

$$dW = Fds = \frac{k\,ds}{s^2}$$

In order to evaluate the work, we must integrate.

$$W = k \int_{.001}^{\infty} \frac{ds}{s^2} = \frac{k}{0.001}$$

$$= \frac{10^{-6}}{0.001} = 0.001\ \text{J}$$

This work represents the energy increase of the magnet pair in the state of Figure 1.3*a* over that of Figure 1.3*c*.

Example 1.3
If the capacitor of Figure 1.4 has a capacitance of one microfarard, how much work does it take to change it from a potential of 0 to 100 volts?

Solution: The work in this case is a product of the voltage and the increase in charge.

$$dW = edC$$

The charge is proportional to the voltage.

$$C = ke$$

$$k = 10^{-6} F$$

Integrating,

$$W = \int_0^{C_1} e\,dC = k \int_0^{100\,V} e\,de = \left.\frac{ke^2}{2}\right|_0^{100}$$

$$W = \frac{10^{-6} \times (100)^2}{2} = 0.005 \text{ J}$$

Heat

When thermal energy is transferred from one system to another, we define the transfer process as heat. There is no organized or macroscopic force involved. The potential is temperature, or the kinetic energy of the microscopic constituents, that is, the electrons, atoms, and so on. The mechanism involves collisions between these constituents (and also the emission of electromagnetic radiation). In these collisions, energy is transferred from particles of higher velocity to those of lower velocity. Therefore, thermal energy always flows in the direction of lower temperature.

We have already stated that thermal energy cannot be transferred quantitatively as work. This fact of experience is formulated as the Second Law of Thermodynamics. It can be stated in many ways, but one of the most useful expressions for our purposes is, "It is impossible to build a system in which there are no other effects than the addition of heat and the production of work."

The question of how much work one could obtain from thermal energy or a heat transfer was pondered by Carnot in the first quarter of the nineteenth century. In a remarkable intuitive leap, he deduced that the fraction of thermal energy that could be converted to work depended on the temperature at which the thermal energy existed (the source temperature) and the temperature of the region to which the thermal energy not converted must be transferred (the sink temperature). The function is

$$\frac{dW_{\text{max}}}{dQ_A} = \frac{T_A - T_R}{T_A} \tag{1.1}$$

where

dW_{max} is the maximum work obtainable from a heat transfer dQ_A
dQ_A is the heat transferred from the source
T_A is the absolute temperature of the source
T_R is the absolute temperature of the sink

The sink temperature is required because, if we cannot convert the thermal energy dQ_A entirely to work (Second Law), and at the same time the unconverted thermal energy cannot be lost (First Law), it follows that thermal energy must appear somewhere. It must be rejected to some heat sink at a temperature T_R that must be lower than T_A. Equation 1.1 is written in differential form because it frequently happens that heat transfer from a source changes its temperature. The use of this expression will be illustrated by some examples.

Example 1.4

Suppose we had liquid water at its atmospheric boiling point, 373.2°K (100°C). We increase its thermal energy by vaporizing it at constant pressure and, therefore, constant temperature. What is the maximum work obtainable from this increase in thermal energy?

Solution: We will assume that thermal energy is withdrawn by heat transfer from the vapor at 373.2°K until it is all condensed. We also assume that the environment is used as a heat sink at a constant temperature of 293°K. The maximum fraction of the heat that is obtainable as work is

$$\frac{dW_{\max}}{dQ} = \frac{T_A - T_R}{T_A}$$

Since the temperatures, T_A and T_R, are constant, this equation becomes

$$\frac{W_{\max}}{Q} = \frac{373.2 - 293}{373.2} = 0.215$$

In other words, only 21.5% of the thermal energy of vaporization can be converted to work. One can do better by increasing T_A. The reason that steam power plants operate at high pressures is to increase the vaporization temperature (T_A) and thereby increase the proportion of thermal energy converted to work.

Example 1.5

Suppose that air is heated to 2000°K from 293°K. What is the maximum amount of work obtainable from this increase in thermal energy?

Solution: We must first ask how much heat can be transferred from the air to cool it back to 293°K. We assume the heat capacity of air is a constant 1050 J/kg · °K.

$$dQ = CdT$$

$$Q = C(T_2 - T_1) = 1050(2000 - 293) = 1.792 \times 10^6 \text{ J/kg}$$

$$\frac{dW_{\max}}{dQ_A} = \frac{T_A - T_R}{T_A}$$

Since T_A varies with heat transfer from the air, we must integrate.

$$W_{max} = \int dQ_A - T_R \int \frac{dQ_A}{T_A}$$

We use a constant sink temperature, T_R, of 293°K.

$$W_{max} = Q_A - 293 \int_{293}^{2000} \frac{C dT_A}{T_A}$$

$$= 1.792 \times 10^6 - 293 \times 1050 \ln \frac{2000}{293}$$

$$= 1.20 \times 10^6 \text{ J}$$

The maximum fraction of the thermal energy obtainable as work is

$$\frac{W_{max}}{Q_A} = \frac{1.20 \times 10^6}{1.792 \times 10^6} = 0.67$$

The most important point to understand here is that thermal energy may be said to have a lower quality than other energy forms, such as those depicted in Figures 1.1-1.4. The basis of quality is the fraction that can be transferred as work. To put it another way: thermal energy cannot be converted quantitatively to other forms, but all other forms can be converted quantitatively to thermal energy. The quality of thermal energy itself varies with temperature; the higher the temperature, the higher the quality. Another conclusion that follows is that whenever we have heat transfer we have a loss in energy quality, because heat transfer occurs only in the direction of lower temperature.

Friction

Since the word "friction" is a part of everyone's vocabulary, we should discuss its role in energy transformations. From a thermodynamic standpoint, friction is any process by which a work transfer produces the same effect as if only heat transfer had occurred. Friction is present in many processes. In Example 1.1, for instance, the work that produced the ball's kinetic energy ends up as thermal energy as the ball bounces on the ground. This is a frictional process. If, in the situation of Figure 1.1, there is air resistance as the ball falls through the air, the air's thermal energy is increased and the ball's kinetic energy correspondingly reduced. This is a frictional process. Any electrical resistance heater (for example, a toaster) is a frictional device. Unfortunately, the reverse of the frictional process—that is, a heat transfer to produce the same effect as a work transfer—does not exist. Such a process would violate the Second Law.

Chemical and Nuclear Energy

Two important forms of energy not yet discussed are chemical and nuclear energy. Together with the electromagnetic radiation that the earth receives from the sun, these energy forms constitute the great bulk of the earth's energy resources. The two forms are closely related but have important differences. Together with thermal energy, chemical and nuclear energy are parts of what the thermodynamicist calls internal energy. Internal energy is dependent on "internal" properties such as temperature, pressure, and composition. Unlike thermal energy, however, chemical and nuclear energy are not transferred because of temperature differences. These energy forms are depicted in Figure 1.6. Chemical energy, for example, is shown as three atoms combining in a chemical reaction to form a molecule. Note the similarity to the magnet in Figure 1.3. In Example 1.2 we showed that the magnet pair in Figure 1.3*a* (magnets separated) is in a higher energy state than in Figure 1.3*c* (magnets together), because energy must be added to the system 1.3*c* to change it to 1.3*a*. Similarly, in Figure 1.6*a*, the energy of the three atoms is different from the energy of the three-atom molecule. It may be higher or lower. Note that the energy of the magnet systems of Figure 1.3 would be reversed if the two north poles faced each other. The point to be made is that if chemical constituents in a high energy state can be changed through a chemical reaction to a lower energy state, the energy difference must appear in another form, and we have the possibility of putting it to other uses. This is what we do when we bring coal and the oxygen in the air together and burn the coal to form carbon dioxide and water. The latter constituents are in a lower energy state; the difference, usually called the heat of combustion, is converted to the thermal energy of the carbon dioxide, water, unburned oxygen, and nitrogen (combustion gases).

There is no simple a priori way to calculate the energy change associated with a given chemical reaction as we did with the magnet pair in Example 1.3. We use data sources that list the energy of formation of various chemical compounds.[1] These values can be combined according to thermodynamic rules to give the energy change for any reaction involving those compounds. Such calculations will not be discussed here.

Nuclear energy is represented in Figure 1.6*b* in much the same way as chemical energy, except that the changes occur in the nucleus of an atom. Where chemical energy is associated with electron and interatomic binding in molecules, nuclear energy is associated with the attraction or binding between the constituents of the nucleus. Although in principle the two forms are similar, the nuclear forces are so much greater than the interatomic forces that the energy changes in nuclear reactions are on the order of one million times greater than in

[1] See, for example, *Handbook of Chemistry and Physics*, 54th edition, CRC Press, Cleveland, Ohio, 1973.

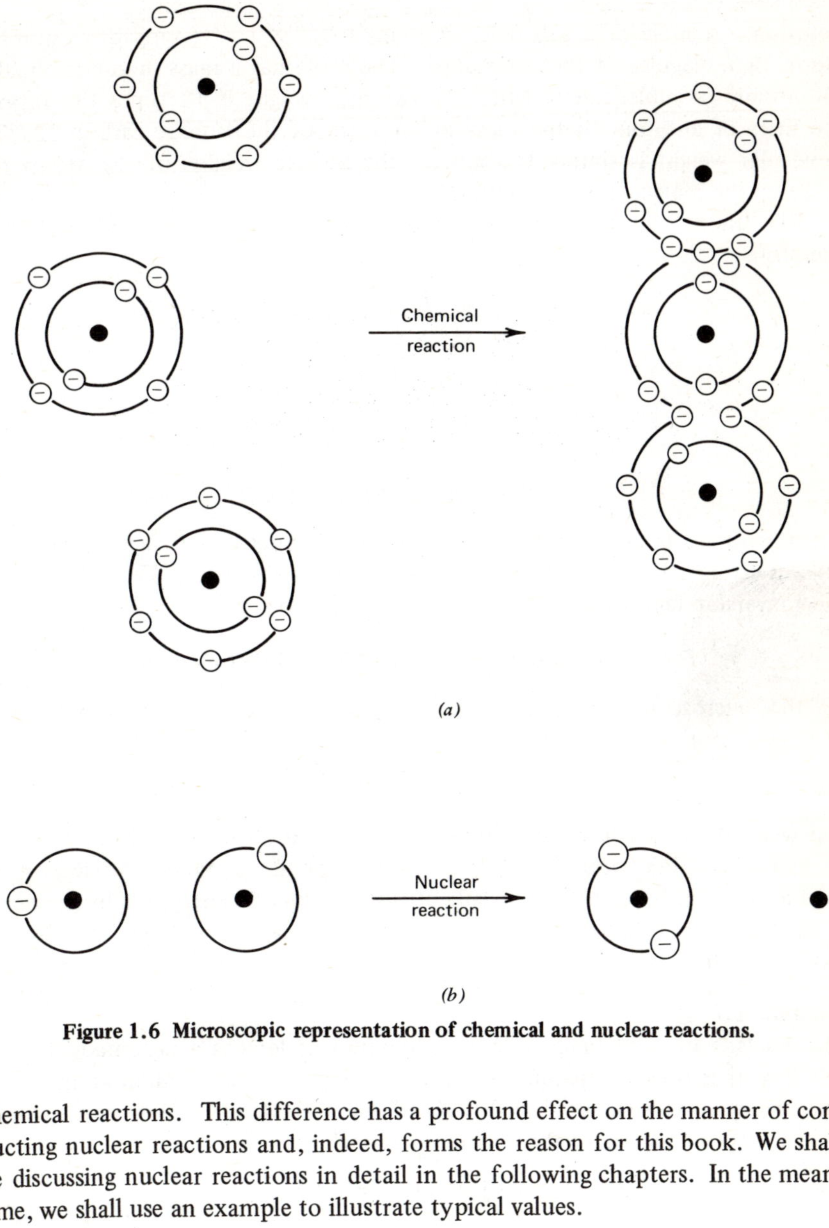

Figure 1.6 Microscopic representation of chemical and nuclear reactions.

chemical reactions. This difference has a profound effect on the manner of conducting nuclear reactions and, indeed, forms the reason for this book. We shall be discussing nuclear reactions in detail in the following chapters. In the meantime, we shall use an example to illustrate typical values.

Example 1.6
Carbon, which is the principal constituent of coal, has a reported energy of combustion of 94,050 calories per gram mole of carbon at 25°C (298°K). For a single atom of carbon, how much energy is this?

Solution: A mole of a substance contains 6.023×10^{23} (Avogadro's number) atoms or molecules of that substance. The mole has a mass in grams equal to the atomic or molecular weight. The atomic weight is 12 times the ratio of the mass of an atom to the mass of an atom of the isotope carbon-12. The molecular weight is simply the sum of the atomic weights of the atoms that make up the molecule in question.

The energy involved in the combustion of a single atom of carbon is, therefore

$$E = 94{,}050 \times \frac{1}{6.023 \times 10^{23}} = 1.562 \times 10^{-19}$$

$$\frac{\text{cal}}{\text{mole}} \qquad \frac{\text{mole}}{\text{atoms}} \qquad \frac{\text{cal}}{\text{atom}}$$

or

$$E = 1.562 \times 10^{-19} \times 4.187 = 6.538 \times 10^{-19}\ \text{J/atom}$$

When working on the level of single atoms or nuclei, it is usual practice in physics to use the electron volt (eV) as the unit of energy. From Table A.1, the conversion factor is 6.242×10^{18} eV/J. Therefore

$$E = 6.538 \times 10^{-19} \times 6.242 \times 10^{18} = 4.081\ \text{eV/atom}$$

On the macroscopic level we can say that when a mole of carbon and a mole of oxygen, both at 25°C, react,

$$C + O_2 \longrightarrow CO_2$$

and when the resulting carbon dioxide is cooled to 25°C, some 94,050 calories of energy have been transferred to the environment. An equivalent view on the atomic level is that a CO_2 molecule at rest (no kinetic energy) is in an energy state 4.081 eV below the system of one atom carbon and one molecule of oxygen, both at rest.

Example 1.7

The nucleus of a particular atom of uranium undergoes a decomposition, or splitting, that is called fission. The atomic weight of the uranium atom is 235. The energy release per atom is 200×10^6 eV or 200 MeV. How much energy is released if a kilogram of this uranium fissions?

Solution: The number of atoms in a kilogram is

$$1000 \times \frac{1}{235} \times 6.023 \times 10^{23} = 2.563 \times 10^{24}$$

$$\frac{\text{g}}{\text{kg}} \qquad \frac{\text{mole}}{\text{g}} \qquad \frac{\text{atoms}}{\text{mole}} \qquad \frac{\text{atoms}}{\text{kg}}$$

The energy release is, therefore

$$2.563 \times 10^{24} \times 200 \times 10^{6} \times 1.6021 \times 10^{-19} = 8.212 \times 10^{13}$$

$$\frac{\text{atoms}}{\text{kg}} \qquad \frac{\text{eV}}{\text{fission}} \qquad \frac{\text{J}}{\text{eV}} \qquad \frac{\text{J}}{\text{kg}}$$

A very important fact about chemical and nuclear energy is that, in theory, they are not subject to the same restriction as thermal energy—that is, not subject to the restriction of Eq. 1.1. A much higher fraction of this energy can be transferred as work. Unfortunately, as will be discussed later, we have so far been unable to take advantage of this fact. Most of our conversions of chemical and nuclear energy involve, first, conversion to thermal energy, and second, conversion to work. Once the energy is in thermal form, however, the fraction that can be converted to work is subject to the rules already discussed (Examples 1.4 and 1.5).

The Ignition Problem

Obtaining an energy source is not as simple as finding two chemical states or two nuclear states with a large energy difference and allowing the reaction to go in the direction of the lower energy state. Often it is difficult to make the reaction go. One concept of the situation is shown in Figure 1.7. The situation is similar to Figure 1.1*a*, except that the ball is held in a groove at 10 m above ground level. We cannot transform the potential energy unless we can first raise the ball 2 m to clear the lip of the groove. We must put in some energy before we can get any out. Another common example is a combustible material and air. We know that we must supply some external energy before the reaction will start. It will sustain itself if some of the energy from the reaction is available to raise some of the unreacted material to the threshold energy—hence the term "ignition problem." It is fortunate, of course, that this

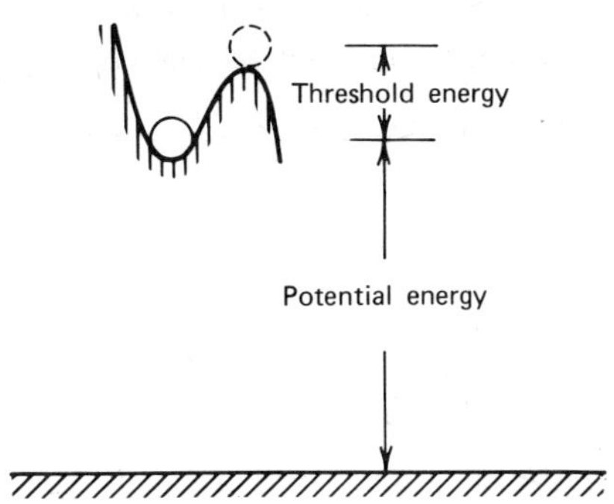

Figure 1.7 An example of threshold energy or input energy needed to start.

is the situation; otherwise, combustible material would not exist. Unfortunately, in the case of a most promising kind of nuclear reaction, fusion, the ignition problem is proving to be an extraordinarily difficult obstacle in the path to the useful application of fusion.

Power Cycles

We have not mentioned power up to this point. Power is the rate of energy conversion or energy transfer. The SI unit is the watt (W), which is one joule per second. Power cycle refers to some process by which we convert an energy form–thermal, chemical, or nuclear–to a form more useful to us, usually mechanical or electrical energy. The automobile engine is an example of mechanical energy output. The utility electric station produces electricity. The process is cyclical because it involves a working fluid such as air or water continuously passing through a repetitive series of states.

We have already mentioned that all large-scale power plants using chemical or nuclear reactants convert their energy first to thermal energy and then to mechanical or electrical energy. The cycles are sometimes rated by their "thermal efficiency," or the fraction of the thermal energy that is converted to work. The conversion of chemical or nuclear energy to thermal energy is close to 100%, so the thermal efficiency is also the efficiency based on the chemical or nuclear energy. There are many power cycles, with countless variations and refinements on each of them, so we shall not discuss them here. We shall be discussing the nuclear power cycles in a later chapter.

Example 1.8
A certain coal has a "heating value" of 12,000 Btu/lb. If the thermal efficiency of an electric generating station using this coal is 38% and the station net power is 1000 MW(e), how many tons of coal per day does the station consume? What is the "heat rate" of the plant?

Solution: Although the mixture of units given here is quite a hodge-podge, it is typical of current usage. First, express the heating value in SI units.

$$\underset{\frac{\text{Btu}}{\text{lb}}}{12{,}000} \times \underset{\frac{\text{J}}{\text{Btu}}}{1055} \times \underset{\frac{\text{lb}}{\text{kg}}}{2.205} = \underset{\frac{\text{J}}{\text{kg}}}{2.792 \times 10^7} = \underset{\frac{\text{MJ}}{\text{kg}}}{27.92}$$

We may translate "heating value" as internal energy, although there are some small theoretical distinctions. In the trade there are standard testing procedures for measuring the heating value. The interpretation of the 38% thermal efficiency is that the electrical energy leaving the station is equal to 38% of the heating value of the coal consumed. In one day the electrical energy leaving the station is

$$\underset{\frac{\text{MJ}}{\text{s}}}{1000} \times \underset{\frac{\text{s}}{\text{d}}}{24 \times 3600} = \underset{\frac{\text{MJ}}{\text{d}}}{8.64 \times 10^7}$$

The total energy or heating value of the coal consumed is, therefore

$$\frac{8.64 \times 10^7}{0.38} = 2.274 \times 10^8 \text{ MJ/d}$$

The mass of coal consumed per day is

$$\underset{\frac{\text{MJ}}{\text{d}}}{2.274 \times 10^8} \times \underset{\frac{\text{kg}}{\text{MJ}}}{\frac{1}{27.92}} = \underset{\frac{\text{kg}}{\text{d}}}{8.144 \times 10^6}$$

In tons, this amounts to

$$\underset{\frac{\text{kg}}{\text{d}}}{8.144 \times 10^6} \times \underset{\frac{\text{lb}}{\text{kg}}}{2.205} \times \underset{\frac{\text{ton}}{\text{lb}}}{\frac{1}{2000}} = \underset{\frac{\text{ton}}{\text{d}}}{8978}$$

The "heat rate" is an expression from the jargon of the electric power business. It is the number of Btu in fuel-heating value required to produce one kilowatt-hour of electricity. The value may be obtained directly from the thermal efficiency.

$$\text{Heat rate} = \underset{\frac{\text{Btu }(t)}{\text{kWh }(t)}}{3412} \times \underset{\frac{\text{kWh }(t)}{\text{kWh }(e)}}{\frac{1}{0.38}} = \underset{\frac{\text{Btu }(t)}{\text{kWh }(e)}}{8979}$$

Here is a case where it is necessary to distinguish between the different forms of energy, that is, thermal (*t*) and electrical (*e*).

Example 1.9
A nuclear electric station produces 1000 MW (*e*) with a thermal efficiency of 32%. The "heating value" of uranium[2] when fissioned is 8.21×10^7 MJ/kg. How many kilograms of uranium does the station consume per day?

[2] For the present we shall use the term uranium to represent nuclear fission fuel in general.

Solution: "Heating value" here simply means the amount of nuclear energy converted to thermal energy in the fission process. (It is not a term used in the trade.) Following Example 1.8:

$$\text{Uranium consumed} = \underset{\frac{\text{MJ}(e)}{\text{s}}}{1000} \times \underset{\frac{\text{s}}{\text{d}}}{24 \times 3600} \times \underset{\frac{\text{MJ}(t)}{\text{MJ}(e)}}{\frac{1}{0.32}} \times \underset{\frac{\text{kg}}{\text{MJ}(t)}}{\frac{1}{8.21 \times 10^7}}$$

$$= 3.29 \text{ kg/d}$$

This is a small quantity; fission reactors do not have to be fueled every day. In the case of U.S. power reactors, one year is a more typical refueling period. The quantity of uranium fissioned, however, does not give an accurate picture of the quantity needed for refueling. Unlike the case of coal, where the fuel is entirely consumed in the combustion process, only a small fraction of the uranium in a reactor is consumed.

Direct Conversion

If thermal energy is a lower-quality form of energy than other forms, why do we first transform chemical or nuclear energy to thermal energy? The principal reason is that we have not found economic ways to convert chemical or nuclear energy directly to an electrical or mechanical form on a large scale. These systems or processes are loosely classified as direct conversion. Desirable as such an achievement would be, the efforts have so far failed except in special cases. The most promising approach is the fuel cell, in which combustible materials react with oxygen to form electrical energy. Storage batteries are very useful, of course, but they do not convert a primary energy resource. While it does appear that the fuel cell can be made economic for more applications than at present, it is still too early to tell if any direct-conversion nuclear devices will ever prove economically feasible.

Another class of direct-conversion systems converts thermal energy to electrical energy without the rotating machinery of the usual power cycle. These systems, of course, are subject to Second Law restrictions on conversion of thermal energy to work. By far the most promising for large-scale applications is the so-called magnetohydrodynamics system, in which electricity is generated in a partially ionized, very high-temperature gas ($>2000°K$) by forcing the gas through a magnetic field. This offers promise of thermal efficiencies much higher than are currently obtainable in power stations.

Electromagnetic Radiation

Some of the energy of most chemical and nuclear reactions appears as electromagnetic radiation. The atomic and nuclear participants in these reactions are often in a high energy state from which they descend by emitting photons of radiation. The visible flame in combustion is one of the more obvious examples. Electromagnetic radiation is in fact given off or emitted by all objects, as a result of the thermal motion of the electrons. This latter is called thermal radiation. Although all of this radiation is part of the electromagnetic spectrum, there are widely different characteristics depending on the energy of the photons. Later in this book we shall discuss the characteristics of the electromagnetic radiation from nuclear reactions—that is, gamma radiation—in considerable detail.

In most cases the energy of electromagnetic radiation is converted to thermal energy when the radiation is absorbed in matter. There are important exceptions, however. A portion of the solar radiation is converted to chemical energy through photosynthesis. This is the process to which we owe our fossil fuels. At the present time there is a great national effort directed to the large-scale conversion of solar radiation to electrical energy through the development of solar cells. In these devices, electrons are moved to a higher potential state by interaction with individual photons. The alternative solar-electric approach is to absorb the solar radiation as thermal energy and convert it to electrical energy by means of a more or less conventional power cycle.

Mass-Energy Relationship

Due to the work of Einstein, it was recognized that mass and energy are not independent properties, even though for many practical purposes they can be treated as such. All energy has mass associated with it, and, conversely, all mass has its energy equivalence. The basic equations are

$$m = \frac{m_0}{\sqrt{1 - (v/c)^2}} \tag{1.2}$$

$$E = mc^2 \tag{1.3}$$

where

m_0 = rest mass (at $v = 0$) of an object

m = mass of same object having speed v

c = velocity of light in a vacuum (2.998×10^8 m/s)

E = relativistic total energy

Energy and mass are in the conventional SI units.

The implication of these equations is that if an object is moving it has greater mass than if it is at rest. The energy does not have to be in the kinetic form, however. If an object or system is changed to a higher or lower energy state by transfers of heat or work, it has a higher or lower mass. If an object has sufficient speed or kinetic energy to significantly increase its rest mass, then we say it is relativistic. Its momentum p follows the same formula, $p = mv$, as the one used for nonrelativistic systems, but the kinetic energy, T, takes a different form. Since kinetic energy is that possessed by an object as a result of its speed

$$T = mc^2 - m_0c^2 \tag{1.4}$$

$$T = m_0c^2\left(\frac{1}{\sqrt{1 - (v^2/c^2)}} - 1\right) \tag{1.5}$$

The term in the parentheses can be expanded to give

$$T = m_0c^2\left(1 + \frac{1}{2}\frac{v^2}{c^2} + \frac{3}{8}\frac{v^4}{c^4} + \cdots - 1\right) \tag{1.6}$$

When $v/c \ll 1$, the kinetic energy takes on its nonrelativistic form

$$T = \frac{m_0v^2}{2}$$

It is common to hear the statement that in nuclear reactions mass is converted to energy. This is misleading for two reasons. In the first place, mass and energy are equivalent. One is not converted to the other. It is not, for example, analogous to the case of Figures 1.1*a* and 1.1*b*, in which we can say potential energy is converted to kinetic energy; one is gone and the other has appeared. Energy and mass go *together*. If an object loses energy it loses an equivalent mass. In the second place, the mass-energy relationship is universal; it is not limited to nuclear reactions. All nature obeys the laws of relativity. In many cases the relativistic effect is so small as to be practically unobservable. The specific energy (energy per unit mass of reactant) of most nuclear reactions, on the other hand, is so large that relativistic effects are easily observed. Even in nuclear reactions, however, the energy change is typically equivalent only to a fraction of 1% of the mass of the material involved.

The fact that the total energy of any material at rest is equal to m_0c^2 means that we can have both energy and mass units in the same conversion table. One kilogram, for example, is equal to $(2.998 \times 10^8)^2$ or 8.988×10^{16} J. Table A.1 gives such conversion factors.

Example 1.10
In previous examples we have used values of 12,000 Btu/lb and 8.21×10^7 MJ/kg for the energy of combustion of coal and the energy of fission of uranium, respectively. In each case, calculate the loss of mass by the fuel when this energy is removed from it.

Solution: The mass equivalence of energy is

$$\frac{m}{E} = \frac{1}{c^2} = \frac{1}{(2.998 \times 10^8)^2} = 1.113 \times 10^{-17} \frac{\text{kg}}{\text{J}}$$

$$12{,}000 \frac{\text{Btu}}{\text{lb}} = 12{,}000 \times 2.205 \times 1055 = 2.792 \times 10^7$$

$$\frac{\text{Btu}}{\text{lb}} \qquad \frac{\text{lb}}{\text{kg}} \qquad \frac{\text{J}}{\text{Btu}} \qquad \frac{\text{J}}{\text{kg}}$$

The mass equivalence in the combustion of coal is, therefore

$$2.792 \times 10^7 \times 1.113 \times 10^{-17} = 3.106 \times 10^{-10} \text{ kg/kg}$$

That is, the mass equivalence of the combustion of 1 kg of coal is 3.106×10^{-10} kg. When the thermal energy is removed from the combustion gases (CO_2, H_2O), this mass goes with it. Obviously, this small a quantity would be difficult to measure.

The same quantity in the case of uranium fission is

$$8.21 \times 10^7 \times 10^6 \times 1.113 \times 10^{-17} = 9.134 \times 10^{-4} \text{ kg/kg}$$

This means that when a kilogram of uranium undergoes the fission reaction, the nuclear energy being converted to thermal energy, the mass equivalence is 9.134×10^{-4} kg, or about 1 g. If the thermal energy is transferred from the uranium, this mass is also transferred. Since it amounts to about 0.1%, it could be measured, although this quantity is usually not considered in mass balances on nuclear power plants.

Example 1.11
In a certain accelerator, 10 GeV (10×10^9 eV) electrons are produced. To put it more carefully, sufficient energy is transferred to the electron to bring its kinetic energy to 10 GeV. An electron volt, one of the few deviations from SI units, is the energy required to raise the potential of an electron by one volt. What is the mass of the electron under this condition?

Solution: We designate the kinetic energy T, and we know that

$$T = 10 \text{ GeV}$$

$$1 \text{ eV} = 1.60219 \times 10^{-19} \text{ J}$$

$$T = 10 \times 10^9 \times 1.60219 \times 10^{-19} = 1.6022 \times 10^{-9} \text{ J}$$

The kinetic energy is also the difference between the total energy and the rest energy.

$$T = mc^2 - m_0c^2$$

$$m_0 = 9.11 \times 10^{-31} \text{ kg (Table A-2)}$$

$$c = 2.998 \times 10^8 \text{ m/s}$$

$$m = \frac{T + m_0c^2}{c^2} = \frac{1.6022 \times 10^{-9} + 9.11 \times 10^{-31} \times (3.00 \times 10^8)^2}{(3.00 \times 10^8)^2}$$

$$= 1.780 \times 10^{-26} \text{ kg}$$

Note that the mass has increased over the rest mass by a factor of almost 20,000. Such an increase has a very profound effect on the trajectory of the electron and on its interaction with other particles.

We can calculate its velocity

$$1 - (v/c)^2 = (m_0/m)^2$$

$$(v/c)^2 = 1 - (m_0/m)^2$$

$$= 1 - \left(\frac{9.11 \times 10^{-31}}{1.780 \times 10^{-26}}\right)^2 = 1 - 2.6 \times 10^{-9}$$

$$v/c = 1 - 1.3 \times 10^{-9}$$

We see that the velocity has approached very closely to that of light. Note also that this approach is 99% of the way when T is only about 3 MeV.

PROBLEMS

1.1 An accelerator increases the energy of electrons uniformly to 10 GeV over a 3000 m path. That means that at 30 m, 300 m, and 3000 m, the energy is 10^8, 10^9, and 10^{10} eV, respectively. At each of these distances, compute the velocity relative to light (v/c), and the mass in atomic mass units.

1.2 In fission reactors one deals with neutrons having kinetic energies as high as 10 MeV. How much error in computing the speed is incurred by using the energy expression $m_0 v^2/2$ rather than the relativistic expression?

1.3 What speed (m/s) and kinetic energy (MeV) would a neutron have if its relativistic mass is 10% greater than its rest mass?

1.4 An electric power plant has an output of 1000 MW(*e*). The efficiency with which it converts thermal energy to electrical energy (that is, the thermal efficiency) is 32%. If the rejected thermal energy is transferred to cooling water in the power plant condensers, what will be the flow of cooling water (kg/s) if the allowed temperature rise is 10°C?

1.5 An accelerator raises the kinetic energy of protons from 0 to 0.2 MeV. If the current represented by the charge on these protons is 1 mA, what rate of energy (W) must be put into the acceleration? If these protons are absorbed on a target and their kinetic energy converted to thermal energy, what would be a reasonable rate of cooling water flow (kg/h) to keep the target from overheating?

Chapter 2

The energetics and mechanics of nuclear reactions

In this chapter we first present the procedures for evaluating the energy changes involved in any nuclear reaction. The calculations are both simple and essential to a quantitative understanding of nuclear energy. By analogy to other fields, this subject might be termed the thermodynamics of nuclear reactions. It is directly analogous to the calculation of heats of reaction in chemical reactions. The reader may or may not find such analogies useful, depending on prior experience. In any event, no prior experience is necessary.

The second subject of this chapter is the mechanics or dynamics of nuclear reactions. Although it is obvious that the energy change in a nuclear reaction is of direct relevance to the subject of nuclear energy, it is not so obvious in the case of the mechanics of the reactions. The subject is relevant, however, because we are basically concerned with the conversion of nuclear energy to thermal energy. The energy released in a nuclear reaction appears initially as the kinetic energy of particles or the energy of photons of electromagnetic radiation, or as both. In order to understand the conversion to thermal energy, it is necessary to understand the origin and life history of these particles.

A MODEL OF THE NUCLEUS

Since we wish to treat nuclear reactions or transformations, we must have a picture of what the nucleus "looks" like. Nuclear structure can be a very complex subject indeed. Fortunately for our purposes, a very simple model will suffice. We start with the conventional atom, consisting of a nucleus surrounded by a cloud of orbiting electrons. The nucleus can be viewed as a dense, tightly bound cluster of protons and neutrons. Protons have a positive electric charge equal in magnitude to that of an electron. Neutrons, as their name implies, have no charge. The number of protons in the nucleus is called the atomic number and designated Z. The number of neutrons in the nucleus is designated N. A value in much more common use is the sum, $N + Z$, which is designated A, the mass number.

The number of electrons orbiting the nucleus is equal to the number of protons in the nucleus, making the atom electrically neutral. The atomic number, Z, determines the chemical characteristics of a material; all atoms of a given Z value belong to the same chemical element.

Single protons and neutrons exist in the free state. The two are about equal in their rest mass and are much heavier than electrons (see Table A.2). The cluster of neutrons and protons that we call the nucleus is bound together very tightly by nuclear forces. These forces are so great that the nucleus should not be viewed as a basket of red and yellow ping pong balls in which each species retains all of its individual characteristics. A neutron in a nucleus, for example, does not behave as a free neutron. A free neutron has a 50% probability of changing spontaneously into a proton and an electron in about 12 minutes. Obviously, the neutrons in nuclei do not exhibit this behavior; otherwise, matter as we know it would not exist. Nevertheless, it is useful to think of the neutrons and protons in the nucleus as retaining some of their individuality. We refer to the constituent neutrons and protons in a nucleus as nucleons.

Representation of Nuclei

The manner in which one designates a given nucleus follows directly from the model we have discussed. That is, one indicates the number of neutrons and protons of which it is formed. For example, we might visualize a nucleus of four neutrons and three protons.

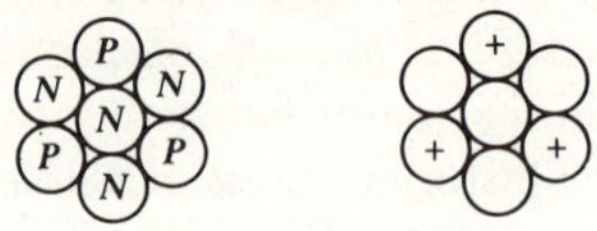

We designate it

$$^{7}_{3}\mathrm{X}$$

where the subscript is Z, the number of protons, and the superscript is the total nucleons, which we term the mass number, A. More generally, therefore, a nucleus is designated

$$^{A}_{Z}\mathrm{X}$$

Formerly, it was convention to represent a nucleus

$$_{Z}X^{A}$$

and many books still follow this convention.

The symbol X here designates the element or elemental species. Note that it conveys no new information, since the atomic number Z defines the element.

Historically, however, the chemical symbol is used. Thus

$$^{12}_{6}X = {}^{12}_{6}C = {}^{12}C = \text{C-12} = \text{carbon-12}$$

Since the atomic number and the symbol for the element convey the same information, they are redundant. Thus, one usually finds ^{12}C.

Data Sources for Nuclear Properties[1]

There are several compilations of data on the characteristics or properties of the different nuclei, but we introduce the reader principally to two: the *Chart of the Nuclides*, prepared by General Electric, and the *Table of Isotopes*,[2] by Lederer, Hollander, and Perlman, which we shall reduce to the simple designation, LHP. First, let us discuss the *Chart of the Nuclides*. Figure 2.1 contains a portion of the chart for illustrative purposes.

Referring to Figure 2.1, we see that each particular nuclear species (or nuclide) is given a box located according to its atomic number Z, and the number of its neutrons, N or $(A - Z)$. To locate carbon, for example, we scan vertically to the row for $Z = 6$ or for the symbol C, depending on what we recall about carbon. In the carbon ($Z = 6$) row, we find one square at the left extreme designated simply C and eight squares designated C-9 to C-16. The left-hand square describes the element carbon as it occurs in nature. The two properties listed are the atomic weight, 12.01115, and a neutron reaction parameter, σ_a, which will be described in a later chapter. The other eight squares describe eight isotopes of carbon that differ from each other in the number of neutrons in the nucleus. These isotopes are divided into two categories: those that are stable, and those that are unstable and are subject to spontaneous change to another nuclide. These latter are called radioisotopes; we shall be discussing them at length in a later chapter. On the *Chart of the Nuclides* the stable isotopes are designated by shading the square. A naturally occurring element is generally made up of one or more stable isotopes (there are some exceptions). In the square of a stable isotope, the number immediately below the isotope designation is the atomic percent (a/o) of that isotope in the natural element. This value is called the abundance. Obviously, there are many other isotope properties in the squares. They apply to subjects yet to be covered. For reference, however, an explanation of all properties is given in the *Chart of the Nuclides*.

[1] The tabulations of data in this book are intended to serve two distinct purposes. One purpose is to provide sufficient data for solutions to examples and exercises in this book so that immediate access to the more complete and more voluminous standard sources is not necessary. A second purpose is to familiarize the reader with standard data sources so that he will be aware of their existence and be confident that he knows how to use them. In presenting samples of data from such references, it is inevitable that certain terms and data will appear before they are discussed in the text.

[2] C. M. Lederer, J. M. Hollander, and I. Perlman, *Table of Isotopes*, Sixth Edition, Wiley, New York, 1967.

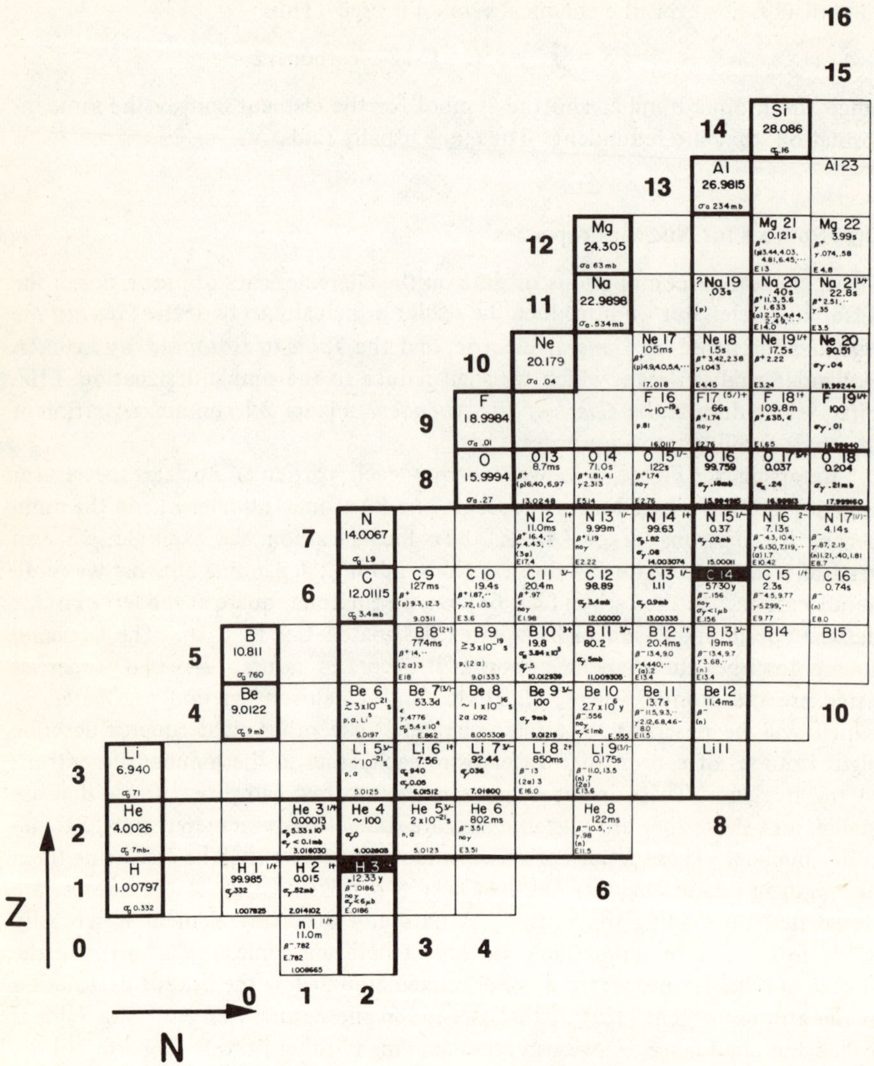

Figure 2.1 Sample of the Chart of the Nuclides.

Summary of the Nomenclature of Nuclear Structure

We have had occasion to introduce certain definitions, which we now summarize:

Isotope: A member of a set of nuclides having the same Z value but different N values

Nuclide: A specific nuclear species with a given number of neutrons (N) and protons (Z)

Nucleon: A neutron or proton that is a constituent of a nucleus

The Mass of a Nucleus

We have already noted that, although the nucleus may be regarded as a dense collection of neutrons and protons, the forces that bind these nucleons together are so strong that the nucleons do not have their free-state characteristics. Just as with the magnets (Figure 1.3), it follows that the energy content of a nucleus is not the same as the energy of an equal number of neutrons and protons. We draw an analogy between the magnet pair and the helium-4 nucleus in Figure 2.2. From our discussion of the magnet systems in Chapter 1 (Figure 1.3, Example 1.2), we conclude that the magnet system (a) of Figure 2.2 is higher in energy than (b) by the work necessary to pull the two apart. The same is true for the helium nucleus and its free nucleons. In nuclear physics this energy is called the "binding energy" of the nucleus. Also, from the discussion of mass-energy equivalence we know that system (a) must have a higher mass in both the magnet case and the nuclear case. In the nuclear case, moreover, the energy difference between states (a) and (b) (that is, the binding energy) is so great that the difference in mass of the two states is a significant fraction of the total mass and is easily measured.

We do in fact know the relative masses of practically all nuclides of interest. A device known as the mass spectrometer makes these measurements relatively straightforward. These masses are expressed in units of atomic mass units, u. By agreement, a carbon-12 atom weighs exactly 12 u. Since there are 6.023×10^{23} (Avogadro's number) atoms in 12 grams of carbon-12, it follows that

$$1\,u = \underset{\frac{\text{g-C}}{\text{mole-C}}}{12} \times \underset{\frac{\text{mole-C}}{\text{atoms-C}}}{\frac{1}{6.023 \times 10^{23}}} \times \underset{\frac{\text{atoms-C}}{u}}{\frac{1}{12}} = 1.660 \times 10^{-24}\ \text{g}$$

The masses of the free neutron and proton as well as the masses of isotopes are tabulated in several references. The *Chart of the Nuclides* (Figure 2.1) gives some mass data. The *Table of Isotopes* (LHP) has much more complete data. Tables A.2 and A.3 in this book give some mass data. There are different conventions in presenting mass data; these require some explanation. In Figure 2.1, for example, the atomic mass is given directly in atomic mass units. In LHP and in Table A.3, the data are presented as the "mass excess" or $m - A$, the mass minus the mass number of the isotope in question. A further difference is that $m - A$ is given in MeV. These differences are not fundamental. One can always

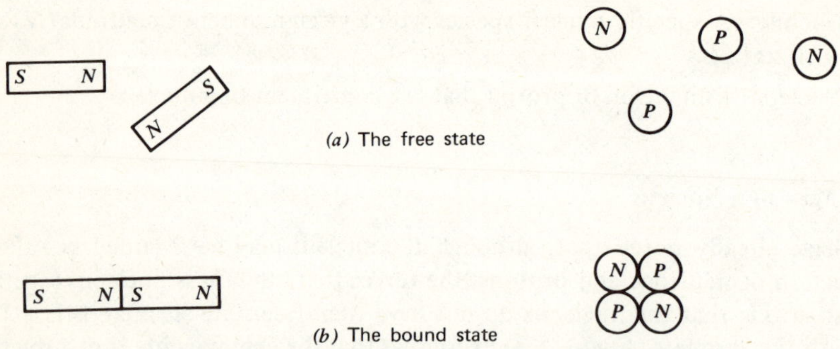

Figure 2.2 The analogy between magnets and nucleons.

convert from one to another. For example, in Table A.3, we read that the mass excess of boron-10 is 12.052 MeV. From Table A.1 we can convert this value to $12.052 \times 1.074 \times 10^{-3}$, or 0.01294 atomic mass units. The mass of boron-10 is, therefore, 10.01294 u, a value that we obtain directly from Figure 2.1.[3]

It will be noted that we are interested in *nuclear* reactions and *nuclear* masses, but are referring to *atomic* masses. The masses which we actually use are atomic masses; these are the values measured experimentally. The atomic mass is the nuclear mass plus the mass of Z electrons minus the binding energy of the orbital electrons. This latter energy is small relative to the energy of nuclear reactions, so that any error introduced by ignoring it is small. We shall also see that in the calculation of mass changes (and, therefore, energy changes) in nuclear reactions, the consistent use of atomic masses properly accounts for the electrons.

Binding Energy

We are now able to calculate the binding energy of any nucleus for which the mass is known. We have already defined binding energy as the energy difference between the nucleons in their free state and the given nucleus. The formula for binding energy of nucleus ${}^A_Z X$ follows very directly:

$$\text{Gross binding energy} = N(m - A)_n + Z(m - A)_{\text{H}} - (m - A)_X \qquad (2.1)$$

Here $(m - A)_n$ and $(m - A)_{\text{H}}$ are the mass excess of the neutron and hydrogen atom, respectively. The unit is usually the MeV. It is also somewhat more common to express binding energy on a "per nucleon" basis. This changes the

[3]It frequently happens that one does not find such perfect agreement in the data from two different references. The student must expect to encounter such differences, because the experiments on which the data are based are rarely in perfect agreement.

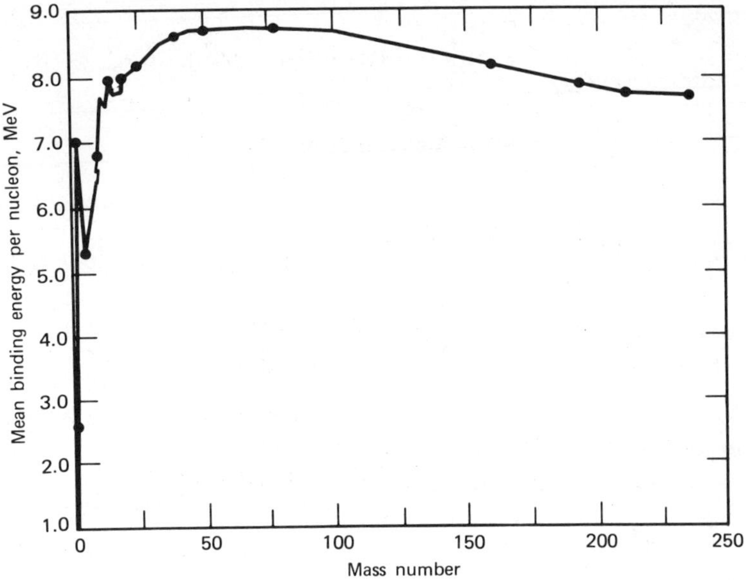

Figure 2.3 The mean binding energy per nucleon for the most stable nucleus at each mass number.

formula to

$$\text{Binding energy} = \frac{N(m - A)_n + Z(m - A)_\text{H} - (m - A)_X}{A} \tag{2.2}$$

A plot of binding energy versus mass number for all nuclei reveals some interesting systematics and tells us something about the universe. Such a curve is given in Figure 2.3. With the exception of the important zig-zags at the lower end, the binding energy increases with mass number, passes through a maximum, and then decreases. The center has the highest binding energy, that is, it is most stable. As we shall learn, reactions that tend toward the center from either end will tend to be energy-releasing reactions.

Example 2.1

Compute the binding energy of cobalt-59. What is this energy value expressed in MJ/kg-^{59}Co?

Solution: From Table A.3 we find that, for ^{59}Co

$$Z = 27$$

$$A = 59$$

$$m - A = -62.233 \text{ MeV}$$

$$(m - A)_n = 8.0714 \text{ MeV}$$

$$(m - A)_\text{H} = 7.2890 \text{ MeV}$$

From Equation 2.2

$$\text{Binding energy} = \frac{(32)(8.0714) + (27)(7.2890) - (-62.233)}{59}$$

$$= 8.768 \text{ MeV/nucleon}$$

Check this value against Figure 2.3. The energy expressed in SI units becomes

$$\underset{\frac{\text{MeV}}{\text{nucleon}}}{8.768} \times \underset{\frac{\text{nucleons}}{\text{atom-Co}}}{59} \times \underset{\frac{\text{atoms-Co}}{\text{kg-Co}}}{\frac{6.023 \times 10^{26}}{59}} \times \underset{\frac{\text{MJ}}{\text{MeV}}}{1.602 \times 10^{-19}} = \underset{\frac{\text{MJ}}{\text{kg-Co}}}{8.460 \times 10^{8}}$$

This is the minimum energy that would be required to separate a kilogram of ^{59}Co into its constituent neutrons and protons. In fact, it would be an almost impossible and, fortunately, not very useful task to achieve.

NUCLEAR REACTIONS

In most nuclear reactions we have two particles or nuclei interacting to form two different nuclei. Thus

$$\underbrace{a + b}_{\text{Reactants}} \longrightarrow \underbrace{c + d}_{\text{Products}}$$

Any reaction must meet the requirement that the sum of the atomic numbers and mass numbers of reactants and products must balance. In other words

$$Z_a + Z_b = Z_c + Z_d$$

$$A_a + A_b = A_c + A_d$$

Mass, however, *is* changed.

Although there is no theoretical limitation on what the nuclides a, b, c and d can be, as a practical matter it often happens that each side of the equation includes a very light nuclide or an electron. These are frequently termed "particles." The most common of these are given in Table 2.1. If we designate the "particle" (keep in mind that the distinction between particle and nucleus is a bit ambiguous) by the lowercase, we can write a nuclear reaction,

$$a + \text{X} \longrightarrow b + \text{Y}$$

In a common shorthand notation, one would write

$$\text{X}(a, b)\text{Y}$$

Table 2.1 Properties of Common Particles in Nuclear Reactions

NAME	FORMAL REPRESENTATION	COMMON REPRESENTATION	REST MASS (*u*)	CHARGE
Electron Beta particle	${}_{-1}^{0}e$	e^- or β^-	0.0005486	-1
Positron Positive beta	${}_{+1}^{0}e$	e^+ or β^+	0.0005486	+1
Neutron	${}_{0}^{1}n$	n	1.008665	0
Proton	—	p	1.007277	+1
Hydrogen	${}_{1}^{1}H$		1.007825[a]	+1
Deuteron	${}_{1}^{2}H$	d	2.014127[a]	+1
Triton	${}_{1}^{3}H$	t	3.016049[a]	+1
Alpha	${}_{2}^{4}He$	α	4.002603[a]	+2

[a]These are masses of the corresponding hydrogen or helium atom, although the charges are those of the nuclei.

This is all much less complicated than it looks formally. An illustration will help. A typical nuclear reaction of interest is

$$ {}_{0}^{1}n + {}_{5}^{10}B \longrightarrow {}_{2}^{4}He + {}_{3}^{7}Li $$

In shorthand,

$$ {}^{10}B(n, \alpha){}^{7}Li $$

or, in speaking, the "*n*-alpha reaction in boron-10." The reactant and product light particles are placed in parentheses and separated by a comma. It is not necessary to be explicit about the product nuclide, since it is determined by balancing charge and mass units.

Example 2.2

Some examples follow of the important classes of reactions we shall be studying. In each case complete the reaction, supplying the identity of any reactant or product not given, as well as the A and Z values of each nucleus and particle involved.

Radioactivity: (a) ${}^{3}H \longrightarrow {}_{-1}e + X$

(b) ${}^{235}U \longrightarrow {}^{4}He + X$

Fission: (c) ${}^{235}U + n \longrightarrow {}^{137}I + X + 3n$

Fusion: (d) ${}^{2}H + {}^{2}H \longrightarrow {}^{1}H + X$

Solution: One must have access to a *Chart of the Nuclides* or *Table of Isotopes* or similar data source in order to relate Z values to the element name and symbol. In each case, add up the A and Z values that are known and obtain the A and Z values of X by balancing.

(a)
$$^{3}_{1}\mathrm{H} \longrightarrow {}^{0}_{-1}e + {}^{3}_{2}\mathrm{X}$$
$$^{3}_{2}\mathrm{X} = {}^{3}\mathrm{He}\ \text{(helium-3)}$$

(b)
$$^{235}_{92}\mathrm{U} \longrightarrow {}^{4}_{2}\mathrm{He} + {}^{231}_{90}\mathrm{X}$$
$$^{231}_{90}\mathrm{X} = {}^{231}\mathrm{Th}\ \text{(thorium-231)}$$

(c)
$$^{235}_{92}\mathrm{U} + {}^{1}_{0}n \longrightarrow {}^{137}_{53}\mathrm{I} + {}^{96}_{39}\mathrm{X} + 3\ {}^{1}_{0}n$$
$$^{96}_{39}\mathrm{X} = {}^{96}\mathrm{Y}\ \text{(yttrium-96)}$$

(d)
$$^{2}_{1}\mathrm{H} + {}^{2}_{1}\mathrm{H} \longrightarrow {}^{1}_{1}\mathrm{H} + {}^{3}_{1}\mathrm{X}$$
$$^{3}_{1}\mathrm{X} = {}^{3}\mathrm{H}\ \text{(hydrogen-3, called tritium)}$$

The Energy (Q-value) of Nuclear Reactions

If we sum the rest masses of the reactants and the products of a nuclear reaction and find a discrepancy, then the law of conservation of energy and the equivalence of energy and mass forces the conclusion that some energy must have been withdrawn from (exothermic, or exoergic) or added to (endothermic, or endoergic) the reacting mixture. We might visualize it thus:

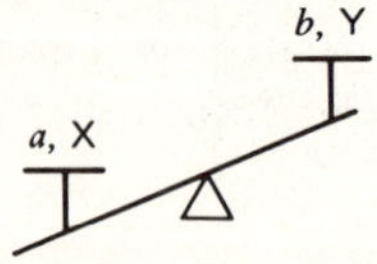

Exothermic (or exoergic)
(Positive Q)

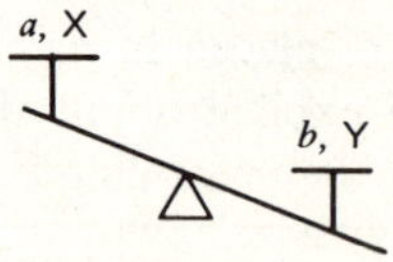

Endothermic (or endoergic)
(Negative Q)

We can measure the quantity thus:

$$Q = m_a + m_X - m_b - m_Y$$

or

$$Q = \sum_{\text{Reactants}} m - \sum_{\text{Products}} m \qquad (2.3)$$

Q (or the Q-value) of a reaction is defined so that Q is positive for exothermic reactions and negative for endothermic reactions. We can use the mass excesses (Table A.3) just as easily.

$$Q = \sum_{\text{Reactants}}(m - A) - \sum_{\text{Products}}(m - A) = \sum_{\text{Reactants}} m - \sum_{\text{Products}} m \tag{2.4}$$

since

$$\sum_{\text{Reactants}} A = \sum_{\text{Products}} A \tag{2.5}$$

Example 2.3

Compute the Q-value of the reaction $^{10}B(n, \alpha)$. What energy does this represent if all the ^{10}B in 1 kg of natural boron undergoes this reaction?

Solution: The reaction more completely represented is

$$^{10}_{5}B + ^{1}_{0}n \longrightarrow ^{4}_{2}He + ^{7}_{3}Li$$

From Table A.3 we obtain the mass excess values to substitute in Eq. 2.4.

	$m - A$ (MeV)	
NUCLIDE	REACTANTS	PRODUCTS
^{10}B	12.052	
n	8.0714	
^{4}He		2.4248
^{7}Li		14.907
	20.123	17.332

$$Q = 20.123 - 17.332 = 2.791 \text{ MeV}$$

This calculation applies to the case of a single ^{10}B nucleus reacting. From Table A.3, we find that natural boron is 19.8 atomic percent (a/o) ^{10}B. In one kilogram of boron, therefore, there is

$$1000 \times 0.198 \times \frac{1}{10.81} \times 6.023 \times 10^{23} = 1.103 \times 10^{25}$$

$$\frac{\text{atoms-}^{10}\text{B}}{\text{atom-B}} \times \frac{\text{mole-B}}{\text{g-B}} \times \frac{\text{atoms-B}}{\text{mole-B}} = \frac{\text{atoms-}^{10}\text{B}}{\text{kg-B}}$$

If all these nuclei reacted, the resulting energy would be

$$1.103 \times 10^{25} \times 2.791 = 3.079 \times 10^{25} \text{ MeV}$$

From Table A.1, this value is equivalent to

$$3.079 \times 10^{25} \times 1.602 \times 10^{-19} = 4.933 \times 10^{6} \text{ MJ}$$

Compare this value to 170 MJ for the combustion of 1 kg of hydrogen.

Example 2.4
Calculate the Q-value of the radioactive decay reaction

$$^{137}\text{Cs} \longrightarrow {}^{137}\text{Ba} + {}_{-1}\text{e}$$

Solution: From Table A.3, we obtain the mass excess values

	$m - A$ (MeV)	
^{137}Cs	-86.9	(reactant)
^{137}Ba	-88.0	(product)

$$Q = -86.9 - (-88.0) = 1.1 \text{ MeV}$$

Note that we did *not* include the mass of the electron (beta particle). The reason is that the mass values are *atomic* masses. $^{137}_{56}$Ba has one more orbital electron than $^{137}_{55}$Cs, so the "extra" electron mass is included in its atomic mass.

Example 2.5
Calculate the Q-value of the $^{59}\text{Co}(p, n)$ reaction.

Solution: From our data sources we obtain information to write the reaction in greater detail.

$$^{59}_{27}\text{Co} + {}^{1}_{1}\text{H} \longrightarrow {}^{59}_{28}\text{Ni} + {}^{1}_{0}n$$

	$m - A$ (MeV)	
NUCLIDE	REACTANTS	PRODUCTS
^{59}Co	-62.233	
^{1}H	7.289	
^{59}Ni		-61.159
n		8.071
	-54.944	-53.088

$$Q = -54.944 - (-53.088) = -1.856 \text{ MeV}$$

The fact that the Q-value is negative means that the products are not as energetically bound as the reactants. Energy must be supplied from the outside to achieve this change. This energy is most commonly obtained by accelerating the proton (hydrogen nucleus) in an electric field to give it the necessary kinetic energy.

ENERGY LEVELS OF THE NUCLEUS

A nucleus can possess energy in two ways: as kinetic energy or as internal energy. We are not concerned with the nature of the internal energy. We need only recognize its existence and that the necessary data can be obtained from a reference such as LHP. The internal energy of a nucleus is not possessed in a continuum of values. Rather, it is quantized at a number of discrete values. Similar behavior is exhibited by the orbital electrons of atoms. Each nucleus has its own particular set of energy levels. The conventional representation of these levels is shown in Figure 2.4. Three energy levels above a datum are shown for magnesium-25. These levels are usually referred to as excited levels or states. The datum is called the ground state. Following the format of LHP, the energy value of each state (in MeV) is designated by the number written on the line to the right.

A nucleus in an excited state will spontaneously descend to a lower state, usually very rapidly. These descents (the excited state is said to decay) are indicated in Figure 2.4 by the vertical arrows. The ground state is simply that at which no further tendency to decay is observed. When an excited state decays to a lower state, the conservation of energy requires that the energy difference in the two states appear somewhere. It appears as a photon of electromagnetic radiation with an energy just equal to the difference in the two states. This particular class of electromagnetic radiation is termed gamma radiation. The

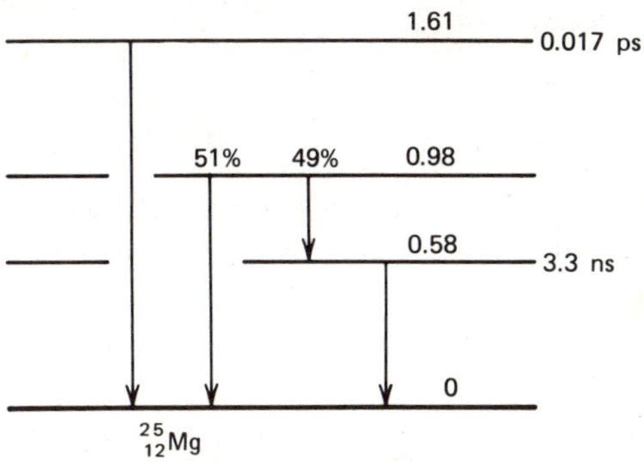

Figure 2.4 Representation of the excited levels of the magnesium-25 nucleus.

vertical arrows in Figure 2.4, therefore, represent emission of gamma photons.[4] When an excited level decays to the ground state in a single step, then the gamma energy is obvious. In Figure 2.4, for example, the decay from the 1.61-MeV level results on each occurrence in the emission of a 1.61-MeV gamma photon, as it also does in the decay from the 0.58-MeV level. In the case of the 0.98-MeV level results on each occurrence in the emission of a 1.61-MeV gamma photon. In the case of the 0.98-MeV level, however, the numbers tell us that, 51% of the time, that state decays directly to the ground state with the emission of a 0.98-MeV photon. The other 49% of the time it decays to the 0.58-MeV level with the emission of a 0.40-MeV photon, followed by the decay from the 0.58-MeV level to ground state and the emission of a 0.58-MeV gamma photon.

How do nuclei get into excited states? In the cases of interest here, it is almost always the result of a nuclear reaction. One or both of the products of a reaction are in one or more excited levels immediately following the reaction. The situation is analogous to the excited electronic levels of the products of the combustion reaction, immediately following the reaction: the decay of such excited states results in the emission of visible radiation, which we call a flame. In the case of nuclear reactions, we have previously represented them generically as

$$a + \mathrm{X} \longrightarrow b + \mathrm{Y}$$

where a and b are "particles" and X and Y are nuclei. A better understanding would be

$$a + \mathrm{X} \longrightarrow c + \mathrm{Y}^*$$

$$\mathrm{Y}^* \longrightarrow \mathrm{Y} + \gamma$$

where Y^* represents the nucleus Y in an excited state. The other product, c, could also be in an excited state, but light particles do not have excited states at energy levels of interest here.

Gamma Radiation

The term "gamma" is applied to radiation emitted by excited nuclei and not to a particular energy band. Gamma radiation generally falls in the 0.01 to 10-MeV range. In the electromagnetic spectrum it overlaps on the lower energy side with X radiation, which is electronic, not nuclear, in origin. On the higher side it overlaps with cosmic radiation.

[4]There is an important exception to the rule that a gamma photon always accompanies a level transition. There is an alternative transition process, called internal conversion, which results in electron emission instead of a photon. It need not concern us here but will be discussed in a later chapter.

As with all forms of electromagnetic radiation, the frequency and wavelength can be related to the photon energy.

$$E = h\nu = hc/\lambda$$

where ν and λ are the frequency and wavelength, respectively, and h is Planck's constant. The 1.61-MeV photon indicated in Figure 2.4, for example, has a wavelength of

$$\lambda = hc/E$$

$$\lambda = \underset{\text{J}\cdot\text{s}}{6.6256 \times 10^{-34}} \times \underset{\text{m/s}}{2.9979 \times 10^{8}} \times \underset{\text{1/MeV}}{\frac{1}{1.61}} \times \underset{\text{MeV/J}}{6.242 \times 10^{12}}$$

$$\lambda = 7.70 \times 10^{-13}\ \text{m}$$

Photons can also exhibit the behavior of particles. Although they do not have any rest mass, they exhibit a momentum equal to E/c. This property is very important in the analysis of the interaction of gamma photons with matter.

Example 2.6
Consider that a ^{25}Mg nucleus is at the 1.61-MeV excited level, as shown in Figure 2.4. Assuming it to be at rest, what kinetic energy is imparted to it by the emission of a gamma photon as it decays to the ground state?

Solution: A momentum balance requires

$$(mv)_{Mg} = E_\gamma/c$$

$$T_{Mg} = \frac{mv^2}{2} = \frac{E_\gamma^2}{2mc^2}$$

The mc^2 in the denominator is the energy equivalence of the mass of the magnesium-25 nucleus.

$$mc^2 = 25 \times 931.5 = 2.33 \times 10^4\ \text{MeV}$$

$$T_{Mg} = \frac{(1.61)^2}{2 \times 2.33 \times 10^4} = 5.56 \times 10^{-5}\ \text{MeV} = 55.6\ \text{eV}$$

This recoil energy of the magnesium nucleus does not significantly affect the 1.61 MeV of the gamma photon.

In some cases gamma photons can also induce a nuclear reaction—that is, appear on the left-hand or reactant side of the equation. Such reactions are termed photonuclear reactions, analogous to photochemical reactions induced by electromagnetic radiation in the ultraviolet or visible part of the spectrum. A well known example of a photonuclear reaction is that between a photon and deuterium.

$$^{2}_{1}\mathrm{H} + ^{0}_{0}\gamma \longrightarrow ^{1}_{1}\mathrm{H} + ^{1}_{0}n$$

The product neutrons are called photoneutrons. Since the Q-value of this reaction is −2.23 MeV, the minimum energy that the gamma photon can have to cause this reaction is 2.23 MeV.

The principal reason for concern with gamma radiation is its serious effect on biological systems and the fact that it can penetrate materials. We shall deal with both these subjects in later chapters.

Isomeric States

Usually, a nucleus exists in an excited level for a very short time, say less than one nanosecond. Its rate of decay or transition follows radioactive decay laws, which we shall discuss in Chapter 3. The decay rate is characterized by the time it takes for 50% of a given amount of a radioactive material to decay, called its half-life. Where experimental data are available, the half-life of an excited state is indicated as in Figure 2.4. For example, the first excited level of ^{25}Mg has a half-life of 3.3 nanoseconds. Occasionally, an excited state has a relatively long half-life. Such states are called isomeric, the nuclide in that state being designated an isomer of the isotope. Isomers are sometimes given the designation ^{Am}X, for example, $^{89\mathrm{m}}$Sr. In the *Chart of the Nuclides*, isomeric states are indicated by dividing the square representing the given values of A and Z. When an isomer decays with the emission of gamma radiation, the process is called isomeric transition, designated IT. The designation of an excited level as isomeric is a bit arbitrary. LHP uses the criterion that the half-life be about one second or greater.

THE MECHANICS OF NUCLEAR REACTIONS

In the design of nuclear reacting systems it is frequently useful, even necessary, to have explicit knowledge of the microscopic details of the various energy transformations that take place during and following a nuclear reaction. Immediately following a positive-Q-value reaction, the energy (the Q-value) is either in the form of excitation energy or in the form of the kinetic energy of the products. We cannot calculate how the Q-value is divided between these two forms; we must obtain that information from experimental data on each reaction. The excitation energy quickly appears as gamma radiation. The kinetic energy of the products is quickly converted to thermal energy[5] through scattering interactions with the atoms and electrons of the surrounding medium. The kinetic energy of one or a few particles is thus distributed to a large number of particles in the medium.

Nonrelativistic Mechanics of Nuclear Reactions[6]

One of the important microscopic details of nuclear reactions is the distribution of kinetic energy between the products of the reaction. As before, we represent the reaction

$$a + \mathrm{X} \longrightarrow b + \mathrm{Y}$$

where a and b are light particles and X and Y are nuclei. For the moment, we assume that Y is formed in its ground state. The simplest case is when T_a and T_X are zero. Then

$$Q = T_b + T_Y = \frac{m_b v_b^2}{2} + \frac{m_Y v_Y^2}{2} \tag{2.6}$$

Since there is no initial momentum, the conservation of momentum requires

$$m_b \mathbf{v}_b + m_Y \mathbf{v}_Y = 0 \tag{2.7}$$

It is obvious that the velocity vectors must be at 180° to each other. In terms of the magnitude of the velocities

$$m_b v_b = m_Y v_Y$$

$$T_b = \frac{Q}{1 + m_b/m_Y} \tag{2.8}$$

$$T_Y = \frac{Q}{1 + m_Y/m_b}$$

$$\frac{T_b}{T_Y} = \frac{m_Y}{m_b} \tag{2.9}$$

The kinetic energy is divided inversely as the masses of the two particles.

[5]In theory and, to a very limited extent, in practice, it is possible to convert the kinetic energy of charged particles to electrical energy by allowing them to give up their kinetic energy working against an electric field and depositing their charge at a high potential. This would be analogous to a baseball thrown upward in a gravitational field and caught near the peak of its trajectory. Its potential energy is then available to do work and can be used *if* a suitable mechanism is provided. Unfortunately, the complexity of such "direct conversion" devices for nuclear reactions, as well as certain inherent difficulties, does not offer much hope that they will produce energy on a large scale in the near future.

[6]The treatment in this section follows very closely that of Walter E. Meyerhof, *Elements of Nuclear Physics*, McGraw-Hill, New York, 1967, Chapter 5.

Example 2.7
In the reaction, $^3H(d, n)^4He$, compute the kinetic energy of the product neutron and helium nucleus, assuming the reactants have no initial energy.

Solution: In our reaction representation the correspondence is

$$\underset{a}{{}^2_1H} + \underset{X}{{}^3_1H} \longrightarrow \underset{b}{{}^1_0n} + \underset{Y}{{}^4_2He}$$

This reaction illustrates the arbitrariness of the distinction between particle and nucleus. Fortunately, consistency is all that is required. We first obtain the Q-value. From the $(m - A)$ values of Table A.3

$$Q = 13.1359 + 14.9500 - (8.0714 + 2.4248) = 17.59 \text{ MeV}$$

All of this energy is in the form of kinetic energy of the particles. From Eq. 2.8, the kinetic energies of the products are

$$\text{neutron:}\quad T_b = \frac{17.59}{1 + 1/4} = 14.07 \text{ MeV}$$

$$\text{helium:}\quad T_Y = \frac{17.59}{1 + 4} = 3.52 \text{ MeV}$$

This particular reaction, often called the DT reaction, is a leading candidate in the effort to obtain energy by nuclear fusion. The fact that 80% of the energy is in the form of kinetic energy of the neutron would have an important effect on the design of a DT reactor.

In a more complex (and more usual) case, the light reactant particle has significant kinetic energy, T_a. Then energy conservation requires

$$Q = T_b + T_Y - T_a \tag{2.10}$$

To represent conservation of linear momentum, we must have a diagram:

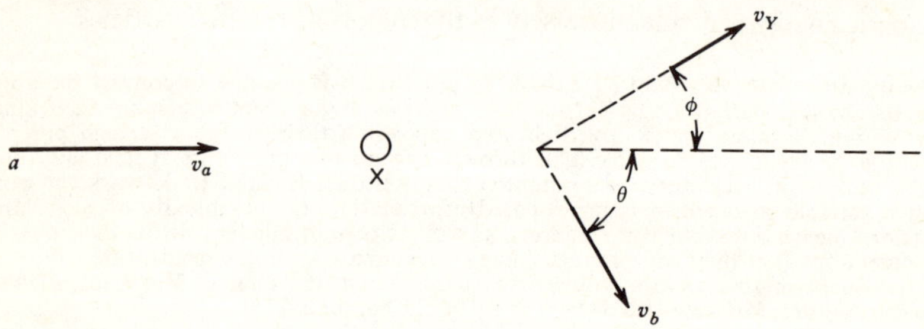

Conservation of momentum requires

$$m_a v_a = m_Y v_Y \cos\phi + m_b v_b \cos\theta$$

$$0 = m_Y v_Y \sin\phi - m_b v_b \sin\theta$$

The simultaneous solution of the energy equation and momentum equations gives

$$Q = T_b(1 + m_b/m_Y) - T_a(1 - m_a/m_Y) - \frac{2}{m_Y}(m_a T_a m_b T_b)^{1/2} \cos\theta \qquad (2.11)$$

We do not know a priori, in any given reaction, what angle θ the particle b will take. One can derive a statistical distribution of direction. Such detail, however, would serve little purpose here.

If it happens that the product Y is in an excited state, Y^*, then Y^* must be treated as having mass equal to the ground state plus the mass equivalent to the energy of the excited state above the ground state. The energy balance then becomes

$$Q = T_b + T_{Y^*} - T_a + E_\gamma \qquad (2.12)$$

As an earlier example, we computed a Q-value of 2.791 MeV for the reaction

$$^{10}\mathrm{B} + n \longrightarrow {}^{4}\mathrm{He} + {}^{7}\mathrm{Li}$$

assuming the ^{7}Li is in the ground state. That is indeed the way the reaction goes sometimes, but it also can lead to an excited state of ^{7}Li, which is 0.4774 MeV above the ground state. In that case we have $T_b + T_{Y^*} = 2.791 - 0.477 = 2.314$ MeV. The decay of the excited state to the gound state with the emission of a gamma photon is a separate process, which we discussed in a previous section. By convention, Q is always calculated on the basis of ground states.

Example 2.8

Consider that a ^{10}B(n, α) reaction is caused by a neutron with 1 MeV energy. Further, the product alpha particle has a path at 90° to that of the incoming neutron. What kinetic energy does the alpha particle possess? At what angle to the neutron's path and to the alpha particle's path is the ^{7}Li nucleus traveling? What speed does it have?

Solution: From the text

$$n + {}^{10}\mathrm{B} \longrightarrow {}^{4}\mathrm{He} + {}^{7}\mathrm{Li}; \quad Q = 2.791 \text{ MeV}$$

$$a + \mathrm{X} \longrightarrow b + \mathrm{Y}$$

First, we designate the value assignments for substitution in Equation 2.11.

$$T_a = 1 \text{ MeV}$$

$$m_b/m_Y = 4/7 = 0.571$$

$$m_a/m_Y = 1/7 = 0.1428$$

$$\cos\theta = 0$$

Solving Eq. 2.11 for T_b

$$T_b = \frac{2.791 + (1 - 0.1428)}{1.571} = 2.322 \text{ MeV}$$

We can obtain ϕ, the angle the ^{7}Li makes with the neutron's path, from the conservation of momentum

$$m_Y v_Y \sin\phi = m_b v_b \sin\theta = m_b v_b$$

$$\sin\phi = \frac{m_b v_b}{m_Y v_Y}$$

or

$$\sin\phi = \left(\frac{T_b}{T_Y} \times \frac{m_b}{m_Y}\right)^{1/2}$$

$$T_Y = Q + T_a - T_b = 2.791 + 1 - 2.322 = 1.469 \text{ MeV}$$

$$\sin\phi = 0.950$$

$$\phi = 71.8^\circ$$

Momentum considerations also require that the original neutron path and the ^{4}He and ^{7}Li paths all lie in the same plane, so the angle between the products must be the sum of θ and ϕ, or 161.8°. The speed of the ^{7}Li nucleus is calculated from its kinetic energy.

$$T_Y = \frac{m_Y v_Y^2}{2} = 1.469 \text{ MeV} = 2.353 \times 10^{-13} \text{ J}$$

The kinetic energy is such a small fraction of the energy equivalence of the rest mass of ^{7}Li that we can treat the problem nonrelativistically.

$$m_Y = 7.016 \text{ u} = 1.165 \times 10^{-26} \text{ kg}$$

$$v = \left(\frac{2 \times 2.353 \times 10^{-13}}{1.165 \times 10^{-26}}\right)^{1/2} = 6.356 \times 10^6 \text{ m/s}$$

Relativistic Mechanics

The reader may observe that the mechanics used in the above derivations looks remarkably like Newtonian mechanics of high school and freshman physics courses. Indeed, it is the same. In the analysis of nuclear reactions, however, relativistic mechanics can be required, so it is important that the reader understand when ordinary mechanics gives a satisfactory approximation.

From Chapter 1, the relativistic total energy is given

$$E = mc^2 = m_0c^2 + T \tag{2.13}$$

where m_0 is the rest mass and T is the kinetic energy. When $T/m_0c^2 \lesssim 0.01$, there is little error introduced by making the assumption of classical mechanics that $T = m_0v^2/2$. Many of the nuclear reactions with which we will be concerned meet this condition. Consider, for example, a neutron with a kinetic energy of 10 MeV. The neutron has a mass of about 1 u or $m_0c^2 = 932$ MeV (see Table A.1). Therefore $T/m_0c^2 \cong 0.01$, and $T = m_0v^2/2$ is a good approximation. Likewise the momentum, p, can be assumed equal to m_0v.

In dealing with electrons, however, such an assumption is not so often justified. The value of m_0c^2 for an electron is about 0.5 MeV. Electrons emitted from a nucleus typically have kinetic energies of 1 MeV. Here $T/m_0c^2 > 1$. Consequently, one must use the more rigorous relationships

$$T = (m - m_0)c^2 \tag{2.14}$$

$$m = \frac{m_0}{\sqrt{1 - v^2/c^2}} \tag{2.15}$$

$$p = mv \tag{2.16}$$

Another instance in which relativistic mechanics must be employed is in the interaction of gamma photons with particles. Example 2.6 illustrated this point.

Example 2.9

A free neutron is itself radioactive and spontaneously "disintegrates" into a proton and an electron.

$${}^1_0n \longrightarrow {}^1_1\text{H} + {}^{\ 0}_{-1}e$$

Make the simplifying assumption that the neutron has zero speed. Calculate the speed of the proton and electron immediately following the disintegration.

Solution: First we must obtain the Q-value, which will represent the sum of the kinetic energy of the two product particles. Taking the mass excess data from Table A.3

$$Q = 8.0714 - 7.2890 = 0.7824 \text{ MeV}$$

Designating the proton and electron by sub-p and sub-e, respectively

$$T_p + T_e = 0.7824$$

$$m_p v_p = m_e v_e$$

$$T_p = \frac{m_p v_p^2}{2}$$

$$T_e = (m_e - m_{0,e})c^2$$

Even though we will treat the electron relativistically and Eq. 2.9 is not quantitatively applicable, we can assume that, as Eq. 2.9 indicates, the preponderance of the energy is with the electron. As a first-iteration guess, assume

$$T_e = 0.7824 \text{ MeV}$$

$$T_e = \frac{m_{0,e}c^2}{\sqrt{1 - v_e^2/c^2}} - m_{0,e}c^2 = 0.7824$$

$$\frac{0.5110}{\sqrt{1 - v_e^2/c^2}} = 1.2934$$

$$\sqrt{1 - v_e^2/c^2} = 0.395$$

$$v_e/c = 0.9186$$

$$v_e = 0.9186 \times 2.998 \times 10^8 = 2.754 \times 10^8 \text{ m/s}$$

$$m_p v_p = \frac{m_{0,e} v_e}{\sqrt{1 - v_e^2/c^2}} = \frac{5.486 \times 10^{-4} \times 2.754 \times 10^8}{0.395} = 3.825 \times 10^5 \text{ u} \cdot \text{m/s}$$

$$v_p = \frac{3.82 \times 10^5}{1.00728} = 3.797 \times 10^5 \text{ m/s}$$

These are first-iteration answers. We can now adjust T_e.

$$T_p = \frac{m_p v_p^2}{2} = \frac{1.00728 \times 1.66 \times 10^{-27} \times (3.797 \times 10^5)^2 \times 6.242 \times 10^{12}}{2}$$

$$\text{u} \quad \times \quad \frac{\text{kg}}{\text{u}} \quad \times \quad \frac{\text{m}^2}{\text{s}^2} \quad \times \quad \frac{\text{MeV}}{\text{J}}$$

$$= 7.53 \times 10^{-4} \text{ MeV}$$

T_e should be adjusted from 0.7824 MeV to 0.7816 MeV. The effect on the values of v_e and v_p would be negligible.

PROBLEMS

2.1 Fluorine-19 bombarded by 4-MeV protons undergoes a (p, α) reaction. What speed would each of the product nuclei have immediately after the reaction, assuming all the energy goes into kinetic energy?

2.2 It is proposed to use a fission reactor as a rocket propulsion device by fissioning only at a back surface (as if that could be done). One half the fission fragments from ^{235}U leave this surface, while the other half are embedded in the rocket. Assume 167 MeV/fission goes into the kinetic energy of a fission fragment pair. What is the thrust of this rocket in newtons per megawatt reactor power?

2.3 For the fusion reaction $^{3}H(d, n)^{4}He$, compute the velocity of each of the two product nuclides.

2.4 Consider that a cesium-137 nucleus at rest decays with the emission of a β^- particle having 0.52 MeV energy (kinetic). Compute the energy and the velocity of the daughter nucleus immediately following this reaction. Treat the beta particle relativistically.

2.5 Complete the following representations of nuclear reactions by adding the nuclide formed in the reaction:

$$^{238}U(d, 3n) \qquad ^{204}Hg(d, p)$$
$$^{232}Th(\alpha, 4n) \qquad ^{20}Ne(d, \alpha)$$

2.6 It is postulated that one of the nuclear reactions occurring in certain stars is the nitrogen-15 (p, α) reaction. Calculate the energy available from this reaction in Btu per pound of nitrogen-15 consumed.

2.7 An excited state of a boron-11 nucleus decays by the reaction

$$^{11}B \longrightarrow {}^{4}He + {}^{7}Li$$

The ^{4}He has a kinetic energy of 3 MeV. What is the kinetic energy of the ^{7}Li?

2.8 For a hypothetical isotope, $^{200}_{80}X$, give the atomic and mass numbers of the product nucleus if it undergoes the following reactions:

β^- decay
β^+ decay
α decay
(n, α)
(n, γ)
(n, p)

2.9 What is the relative value of the momentum of a particle of rest mass m_0 and kinetic energy T and a photon of $E = h\nu$?

2.10 A neutron of negligible energy reacts with a nucleus of boron-10 in a (n, α) reaction. What is the kinetic energy of each resulting particle?

2.11 Immediately following the radioactive decay of a ^{14}C nucleus, what are the speeds of the beta electron and the ^{14}N nucleus?

2.12 Following the $^{1}H(n, \gamma)$ reaction, what is the kinetic energy of the product deuterium nucleus if the proton has a kinetic energy of 1 eV?

2.13 A 1-MeV neutron reacts with a stationary ^{10}B nucleus in the $^{10}B(n, \alpha)$ reaction. The alpha particle leaves the reaction site at 90° (in the lab system) to the line of direction of the incoming neutron. Calculate the direction of the other particle.

2.14 Calculate the binding energy per nucleon of ^{11}B.

2.15 A neutron with an energy of 10 MeV decays into a proton and an electron. What range of energies will the electrons and protons have?

2.16 A 1-MeV proton strikes a ^{7}Li nucleus, and the (p, α) reaction occurs. What will be the range of kinetic energy of the product particles?

2.17 It is proposed that in a laser fusion process it might be possible to cause two lithium-6 nuclei to react to form 3 alpha particles. How many MJ per kg of *natural* lithium would be obtained if one could react all the lithium-6 that way?

2.18 The $^{7}Li(p, n)$ reaction can be used for the production of monoenergetic neutrons. What is the maximum energy (MeV) neutron that can be produced with a 3-MeV proton accelerator? What is the speed (m/s) of those neutrons? If 3-MeV protons are used, what is the angle with respect to the proton beam axis at which neutrons of 1 MeV are emitted?

Chapter 3

Energy-producing nuclear reactions

Fossil fuels provided the United States with about 95% of the energy it consumed in the early 1970s. A single class of energy-producing chemical reactions, namely combustion, forms the basis of this energy production.[1] Hydrogen and carbon in the variety of chemical compounds found in coal, oil, natural gas, and wood combine with oxygen of the air to form water and carbon dioxide. There are, of course, many other exothermic chemical reactions, but for one reason or another they cannot serve as the basis of large-scale energy production.

Among the large number of exothermic (positive Q-value) nuclear reactions, there are three classes that deserve consideration. These are: radioactivity, the spontaneous change of certain nuclides into a different isotope or isomer; fission, the neutron-induced splitting of certain heavy isotopes; and fusion, the reaction of two light nuclei or particles to form nuclides of higher total binding energy. These reactions and their associated reacting systems are the principal subject of this book. It would appear logical, therefore, to treat each of them in turn. For a variety of reasons, however, it appeared more useful to introduce all three classes of reactions in this chapter, reserving a more detailed treatment of each for later. The most compelling reason for this approach is that the reactions and reacting systems are so interrelated. The only sizable source of radioactive materials, for example, is the fission reactor; all fission fuels and some fusion fuels are themselves radioactive; fission is one mode of radioactive decay. It is not possible, therefore, to have a good understanding of any one of these classes of reaction without knowing something about the others.

Our primary interest is in reactions that can form the basis of a large-scale energy resource. In order to determine the suitability of any given reaction, we need some requirements that a large-scale energy source must meet. We shall use the following:

1. The energy change between reactant and product must be negative and of

[1] We know from the First Law of Thermodynamics that energy can neither be produced nor destroyed. It is hoped that the reader is sufficiently grounded in this physical fact not to be misled by the use of a term, such as "production," which is in such common use in the vernacular.

substantial magnitude. In other words, the Q-value should be large and positive.

2. A satisfactory mechanism or system for conducting the reaction and for converting the nuclear energy released to suitable forms (such as thermal or electrical energy) must exist or be capable of development. To conduct the reaction it is desirable to be able to initiate it, sustain it at a controlled rate, and terminate it when desired. The system for conducting the reaction is most often called a reactor.
3. The reactant materials must be available in suitable quantity and at a cost competitive with alternative energy resources.
4. The environmental impacts of acquiring the natural materials and processing them, of conducting the reaction, and of disposing of the by-products can be made acceptably low.

In the remainder of this chapter the three classes of nuclear reactions will be described and discussed in the light of these criteria.

RADIOACTIVITY

In Chapter 2 we discussed a simplified model of a nucleus as a cluster of neutrons and protons held together by forces strong enough to make these nucleons lose some of the characteristics which they possess as free particles. The world as we know it is largely made up of atoms whose nuclei are stable. They do not show any tendency to change. There are a limited number—about 270—of these stable nuclei. The locus of these nuclei on a plot of N versus Z (the number of neutrons and protons, respectively) is shown by the black squares in Figure 3.1. Figure 3.1 is simply a chart of the nuclides, reduced in size to show trends. It should be noted that for light elements, stability is achieved with approximately $N = Z$ (for example, ${}^{12}_{6}C$, ${}^{16}_{8}O$). As the mass increases, the locus of stable nuclei rises above a 45° line in Figure 3.1, indicating that stability requires that $N > Z$. The heaviest stable nucleus is ${}^{209}_{83}Bi$.

Another class of nuclide or isotope, called radionuclides or radioisotopes, does tend to change. Perhaps a fair concept is of a cluster of nucleons that possesses a certain amount of binding energy. Under most conditions of distribution of this energy the cluster or nucleus hangs together and, to an outside observer, would appear stable. At some time, however, an energy distribution occurs that creates a local weakness, and a spontaneous change occurs. The resulting nucleus is called the daughter, the original being called the parent. The daughter always has a lower ground-state energy (that is, it has a higher binding energy) than the parent. The daughter may be stable or radioactive. If the daughter is radioactive, the series of decays that results is called a radioactive decay chain. The location of radionuclides on an N versus Z plot is shown in

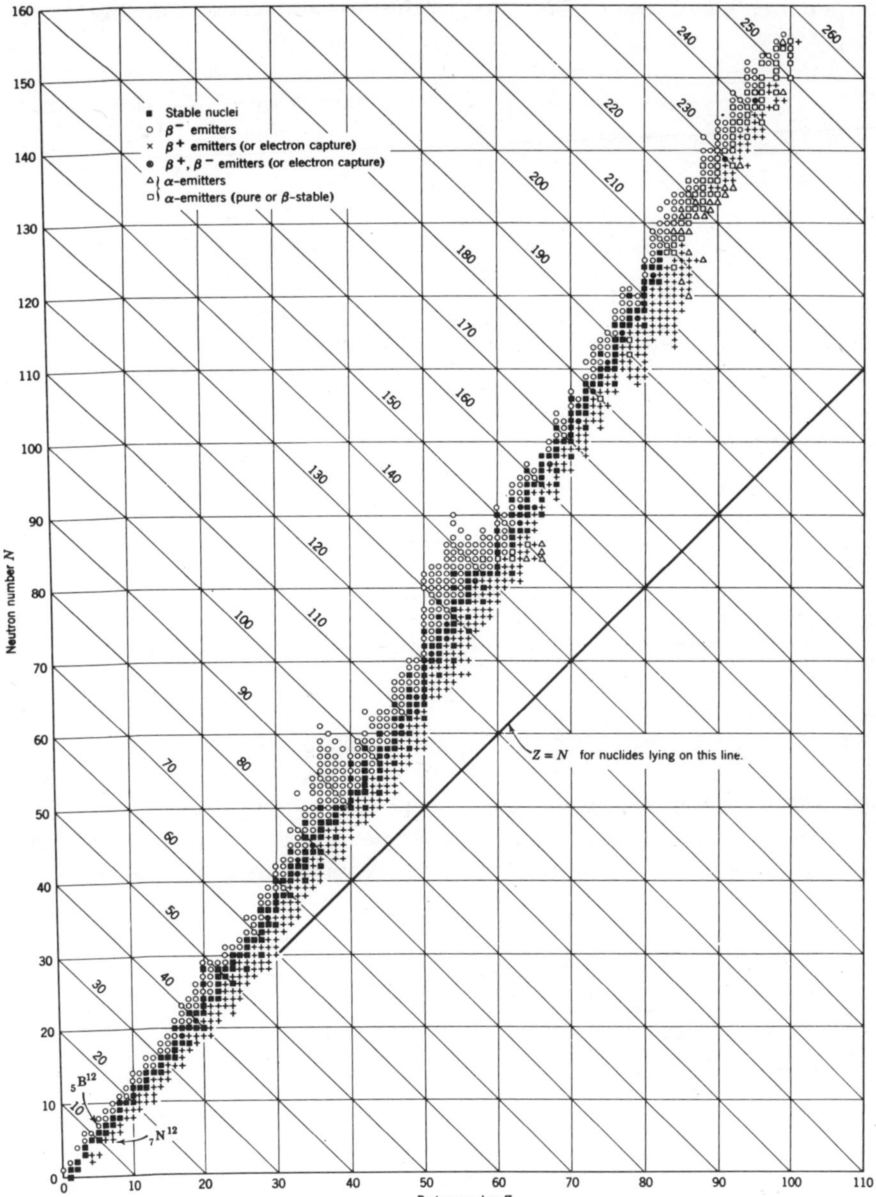

Figure 3.1 A chart of the nuclides, showing stable and radioactive nuclides. From S. E. Liverhant, *Elementary Introduction to Nuclear Reactor Physics*, Wiley, New York, 1960.

Figure 3.1. These radionuclides are identified by the mode of their radioactive decay.

Modes of Radioactive Decay

The spontaneous change of a radionuclide is called decay or disintegration. The latter term is a bit misleading, implying as it does a more drastic change than usually occurs. There are perhaps ten or so ways in which various radioisotopes decay. Some isotopes, in fact, exhibit different modes of decay. We must learn from laboratory experiments the percent of such nuclei that decays by each mode. Our immediate concern is limited to two modes, alpha emission and negative beta emission, because these are the only two that we can produce in sufficient quantities to consider as energy sources. The alpha decay reaction can be written

$$ {}^{A}_{Z}X \text{ (parent)} \longrightarrow {}^{A-4}_{Z-2}X^* + {}^{4}_{2}\text{He (alpha, } \alpha) $$

$$ {}^{A-4}_{Z-2}X^* \longrightarrow {}^{A-4}_{Z-2}X \text{ (daughter)} + \gamma \text{ photon(s)} $$

Similarly, the negative beta decay process can be written

$$ {}^{A}_{Z}X \text{ (parent)} \longrightarrow {}_{Z+1}^{\;\;A}X^* + {}_{-1}^{\;\;0}e \text{ (beta, } \beta^-) + \bar{\nu} \text{ (antineutrino)} $$

$$ {}_{Z+1}^{\;\;A}X^* \longrightarrow {}_{Z+1}^{\;\;A}X \text{ (daughter)} + \gamma \text{ photon(s)} $$

Figure 3.2 is a pictorial representation of the beta decay process in nitrogen-17.

The photons and particles produced in radioactive decay are sometimes referred to collectively as radiation. One can still find the expressions "alpha rays" and "beta rays" in the literature. These different forms of radiation vary widely in the distance they travel in solid or dense material. To treat the matter very briefly:

alpha particles travel a few micrometers;

beta particles travel on the order of one or two millimeters;

gamma photons travel centimeters to meters;

neutrinos travel great distances, many kilometers.

Note that it is the charged particles that do not get very far. Note also that we are talking about distances in normal materials: water, concrete, steel, and so on. In a highly rarefied atmosphere, as in outer space, alpha particles and electrons can travel long distances. We shall treat the subject of the travel or interactions of gamma photons in materials in some detail in Chapter 4.

We are interested in the travel of radiation in matter for two reasons: first, the conversion of the energy of the radiation to thermal energy takes place

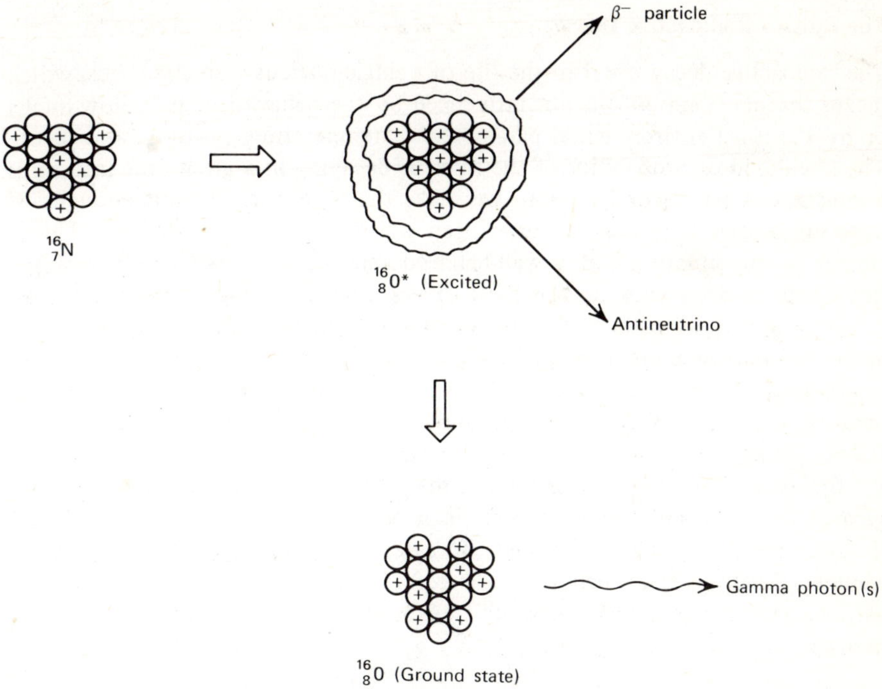

Figure 3.2 Schematic representation of the β^- decay process.

where the radiation is impeded and stopped, *not* where the radiation originates; second, because radiation is harmful to biological systems, we need to know how to shield them from sources of radiation, particularly gamma photons and neutrons. The neutrinos do not present a problem, because their penetrability is so great that they travel through a biological system without any interactions.

As indicated in the reaction equations, the alpha particle is a helium nucleus. Following emission, it rapidly picks up two electrons and becomes an atom of helium. The negative beta particle is simply an electron. The designation "beta" tells us it came from the nucleus. Once it is emitted it is just like any other electron, being distinguished only by the kinetic energy that it may possess. The antineutrino is distinguished by its unobtrusiveness. Having neither charge nor rest mass, it has little tendency to react and, as a result, travels great distances through matter. We would not even take note of it except for the fact that it can carry off a sizable fraction of the energy released in the decay. The neutrino energy must be totally discounted, because it cannot be captured. Gamma photons are emitted in the radioactive decay process whenever the daughter nucleus is formed in an excited state. This is most often the case. Gamma emission was discussed in Chapter 2.

The Rate of Radioactive Decay

The radioactive decay event in the life of a single nucleus is spontaneous, which means that its occurrence is not influenced by the behavior of its fellow nuclei or by the usual environmental parameters of temperature, pressure, and so on. The likelihood or probability of the nucleus decaying in a given time interval is a constant, λ per second, if only we make the time interval short enough. It does not matter how long we have been waiting for the decay event. This is similar to the situation with a well-behaved coin. The probability of tossing a head with a given coin on the next toss is 0.5, regardless of what has been thrown on previous tosses and regardless of what comes up on neighboring coins. With the coin we express the probability on a "per toss" basis; with the radioactive nucleus we express it on a "per second" basis. To continue the coin analogy, if we have N coins and toss them all, the "expected value" of heads is $0.5N$. We know that in fact there will be variations around this value. Probability theory tells us that as N increases, the *percentage* deviation from $0.5N$ decreases. So also with N nuclei of a single species. The expected value of the decay rate is λN decays per second. In cases of interest to us, N is a very large number, usually greater than 10^{15}. Therefore, we can treat the expected value as the actual frequency with negligible error. We call λN the decay rate or the *activity*.

$$\mathcal{A} \equiv \lambda N \text{ decays/s} \tag{3.1}$$

If at a given time we have a population $N(t)$ of a single radioactive species, and if no other members are being added to this population, then the population must be varying with time according to the equation

$$\frac{dN(t)}{dt} = -\lambda N(t) \tag{3.2}$$

which we can integrate to obtain

$$\int_{N(0)}^{N(t)} \frac{dN(t)}{N(t)} = -\lambda \int_0^t dt$$

$$\frac{N(t)}{N(0)} = e^{-\lambda t} \tag{3.3}$$

Likewise, the decay rate or activity varies with time:

$$\frac{\lambda N(t)}{\lambda N(0)} = e^{-\lambda t} \tag{3.4}$$

The activity can be expressed in decays or disintegrations per second (dps) or per minute (dpm). Most often, however, activity is expressed in curies, Ci.

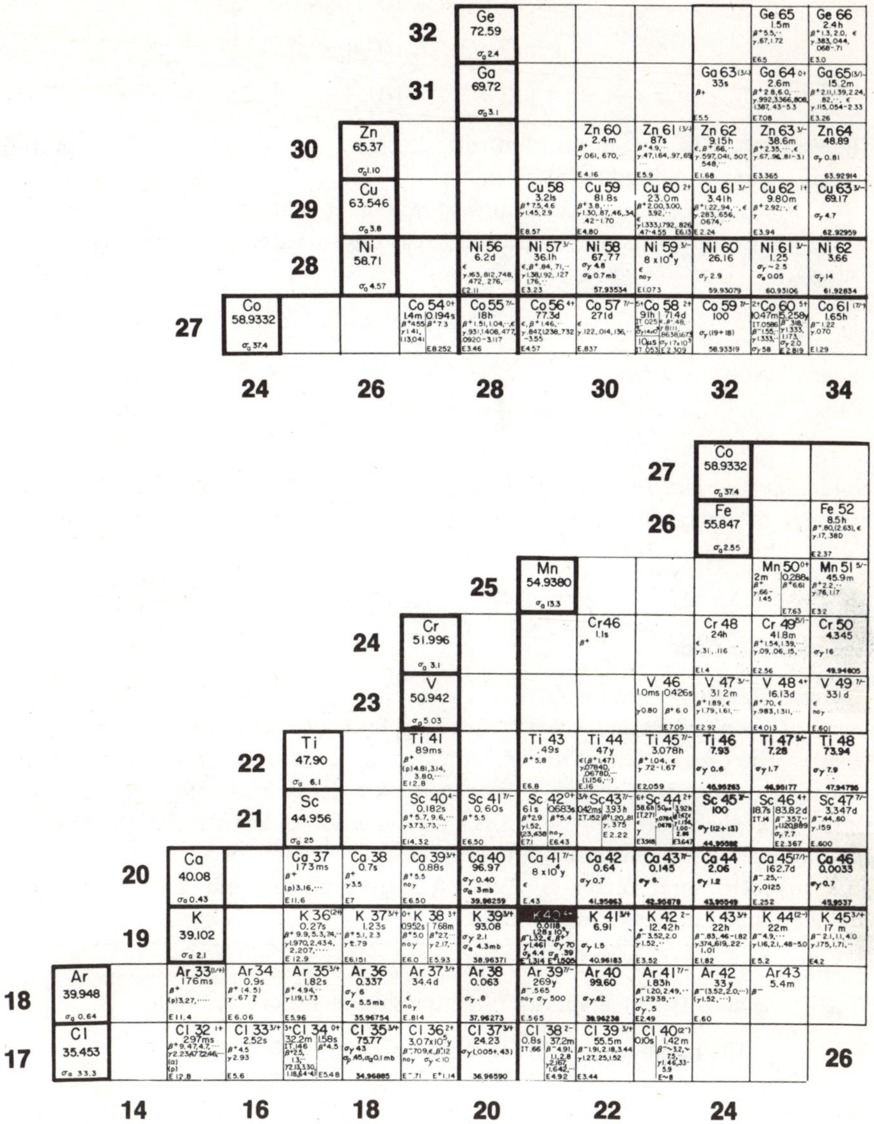

Figure 3.3 A portion of the Chart of the Nuclides. Courtesy of General Electric Co.

$$1\ \text{Ci} \equiv 3.7 \times 10^{10}\ \text{dps}$$

$$1\ \text{mCi} = 3.7 \times 10^{7}\ \text{dps}$$

$$1\ \mu\text{Ci} = 3.7 \times 10^{4}\ \text{dps}$$

The value of 3.7×10^{10} dps is of historical origin. It is close to the disintegration rate of 1 gram of radium (^{226}Ra), one of the earliest known radioisotopes.

Although the decay constant provides an adequate description of the decay rate of an isotope, it is not used as much as the half-life, $t_{1/2}$, which is derived as follows:

$$\frac{N(t)}{N(0)} = 0.5 \equiv e^{-\lambda t_{1/2}}$$

$$t_{1/2} = \frac{\ln 2}{\lambda} \qquad (3.5)$$

Note that $t_{1/2}$, like λ, is independent of time and the population, N. It is a property of a particular radioisotope. In the various data sources it is always the half-life that is reported.

Radioisotope Data

We have already introduced the two data sources which we shall be using, the *Chart of the Nuclides* and *Table of Isotopes*, by Lederer, Hollander, and Perlman (LHP). The *Chart of the Nuclides* provides rapid information on half-lives and modes of decay. It is also very easy to track a decay chain through a series of isotopes on the Chart. Figure 3.3 is a reproduction of a portion of the *Chart of the Nuclides*. For quantitative detail on the decay scheme of a particular isotope, LHP is much more complete. Table A.3 gives the half-life, decay mode, and decay Q-value for many radioisotopes.

Example 3.1
What is the activity of a gram of ^{90}Sr, expressed in curies?

Solution: The number of atoms of ^{90}Sr in a gram is

$$N = \frac{1}{90} \times 6.023 \times 10^{23} = 6.69 \times 10^{21}\ \text{nuclei/g}$$

(Again, the mass number, 90, is used in place of the true atomic mass of ^{90}Sr. This is a matter of convenience and does not introduce a significant error.)
The half-life of ^{90}Sr (Table A.3) is 28.1 years.

$$\lambda = \frac{\ln 2}{28.1 \times 3.154 \times 10^{7}} = 7.82 \times 10^{-10}\ \text{s}^{-1}$$

The decay rate is

$$\lambda N = 6.69 \times 10^{21} \times 7.82 \times 10^{-10} = 5.23 \times 10^{12} \text{ dps/g}$$

Expressed in curies, the activity is

$$\mathcal{A} = \frac{5.23 \times 10^{12}}{3.7 \times 10^{10}} = 141.5 \text{ Ci/g}$$

Example 3.2
Because the radiation emitted by radioisotopes represents a health hazard, there are standards for quantities that can be released to the environment. These standards are often expressed in terms of maximum permissible concentrations (MPCs) in air or water or both. As an example, the MPC for krypton-85 in air is 3×10^{-7} μCi/ml. Suppose one obtained a liter sample of air that had a concentration of ^{85}Kr equal to 1% of MPC. How many disintegrations would occur in one hour?

Solution: The half-life of ^{85}Kr is 10.76 years (Table A.3). This value is not needed explicitly here, but it is relevant because it tells us that over a one-hour period there is no significant decrease in the population, N, and that the activity is constant.

$$\text{Activity} = 3 \times 10^{-7} \times 0.01 \times 1000 \times 3.7 \times 10^{4}$$

$$\frac{\mu\text{Ci}}{\text{ml}} \qquad\qquad \frac{\text{ml}}{\text{l}} \qquad \frac{\text{dps}}{\mu\text{Ci}}$$

$$= 0.111 \text{ dps/l}$$

In one hour the expected number of disintegrations would be 0.111×3600 or 400.

Example 3.3
The excited isomer of ^{137}Ba is obtained in pure form by chemical separation from its parent, ^{137}Cs. If at a given instant there is one μCi of ^{137m}Ba present, how many disintegrations will occur over the next five minutes?

Solution: The half-life of ^{137m}Ba is 2.55 minutes (LHP). This value is short compared to the observation period, so that the activity cannot be treated as constant.

$$\lambda N(0) = 3.7 \times 10^{4} \text{ dps}$$

$$\lambda = \frac{\ln 2}{2.55} = 0.2718 \text{ min}^{-1}$$

$$N(0) = \frac{\lambda N(0)}{\lambda} = \frac{3.7 \times 10^{4}}{0.2718} \times 60 = 8.167 \times 10^{6} \text{ nuclei}$$

$$N\,(5 \text{ minutes}) = 8.167 \times 10^6 \times e^{-5 \times 0.2718}$$
$$= 2.098 \times 10^6$$
$$N(0) - N(5) = 6.07 \times 10^6 \text{ nuclei}$$

So in five minutes there are 6.07×10^6 disintegrations. We can approach this a different way.

$$\lambda N(t) = 3.7 \times 10^4 \times 60e^{-\lambda t} \text{ disintegrations/minute}$$
$$= 2.22 \times 10^6 e^{-\lambda t}$$

Over five minutes the disintegrations will be

$$2.22 \times 10^6 \int_0^5 e^{-\lambda t}\, dt = \frac{-2.22 \times 10^6}{\lambda}\, e^{-\lambda t}\Big|_0^5$$
$$= \frac{2.22 \times 10^6}{0.2718}\,[1 - e^{-5 \times 0.2718}]$$
$$= 6.07 \times 10^6$$

Figure 3.4 shows the decay scheme of a beta-emitter, cobalt-60.

$${}^{60}_{27}\mathrm{Co} \longrightarrow {}^{60}_{28}\mathrm{Ni} + {}^{\;0}_{-1}e\ (\beta^-)$$

This decay scheme is adapted from LHP, with much information irrelevant to the present discussion omitted. The diagram shows that the half-life of ^{60}Co is 5.26 years. We are also informed that beta emission always leads to an excited state of nickel-60 and, therefore, always involves gamma emission. Moreover, the diagram indicates the percent of decays in which each level of ^{60}Ni is reached. In the great majority (>99%) of the decays, the 2.506-MeV level is reached. The vertical arrows represent level transitions and, therefore, gamma photon emissions. The energy of each gamma photon is proportional to the length of the arrow, that is, the difference in energy between the two states. Note that in the ^{60}Co case, a transition from the 2.506-MeV level to the ground state does not occur. Therefore, we expect no gamma photon with an energy of 2.506 MeV. Rather, the 2.506-MeV level decays to the 1.333 level with the emission of a 1.173-MeV (2.506 - 1.333) photon. The 1.333-MeV level decays to the ground state with the emission of a 1.333-MeV photon. On nearly every decay, therefore, we expect a 1.173-MeV and a 1.333-MeV photon. This deduction is supported by the tabular data from LHP's Table I, which is also displayed in Figure 3.4. We note that a 1.173-MeV γ and a 1.333-MeV γ are each emitted on 100% of the decays. We also see that over 99% of the time, a β with a maximum energy of 0.314 MeV is emitted. We could obtain this energy value

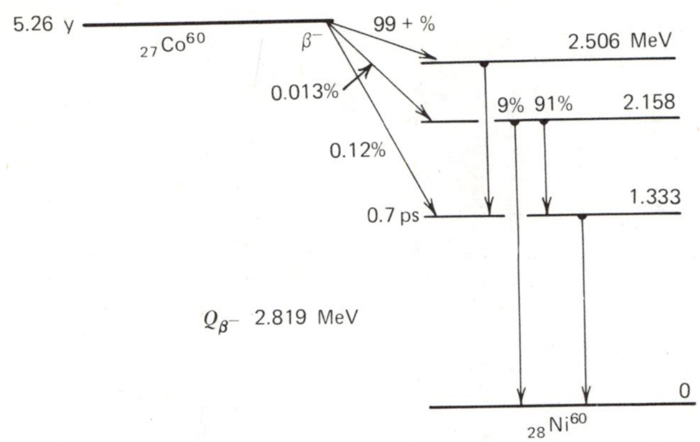

ISOTOPE Z A	HALF-LIFE	TYPE OF DECAY (Ψ); % ABUNDANCE Mass Excess ($\Delta \equiv M - A$) MeV ($C' = 0$) THERMAL NEUTRON CROSS SECTION (σ), BARNS	CLASS; IDENTIFICATION; GENETIC RELATIONSHIPS	MAJOR RADIATIONS; APPROXIMATE ENERGIES (MeV) AND INTENSITIES	PRINCIPAL MEANS OF PRODUCTION
Co^{60}	5.263 y (Gorb563) 5.24 y (GeiKW57) 5.20 y (LocE56) 5.21 y (KasJ53a) 5.27 y (TobJ55, TobJ51) others (LocE53, LivJ41, BrowG50, SinW51)	Ψ β^- (RisJ37) Δ −61.651 (MTW) σ_c −6 (GoldmDT64)	A n-capt (SamM36) chem, excit, cross bomb (LivJ41)	β^- 1.48 max (0.12%), 0.314 max (99+%) γ 1.173 (100%), 1.332 (100%)	Co^{59} (n, γ) (RisJ37, LivJ38a, LivJ41, SerI47b, YafL51)

Figure 3.4 **Decay scheme data on β^- emitter, adapted from LHP. Adapted from C. M. Lederer, J. M. Hollander, and I. Perlman, *Table of Isotopes*, Sixth Edition, Wiley, New York, 1967.**

from the decay diagram by subtracting 2.506 MeV (of the excited level to which the beta emission leads) from the total difference in the ground states, namely, the Q-value or Q_{β^-} shown on the diagram. We get 2.819 - 2.506 or 0.313 MeV. Note this value is designated the *maximum* (max) beta energy. The beta shares the energy of the level change with the antineutrino. The energy division between the two varies from decay to decay, but, on the average, the beta particle gets only about one third of the maximum value.

The Specific Energy of Radioactive Decay Reactions

The energy of the decay reaction is the Q-value. These are tabulated in Table A.3 and also in LHP (see Figure 3.4). In the case of alpha decay the energy is divided between the alpha particle and the gamma photons. In beta decay it is divided between beta electrons and neutrinos and gamma photons. The division must be analyzed for each radioisotope from the decay scheme data (Figure 3.4).

Since a sample of any radioisotope is always reacting at the rate λN, the total rate of energy conversion is simply λNQ. The analysis is illustrated in the following example.

Example 3.4
From the data in Figure 3.4, compute the beta and gamma energy release rates in one gram of ^{60}Co.

Solution: The decay rate is

$$\lambda = \frac{\ln 2}{5.26 \times 3.154 \times 10^7} = 4.178 \times 10^{-9}\ \text{s}^{-1}$$

$$N = \frac{6.023 \times 10^{23}}{60} = 1.004 \times 10^{22}\ \text{nuclei/g}$$

$$\lambda N = 4.194 \times 10^{13}\ \text{dps/g}$$

The principal beta particle is the 0.314-MeV (max), which is emitted practically 100% of the time. Assume that the average energy of the betas is one third of the maximum, or 0.105 MeV.

$$\dot{q}_\beta = 4.194 \times 10^{13} \times 0.105 \times 1.602 \times 10^{-13} = 0.705\ \text{W/g}$$

Since the decay is almost 100% to the 2.506-MeV level of ^{60}Ni, the gamma energy per decay has this value. The decay is in fact in two steps of 1.173 and 1.333 MeV, but the sum is the value of interest here.

$$\dot{q}_\gamma = 4.194 \times 10^{13} \times 2.506 \times 1.602 \times 10^{-13} = 16.84\ \text{W/g}$$

The preponderance of high-energy gamma photons in the decay energy of cobalt-60 limits its usefulness as an isotope energy source. Note that by the time the entire gram of ^{60}Co has decayed, the total energy release would be NQ.

$$NQ = 1.004 \times 10^{22} \times 2.819 \times 1.602 \times 10^{-13} = 4.53 \times 10^{9}\ \mathrm{J/g}$$
$$(4.30 \times 10^{6}\ \mathrm{Btu})$$

It would take 5.26 years to release half this value.

The Production of Radioisotopes

Since our principal concern is with the possibilities of radioisotopes as energy sources, we must know their availability. While there are radioisotopes found in nature, they are far too dilute to permit their use as energy sources. Such radioisotopes must be produced. At present the only practicable source is the fission reactor. Radioisotopes are formed when the fission reaction takes place. The fissioning nucleus splits into two smaller nuclei, both of which tend to be radioactive. These are called fission products. Radioisotopes are also formed when a neutron produced in fission reacts with a nucleus. These are called (n, γ) (n-gamma) reactions, because gamma photons accompany the reaction. The product nucleus is formed in an excited level. For example:

$$^{59}\mathrm{Co} + n \longrightarrow {}^{60}\mathrm{Co}^*$$
$$^{60}\mathrm{Co}^* \longrightarrow {}^{60}\mathrm{Co} + \gamma$$

While a radioisotope is being produced, it is, of course, decaying. The net rate of growth of the isotope is

$$\frac{dN}{dt} = \dot{P} - \lambda N \tag{3.6}$$

where $\dot{P}$ is the production rate in the same units as λN. If $\dot{P}$ is a constant, this differential equation has the solution

$$N(t) = \frac{\dot{P}}{\lambda}(1 - e^{-\lambda t}) + N(0)e^{-\lambda t} \tag{3.7}$$

Usually, we are more interested in the activity of the isotope.

$$\mathcal{A}(t) = \lambda N(t) = \dot{P}(1 - e^{-\lambda t}) + \lambda N(0)e^{-\lambda t} \tag{3.8}$$

Independent of the radioisotope in question, this equation has some characteristics that are worth understanding. These are displayed in Figure 3.5. As t grows large, $\mathcal{A}(t) \to \dot{P}$. This is called the saturation activity. Starting at any arbitrary time with any arbitrary activity, the activity will gain (or lose) 50% of the remaining distance to saturation in one half-life ($\ln 2/\lambda$). These facts are handy for rapid reckoning. If, for example, gold-198, with a 2.7-day half-life, is being produced in a reactor, we can say quickly that if it has been in for a week, or about three half-lives, it has 85% of the activity it would attain if it was left in indefinitely.

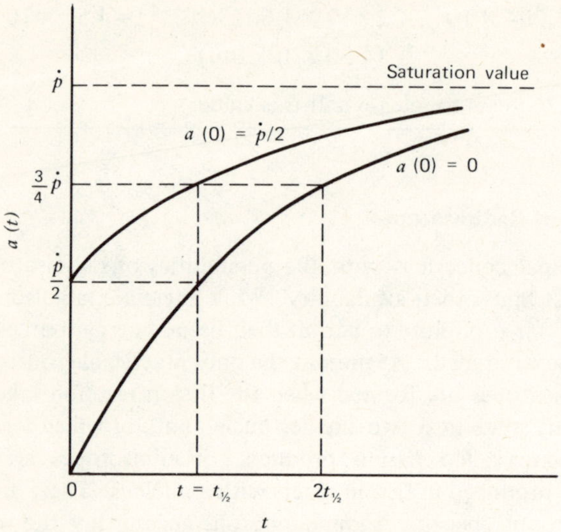

Figure 3.5 Characteristic growth of activity of a radioisotope being produced at a constant rate.

In the case of long half-life isotopes where the half-life is long compared to the production time, we can expand $e^{-\lambda t}$ and take only the first two terms.

$$e^{-\lambda t} \cong 1 - \lambda t \quad (\lambda t \gtrsim 0.1) \tag{3.9}$$

Equation 3.8 then becomes

$$\mathcal{A}(t) = \dot{P}\lambda t + \mathcal{A}(0)(1 - \lambda t) \tag{3.10}$$

The activity grows linearly with the production time.

Example 3.5

A piece of manganese is placed in a neutron beam, where it undergoes 10^{10} (n, γ) reactions per second. What activity (curies) would be reached if the manganese is left in the beam for five hours?

Solution: First we must find out what reaction(s) may be taking place. The *Chart of the Nuclides* is by far the most convenient source of this information, although it can be obtained from LHP. We find that manganese has only a single naturally occurring isotope, ^{55}Mn. The (n, γ) reaction, therefore, leads to

$$^{55}\text{Mn} + n \longrightarrow {}^{56}\text{Mn}$$

The chart tells us that ^{56}Mn undergoes β^- decay with a 2.58-h half-life.

$$^{56}\text{Mn} \longrightarrow {}^{56}\text{Fe} + \beta^-$$

The ^{56}Fe is stable, so it does not contribute to the activity. The ^{56}Mn is being produced at a rate

$$\dot{P} = 10^{10} \text{ nuclei/s}$$

Its activity is, therefore

$$\lambda N = 10^{10} (1 - e^{-\lambda t})$$

$$\lambda = \frac{\ln 2}{2.58} = 0.269 \text{ h}^{-1}$$

$$\lambda N(5 \text{ h}) = 7.39 \times 10^{9} \text{ decays/s}$$

$$= \frac{7.39 \times 10^{9}}{3.7 \times 10^{10}} = 0.200 \text{ Ci}$$

Some important radioisotopes are formed in reactors through a succession of (n, γ) reactions and radioactive decays. Plutonium-238, which is used in heart pacemakers, is an example.

$$^{235}_{92}\text{U} + n \longrightarrow {}^{236}_{92}\text{U}$$

$$^{236}_{92}\text{U} + n \longrightarrow {}^{237}_{92}\text{U}$$

$$^{237}_{92}\text{U} \longrightarrow {}^{237}_{93}\text{Np} + \beta^-$$

$$^{237}_{93}\text{Np} + n \longrightarrow {}^{238}_{93}\text{Np}$$

$$^{238}_{93}\text{Np} \longrightarrow {}^{238}_{94}\text{Pu} + \beta^-$$

Radioisotopes as Energy Sources

We now address the question of how much energy we can obtain from the quantities of radioisotopes that we have the capacity to produce. The preponderant source is the fission products, by-product of the fission reaction. In the following examples, we analyze the thermal power that could be obtained from three of the most energetic of the fission products. In Example 3.6, two long-lived isotopes are analyzed, and in Example 3.7 a relatively short-lived isotope is analyzed.

Example 3.6

Strontium-90 and cesium-137 are produced in 6% of fissions. Compute the maximum energy that could be obtained from the quantity that would be produced in a power reactor that operates at an average thermal power of 2000 MW(t) in 30 years of operation.

Solution: A precise answer to this question depends on certain assumptions. A fission results in a release of 200 MeV energy. Therefore, a 2000 MW(t) reactor has a fission rate of

$$\underset{\frac{\text{J}}{\text{s}}}{2000 \times 10^6} \times \underset{\frac{\text{fissions}}{\text{MeV}}}{\frac{1}{200}} \times \underset{\frac{\text{MeV}}{\text{J}}}{6.242 \times 10^{12}} = \underset{\frac{\text{fissions}}{\text{s}}}{6.242 \times 10^{19}}$$

The production rate of both ^{90}Sr and ^{137}Cs is, therefore

$$\dot{P} = 0.06 \times 6.242 \times 10^{19} = 3.745 \times 10^{18} \text{ nuclei/s}$$

Let us assume that we only obtain the radioisotopes at the end of the 30 years' operation of the reactor. First, we calculate ^{90}Sr. From Table A.3:

$$t_{1/2} = 28.1 \text{ y}$$

$$Q = 0.546 + 2.27 = 2.82 \text{ MeV}$$

This includes the Q-value of the 64-hour daughter of ^{90}Sr, yttrium-90. At the end of 30 years

$$\lambda N = \dot{P}(1 - e^{-\lambda t})$$

$$\lambda = \frac{\ln 2}{28.1} = 0.02467 \text{ y}^{-1}$$

$$\lambda N = 3.745 \times 10^{18} \, [1 - \exp - (0.02467 \times 30)]$$

$$= 1.958 \times 10^{18} \text{ decays/s}$$

$$\dot{q} = \lambda N Q = 1.958 \times 10^{18} \times \frac{2.82}{6.242 \times 10^{12}} = 8.85 \times 10^5 \text{ W}$$

Note that about two thirds of this energy is carried by the neutrinos and is not available. The calculation for ^{137}Cs:

$$t_{1/2} = 30 \text{ y}$$

$$\lambda = 0.02311 \text{ y}^{-1}$$

$$Q = 1.176 \text{ MeV}$$

$$\lambda N = 3.745 \times 10^{18} \, [1 - \exp - (0.02311 \times 30)]$$

$$= 1.873 \times 10^{18} \text{ decays/s}$$

$$\dot{q} = 1.873 \times 10^{18} \, \frac{1.176}{6.242 \times 10^{12}} = 3.53 \times 10^5 \text{ W}$$

Combining the two isotopes and correcting for neutrinos, we would get a thermal power of about 600 kW(t). Considering that it took 30 years of fission reactor operation at 2,000,000 kW(t) to produce this quantity of strontium and cesium, it is apparent that the radioisotopes will not add significantly to the energy supply. The thermal power of the isotopes is significant in absolute magnitude and could conceivably serve some valuable functions. On the large scale, however, they do not have a significant contribution to make.

Example 3.7

A fission product radioisotope, cerium-144, is considered as a possible energy source. It has a relatively short half-life, 284 days. Assume that it is produced in a fission reactor in 6% of fissions for three years at a steady reactor power of 2000 MW(t). Assume further that, following production in the reactor, one year is required to process the cerium to use in a cerium-fueled isotope power plant. What thermal power could be achieved?

Solution: At 200 MeV per fission, the fission rate is (see Example 3.6)

$$2000 \times 10^6 \times \frac{1}{200} \times 6.242 \times 10^{12} = 6.242 \times 10^{19}$$

The production rate of ^{144}Ce is

$$\dot{P} = 0.06 \times 6.242 \times 10^{19} = 3.745 \times 10^{18} \text{ nuclei/s}$$

$$\lambda = \frac{\ln 2}{284/365} = 0.8908 \text{ y}^{-1}$$

$$\lambda N(3y) = 3.745 \times 10^{18}\,[1 - \exp - (0.8908 \times 3)]$$

$$= 3.486 \times 10^{18} \text{ decays/s}$$

$$\dot{q} = \lambda N Q$$

$$Q = 0.3 + 2.989 = 3.309 \text{ MeV}$$

The Q-value of the daughter nuclide, ^{144}Pr, is also included, because ^{144}Pr has a 17.3-minute half-life and will, therefore, decay at the same rate as ^{144}Ce.

$$\dot{q} = 3.486 \times 10^{18} \times \frac{3.309}{6.242 \times 10^{12}} = 1.848 \times 10^6 \text{ W}$$

This is the energy rate at the time of removal from the fission reactor. After one year of processing it will be reduced.

$$\dot{q} = 1.848 \times 10^6 \times \exp - (0.8908 \times 1)$$

$$= 7.584 \times 10^5 \text{ W}$$

The actual thermal energy recoverable is about one third of this value when the neutrino energy is discounted. So we have about 250 kW(t), or about 0.01% of the thermal power of the fission reactor which produced the cerium-144. It obviously is not this simple, because this power will rapidly decline unless the decaying fuel is replenished. The object of this calculation is to show the magnitude of the radioisotope power relative to the power of the fission reactor that produced it.

We are now in a position to apply the criteria presented at the start of this chapter to evaluate radioisotopes as potential large-scale energy sources.

1. *The specific energy of decay reactions*. A typical Q-value of a radioactive decay is 1 MeV. For an isotope of mass number 100, the energy obtainable from the decay of a kilogram is

$$\frac{1000}{100} \times 6.023 \times 10^{23} \times \frac{1}{6.242 \times 10^{12}} = 9.65 \times 10^{11} \text{ J/kg}$$

   or about one million MJ per kg. This is about 400 million Btu/lb. The specific energy is high enough.
2. *A mechanism for conducting the reaction must exist*. There is no mechanism for either starting or stopping radioactive decay, once the isotope is produced. It decays with a release of nuclear energy at a rate dependent only on its quantity, its half-life, and its decay scheme. This statement must be modified to admit of the technical possibility of changing the radioisotope by neutron bombardment or other forms of nuclear radiation. Such a mechanism would be complex and costly and would probably make a radioisotope power source totally impracticable. One could, of course, control the power to some extent by adding and removing material periodically or even continuously. Nevertheless, the inflexibility of a radioisotope power source is a major disadvantage.
3. *The availability of the reactant materials*. It is on this point that radioisotopes fail as large-scale energy sources. Examples 3.6 and 3.7 show that they can never produce power at more than a fraction of 1% of the fission power needed to produce them. While in absolute numbers this power is not negligible, it should always be more economical to increase the fission power level by 1% than to use the by-product isotopes to produce power. This fact relegates radioisotopes to special-purpose, high-unit-cost power supplies. The plutonium-238 power supplies used in the moon landings of the Apollo project are an example.
4. *Environmental impacts*. The principal impact of radioisotopes may well be the impact of the fission reactors that produce them. (Radioisotopes are, of course, produced in such devices as accelerators, but the quantities are relatively small.) The use of radioisotope power sources would, of course, have some incremental impact, because they would increase the transportation of

isotopes and increase the possibilities for escape into the environment. This limitation of their use, however, appears to be much less severe than item 3 above.

FISSION

Of the three classes of nuclear reactions offering energy-source possibilities, fission is the one that currently holds first place in quantity of nuclear energy actually being converted. The word "fission" applies to a class of nuclear reactions in which a nucleus breaks up into two smaller nuclei. Neutron and gamma photon emission also accompany the process. We are interested in a subclass of fission reactions, those caused by neutron interaction with a nucleus. Fission also occurs spontaneously in a few nuclides (for example, ^{240}Pu), following the radioactive decay laws. The nuclides are so rare and the decay rate so low that these reactions are not important as energy sources, however.

The Fission Reaction

We can represent the fission reaction in words.

Neutron + Nucleus ⟶ Fission-product nuclei

+ Neutrons + Beta electrons + Gamma photons

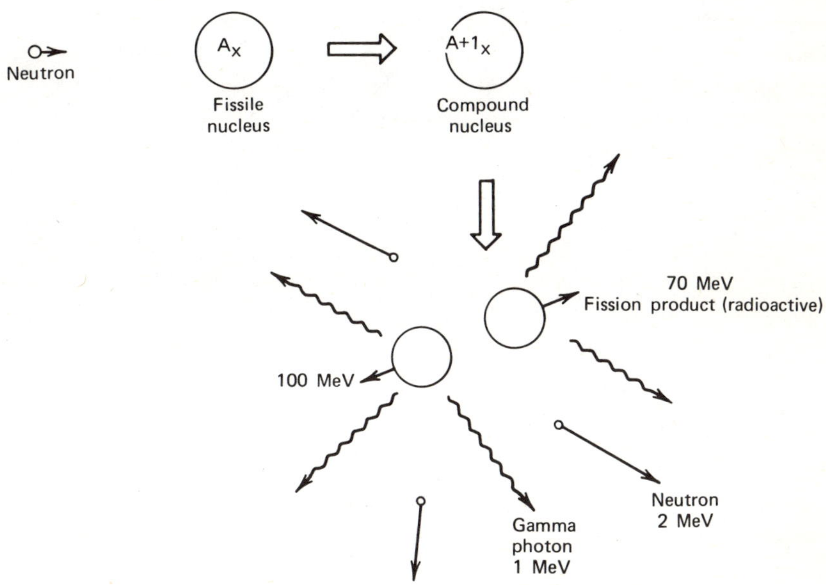

Figure 3.6 Schematic of fission reaction.

The reaction is shown schematically in Figure 3.6. The neutron enters the nucleus and, momentarily, a compound nucleus exists. One possible outcome at this point is for the excited nucleus to decay to its ground state with the emission of gamma radiation. This is the (n, γ) reaction. When fission occurs, however, the compound nucleus splits, usually into a pair of nuclei. These nuclei are called fission products. They are usually radioactive and decay by β^- emission. The beta electrons indicated in the above equation are probably from the decay of fission-product nuclei, but some decays follow the fission event so rapidly that we can regard the sequence as a single event.

Fissionable Materials, Fissile Materials, and Fertile Materials

For our purposes, we define a fissionable isotope as one in which the fission reaction can be caused by neutrons. A fissile isotope is one in which the fission can be caused by low-energy neutrons. There is no minimum neutron energy required. Fissile isotopes are a subclass of fissionable isotopes.

Very few isotopes are fissile. The most important ones are uranium-233, uranium-235, and plutonium-239. Of these, only ^{235}U is found in nature, as 0.71% of natural uranium. ^{238}U forms the other 99.3%. The other two isotopes are themselves products of neutron reactions plus beta decays, as follows:

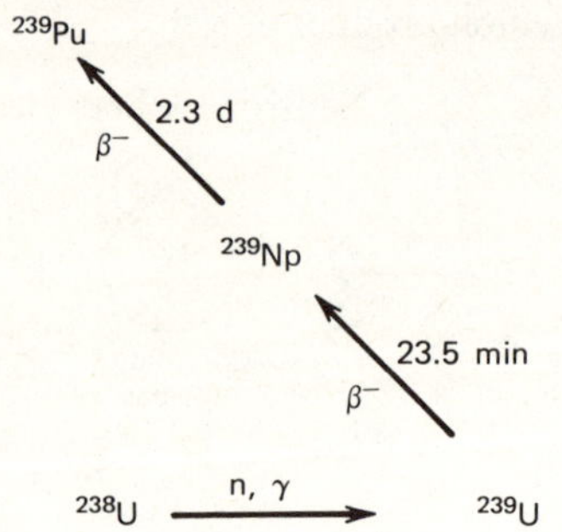

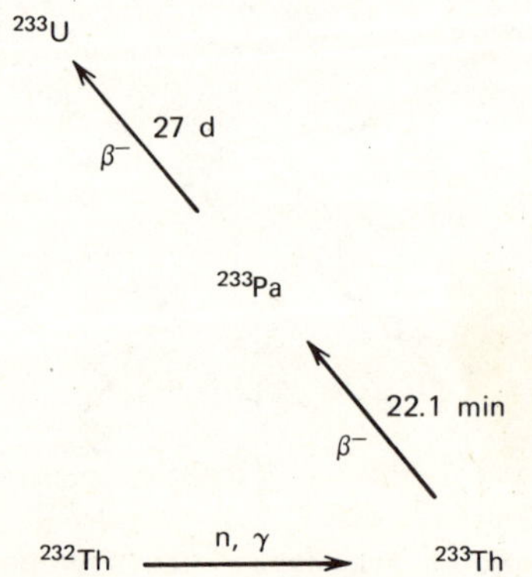

The representation scheme for series of reactions as shown here is to position the various isotopes in the relative position they would occupy in the *Chart of the Nuclides*, a plot of Z versus N. In each of the two reactions shown, the original (n, γ) reaction leads to an isotope with a short half-life that quickly decays to the fissile isotope. Isotopes ^{239}Pu and ^{233}U are also radioactive but have such long half-lives that there is no significant disappearance by decay in times of interest to us.

The two reactions shown above are called conversion reactions. Isotopes ^{238}U and ^{232}Th are called fertile materials. Since ^{238}U forms the great bulk of natural uranium and ^{232}Th is the only isotope of natural thorium, the potential quantities of ^{233}U and ^{239}Pu far exceed those of ^{235}U. In these early days of nuclear energy, however, ^{235}U is the predominant isotope. The expectation that fission energy will substitute for fossil fuel energy depends strongly on the conversion of ^{238}U and ^{232}Th.

Fission Products

In a single fission, two product nuclei are usually formed. Occasionally there are three. The most notable product of a three-way split is tritium, ^{3}H, which is formed about once in 10,000 fissions. Since the particular pair formed varies from fission to fission, one finds a broad distribution of isotopes in a mixture of fission products. Some of the more prominent fission-product isotopes are shown in Table 3.1.

The yield is the percent of fissions in which the isotope is formed. The maximum yield for any isotope is less than 7%. Fission products represent the most abundant source of radioisotopes we have. As such, they represent a by-product asset. Unfortunately, the problem of safely managing this quantity of radioactivity represents a responsibility of serious proportions.

Table 3.1 Some Fission Product Isotopes (from slow neutron fission in ^{235}U)

ISOTOPE				ISOTOPE			
ELEMENT	A	$t_{1/2}$	YIELD (%)	ELEMENT	A	$t_{1/2}$	YIELD (%)
Strontium	89	51 d	4.8	Tellurium	127	105 d	0.04
Strontium	90	28 y	5.8	Tellurium	129	37 d	0.35
Yttrium	91	58 d	5.4	Cesium	137	30 y	6.2
Zirconium	95	65 d	6.3	Cerium	141	33 d	6.0
Ruthenium	103	40 d	3.0	Cerium	144	280 d	6.0
Ruthenium	106	1.0 y	0.4	Promethium	147	2.6 y	2.4
Antimony	125	2 y	0.02	Samarium	151	80 y	0.44

Fission Neutrons

A unique feature of the fission reaction is that among the products of the reaction are neutrons, one of the necessary reactants. It is as if combustion produced oxygen. The neutrons produced in the reaction are called fission neutrons. Since the fission reaction occurs in many different ways, the number of neutrons produced varies from one reaction to another. The number is usually two, three, or four. The *average* number of neutrons produced per fission for any given fissioning isotope is designated $\bar{\nu}$. It is about 2.5 for ^{235}U and 3.0 for ^{239}Pu.

The energy of fission neutrons is of definite interest. As with other products of the fission reaction, these neutrons show an energy spectrum. That means the individual neutrons have different energies; the population as a whole exhibits a distribution or spectrum. Distribution functions are treated in Chapter 4. In the case of the fission neutron population, the spectrum follows that of a normal distribution, with some distortion in the wings. The spectrum shows a peak value (most probable) at about 1 MeV and an average value of 2 MeV. These energies are important, because it will be recalled that there are fissionable isotopes that are not fissile. These isotopes, which include the abundant ^{238}U and ^{232}Th, can be fissioned by neutrons only if the neutron has some minimum kinetic energy. This minimum energy is sometimes called the threshold energy, and such nuclear reactions are called threshold reactions. For the fission process in ^{238}U and ^{232}Th, this threshold is around 0.1 to 0.5 MeV. These values are important, because they indicate that the average neutron produced in fission is capable of causing fission directly in the fertile isotopes ^{238}U and ^{232}Th. (Remember that if a neutron has less than the energy required to cause a given reaction, there is nothing that can be done because, lacking any charge, a neutron has no "handle" by which we can accelerate it.) The direct fission of ^{238}U and ^{232}Th can be important, but in actual reactors it does not occur so often (maybe 1 to 10% of fissions), because the neutrons lose the necessary energy in the more probable scattering reactions.

Fission Gammas

As might be expected, fission-product nuclei are in an excited state immediately following the fission reaction. They decay with the emission of gamma photons, referred to as prompt gammas (Figure 3.6). These photons have an average energy around 1 MeV. In a reactor there is also thermal energy produced by the beta decay of fission products. Gamma photons coming from these decay reactions are called delayed gammas.

Energy Release in Fission

We can calculate a Q-value for a fission reaction just as we did for nuclear reactions in Chapter 2 (see Example 3.8). Fission is a whole series of reactions, however, so some averaging is necessary. Also, there is variation from one fissionable isotope to another. For present purposes we shall take the average energy per fission as

	MeV
Fission fragments	168
Beta particles	8
Gamma photons	19
Neutrons	5
Neutrinos	(12)
	200

The neutrino energy is not counted because it is not recoverable. The bulk of the energy is in the form of the kinetic energy of the two fission fragments (that is, fission-product nuclei). These fragments emerge from the fission reaction as ions, with an initial charge of about +20. They slow down very quickly, picking up orbital electrons in the process. Their energy is converted to thermal energy within micrometers of the fission site. The energy of the gammas and neutrons is deposited over a much wider range.

Example 3.8

One example of a fission reaction is

$$n + {}^{239}\text{Pu} \longrightarrow {}^{137}\text{Cs} + \text{X} + 3n + 7e^{-}(\beta^{-})$$

Compute the Q-value of this reaction from the rest mass values in Table A.3.

Solution: First, complete the reaction by balancing mass numbers and charges.

$${}^{1}_{0}n + {}^{239}_{94}\text{Pu} \longrightarrow {}^{137}_{55}\text{Cs} + {}^{100}_{46}\text{X} + 3\,{}^{1}_{0}n + 7\,{}^{0}_{-1}e$$

From the *Chart of the Nuclides* we determine that ${}^{100}_{46}\text{X}$ is ${}^{100}\text{Pd}$, or palladium-100. From Eq. 2.4

$$Q = \sum_{\text{Reactants}} (m - A) - \sum_{\text{Products}} (m - A)$$

Using the $(m - A)$ values from LHP

$$Q = 8.0714 + 48.60 - (-86.9) - (-85) - 3 \times 8.0714$$

$$Q = 204 \text{ MeV}$$

Note that the mass of the electrons does not have to be included explicitly, because they are accounted for in the atomic masses of the products.

Example 3.9

Uranium-235 constitutes 0.71 a/o of natural uranium. How much fission energy does this isotope represent per kilogram of U_3O_8? The compound U_3O_8, called "yellow cake" in the uranium business, is the chemical form in which uranium moves from the mining and milling operations into the next stage of the fuel cycle.

Solution: Since natural uranium is 99.3% ^{238}U, we can use 238 as the value of its atomic mass. A kilogram of U_3O_8, therefore, contains

$$1 \times \frac{(3 \times 238)}{(3 \times 238) + (8 \times 16)} = 0.848 \text{ kg-U/kg-}U_3O_8$$

The number of ^{235}U atoms is

$$\frac{0.848 \times 1000}{238} \times 6.023 \times 10^{23} \times 0.0071 = 1.524 \times 10^{22} \text{ atoms } ^{235}\text{U/kg-}U_3O_8$$

The fission energy represented is

$$1.524 \times 10^{22} \times 200 \times 1.602 \times 10^{-13} = 4.822 \times 10^{11} \text{ J/kg-}U_3O_8$$

The Fission (or Nuclear) Reactor

A fission reactor is a structure within which the fission reaction is conducted at a controlled rate for various research and energy conversion purposes. In the latter case the reactor is called a power reactor. Fission energy is converted to thermal energy, which is transported out of the reactor by a fluid whose energy is converted to mechanical or electrical energy by a power cycle (Chapter 1).

Occasionally one will see a fission reactor described as a chain reactor. The reason derives from the fact that a unique feature of the fission reaction is that the only large-scale source of one essential reactant, namely the neutron, is the reaction itself. In order to have a sustained reaction rate at a high level, it is necessary that, on the average, one neutron produced in each fission causes one more fission. When this is accomplished, we have a chain reaction. Figures 3.7 and 3.8 are representations of the fission chain.

It should be mentioned that we have low-level sources of neutrons (10^6 to 10^{10} neutrons/s) that can be used to cause fission reactions in fissile material. That process would not be a chain reaction. Californium-252, which emits spontaneous fission neutrons, is an example of such a source. The fission power developed by such a source is too low to be of interest, however.

A term that one hears very often in connection with nuclear reactors is the

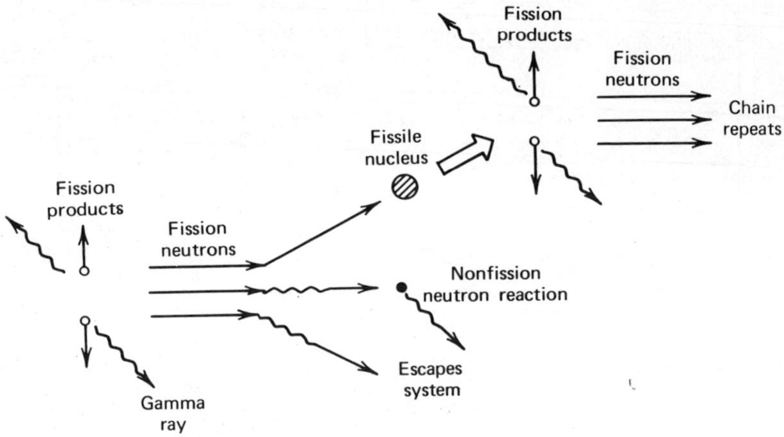

Figure 3.7 Schematic of fission chain.

"breeder reactor." A breeder reactor is shown schematically in Figure 3.9. The fission chain reaction goes on as before. However, some of the neutrons left over react (the conversion reaction) with uranium-238 to form plutonium-239. In this way the ^{239}Pu being consumed is, to some extent, being replaced. If this process occurs to the extent that the ^{239}Pu formed exceeds that consumed, we have a breeder reactor.

Let us say that for every atom of ^{239}Pu consumed, we produce C atoms of new ^{239}Pu via the conversion reaction. We start with N_0 atoms and consume them all. We will then have N_0C atoms left. We now put those in a reactor and consume them; we will have N_0C^2 left. We repeat the process indefinitely. The total amount finally consumed will be

$$N_0 + N_0C + N_0C^2 + N_0C^3 + \cdots$$

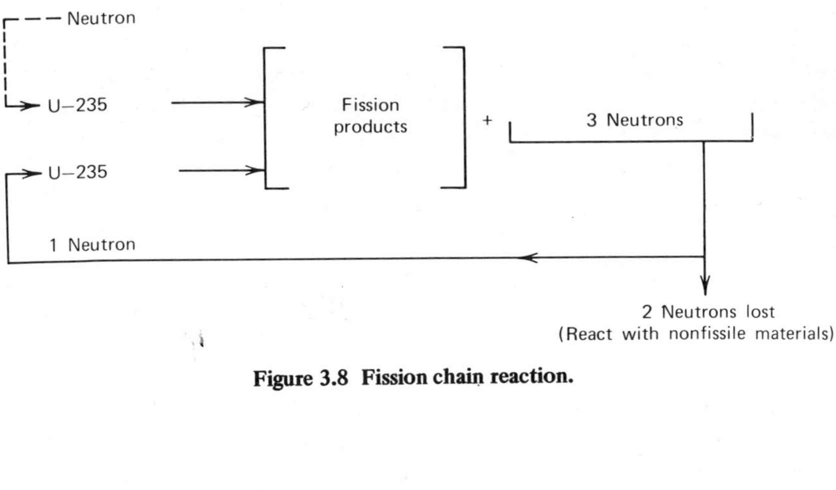

Figure 3.8 Fission chain reaction.

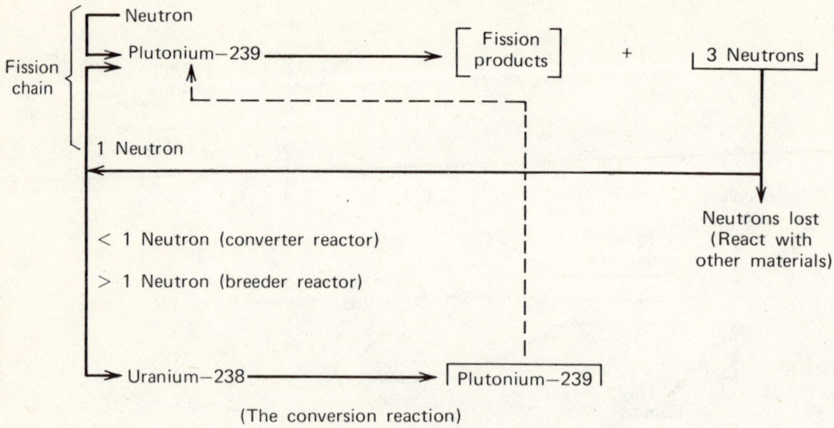

Figure 3.9 Converter/breeder reactor schemes.

If C is less than one, this series converges and the sum is

$$\frac{N_0}{1 - C}$$

If the conversion ratio is 0.5 (about what our present-day U.S. utility reactors achieve), then we have doubled the original amount, N_0. Such a reactor is called a converter reactor. As the conversion ratio approaches one, the gain would appear to grow indefinitely large. Our simple analysis admits of that impossibility, because it does not include the fact that when $N_0 C$ atoms of ^{239}Pu are formed, a like number of atoms of ^{238}U disappears. We cannot form more ^{239}Pu than we have ^{238}U atoms in the first place. Consequently, if the conversion ratio, C, can be made larger than one, the quantity of fissile material does not grow indefinitely, but it *does* approach the quantity of fertile material. Given the high relative quantity of fertile to fissile material in nature, this is an enormous effect.

A breeder reactor is one in which C is greater than one, meaning that more *fissile* material is formed than is consumed. In an economy of our present converter reactors, it is the *fissile* material in nature (^{235}U) that is the limiting resource. In a breeder economy, it is the *fertile* material in nature (^{238}U, ^{232}Th) that is limiting. The difference is shown graphically in Figure 3.10, which gives a resource estimate of the U. S. Energy Research and Development Administration (ERDA) as of 1975. The energy obtainable from the estimated U. S. uranium deposits is 1800 quads (1 quad = 10^{15} Btu). (The total U. S. consumption of energy in all forms in 1974 was 73 quads.) If breeders become economically feasible, however, this same uranium is estimated to yield 130,000 quads. This dramatic gain provides the incentive for the development of breeder reactors.

Although this discussion has been limited primarily to the ^{239}Pu–^{238}U fissile-

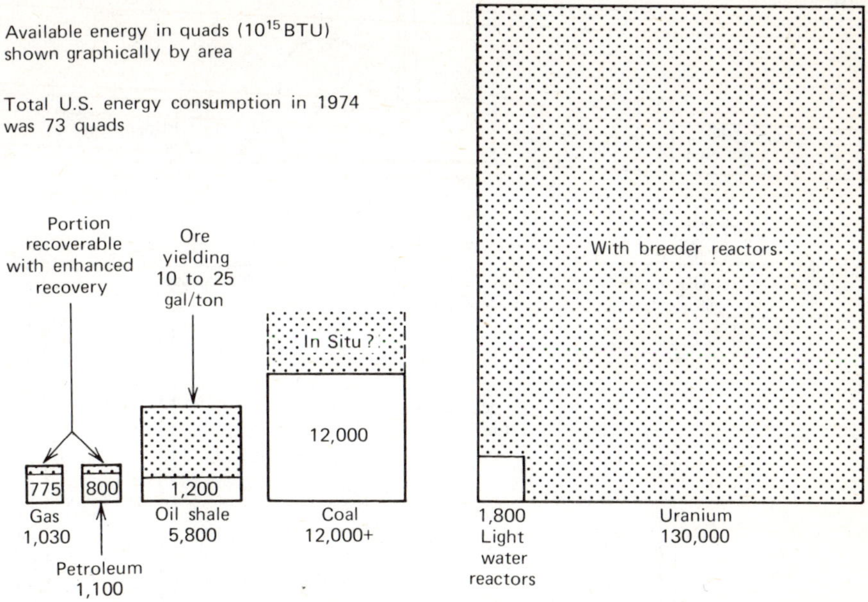

Figure 3.10 Available energy from recoverable domestic energy resources. From "A National Plan for Energy Research, Development and Demonstration," ERDA-48, 1975.

fertile combination, sometimes called the uranium fuel cycle, most of it also applies to the ^{233}U–^{232}Th fissile-fertile combination (the thorium fuel cycle). There are some differences. Unlike uranium, natural thorium contains no fissile material. As a result, a thorium fuel cycle must rely on help from uranium fuel to get started. There also are significant physics differences in the isotopes of the two cycles.

The Nuclear Fission Fuel Cycle

A system of nuclear power reactors must be supported by a series of industrial operations that prepare uranium (or thorium) for insertion in the reactor and that process this material after its removal from the reactor. The entire series of operations is called the fuel cycle. The exact nature of the fuel cycle is dependent on the reactor type that it serves. Figure 3.11 shows the fuel cycle that serves the predominant U. S. utility reactor type, the light-water reactor or LWR. Natural uranium is mined and purified to U_3O_8. It is then chemically converted to uranium hexafluoride, UF_6, a volatile form necessary for the next step: enrichment. This term refers to increasing the concentration of ^{235}U in the uranium from the natural 0.71% to perhaps 3%. At the enrichment plant the natural uranium is split into two portions, a depleted or "tails" portion of 0.2–0.3% ^{235}U, and the enriched portion. The tails portion is stored. The enriched

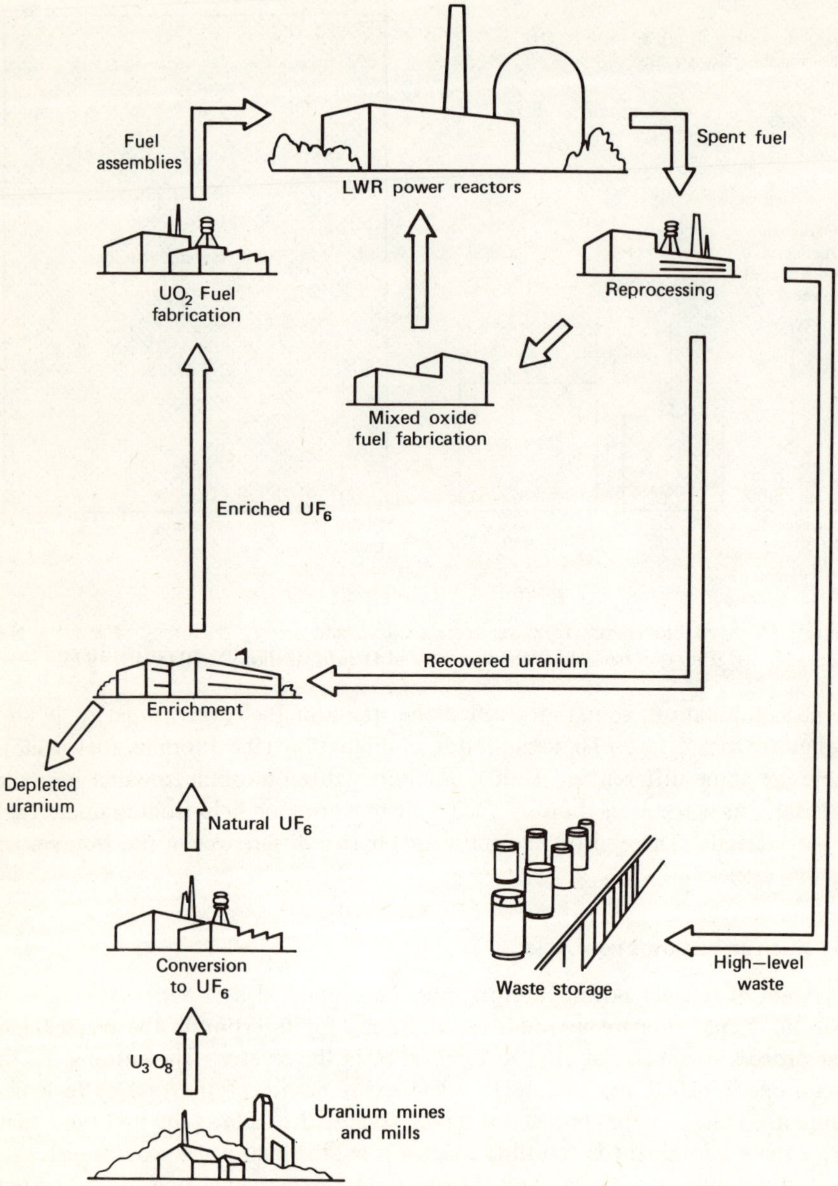

Figure 3.11 Fission fuel cycle.

portion is sent to a fabrication plant, where it is converted chemically to uranium dioxide, UO_2, the usual form for reactor use. The UO_2 is placed inside tubes, and these tubes are fabricated into fuel assemblies for insertion into the reactor. When the assemblies are discharged from the reactor, they are called "spent fuel." There are some possible variations in its subsequent history, but Figure 3.11 indicates that it proceeds to a reprocessing plant where it is chemically separated into three portions: uranium, which may be returned to the enrichment plant, the ^{235}U concentration being low following the fissioning process; fission products, or high-level waste, which moves to storage or disposal; and plutonium, which is recycled to the reactor via mixed oxide fabrication. Mixed oxide refers to a mixture of UO_2 and PuO_2.

Fissionable Materials as Large-Scale Energy Sources

We shall now match the fission reaction against the four criteria of a large-scale energy source given earlier in this chapter. On the first requirement, that of a high specific energy (Q-value), there is no question that the fission reaction meets it. Likewise, the requirement of a satisfactory mechanism for conducting the reaction at a controlled rate exists in nuclear reactors. The difficult questions on the fission reaction arise in the last two criteria: the magnitude and cost of the natural resources and the environmental effects. We shall discuss environmental effects in a later chapter. To judge the adequacy of uranium supply, we must look at future projections. Table 3.2 gives an ERDA 1975 projection of installed nuclear electric capacity and Table 3.3 gives an estimate of uranium resources.

Not shown in Table 3.3 are large quantities of very low-grade uranium in shale and ocean water, which could only be recovered at high cost (\$100–\$500/lb-U_3O_8). Even the U. S. uranium resource estimates given in Table 3.3 are the subject of considerable disagreement among experts. One can find other esti-

Table 3.2 A Projection of Total Electric and Nuclear Electric Capacity in the United States

YEAR	TOTAL CAPACITY (MW(e))	NUCLEAR CAPACITY (MW(e))	CUMULATIVE NATURAL URANIUM (METRIC TON)	CUMULATIVE NATURAL URANIUM (SHORT TON U_3O_8)
1975	496	38		
1980	620	75	73,700	95,800
1985	800	185	218,700	284,300
1990	1040	340	485,400	631,000
1995	1350	545	874,000	1,136,100
2000	1750	800	1,360,900	1,769,000

Table 3.3 An Estimate of Uranium Resources in Explored Areas of United States (Short tons of U_3O_8, cumulative)

COST OF PRODUCTION ($/lb – U_3O_8)	REASONABLY ASSURED RESERVES	ESTIMATED ADDITIONAL RESOURCES	TOTAL
<$8	200,000	530,000	730,000
<$10	315,000	960,000	1,275,000
<$15	420,000	1,530,000	1,950,000
<$30	600,000	2,900,000	3,500,000

mates ranging from 50% to 200% of these values. Because important policy questions depend on this information, there will be a large effort to narrow this range in the next decade.

We can compare the values in Tables 3.2 and 3.3 to see if the uranium resource estimates match with the reactor installation projections. For example, it is estimated that by the year 2000 we will need 1,769,000 short tons (2000 pounds per ton) of U_3O_8. From Table 3.3, we can see that this quantity has not yet been identified in higher grade ores, that is, those yielding U_3O_8 at a cost below $30/lb-$U_3O_8$. It is estimated that this quantity will be discovered by additional exploration. Table 3.2 does not tell the whole story of uranium requirements, however. Many of the reactors of the 800,000 MW(*e*) capacity as of year 2000 will have additional uranium requirements over their lifetime. Approximately 640,000 MW(*e*) of that capacity are present-generation converter reactors. Each 1000 MW(*e*) capacity of these reactors requires 3500 to 5000 short tons of U_3O_8 over its lifetime, depending on how it is operated. That means that their construction represents a committed uranium requirement of 2.25 to 3.2 million tons of U_3O_8. From our current information, therefore, it would appear that our uranium resources are by no means excessive relative to our plans. Current thinking is that installation of breeder reactors, which have a very low demand for natural uranium, will have to begin in the latter part of this century if fission energy is to provide an important share of U. S. energy in the next century.

FUSION

The third class of nuclear reactions that offers possibilities as a large-scale energy source is fusion. In fusion reactions, a more stable pair of nuclei is formed from the reaction of two light nuclei. The pair that is formed possesses a higher binding energy. Figure 2.3 indicates the rapid rise in binding energy with mass num- for the very light nuclides. This physical fact forms the basis for a series of ne reactions known as fusion reactions.

The Fusion Reaction

The first requirement of a fusion reaction, if it is to be an energy-source candidate, is a positive Q-value. On that basis the reaction

$$^2_1\text{H} + ^2_1\text{H} \longrightarrow ^4_2\text{He}$$

$$Q = 2 \times 13.1359 - 2.4248 = 23.85 \text{ MeV}$$

would look very good. Unfortunately, as we have learned from experiments, the reaction does not occur. One surmises that the $^4_2$He nucleus has no way of accommodating the 23.85 MeV of energy, internally or externally. Two deuterium nuclei do in fact react, however. The reaction can go to two different product pairs.

$$^2_1\text{H} + ^2_1\text{H} \longrightarrow ^3_2\text{He} + n; \qquad Q = 3.25 \text{ MeV}$$

$$^2_1\text{H} + ^2_1\text{H} \longrightarrow ^3_1\text{H} + ^1_1\text{H}; \qquad Q = 4.02 \text{ MeV}$$

These are called the DD (for deuterium) reactions. Each occurs about 50% of the time. Another candidate reaction is the DT reaction between deuterium and tritium.

$$^2_1\text{H} + ^3_1\text{H} \longrightarrow ^4_2\text{He} + n; \qquad Q = 17.59 \text{ MeV}$$

Still another is

$$^2_1\text{H} + ^3_2\text{He} \longrightarrow ^4_2\text{He} + ^1_1\text{H}; \qquad Q = 18.35 \text{ MeV}$$

We could go on finding positive-Q-value reactions between light nuclei. However, it will be more instructive if we stop to look at the problem of conducting these reactions.

Figure 3.12 gives a picture of how two deuterium nuclei might interact. Both nuclei are positively charged. Consequently, most of the time when they approach each other, they are mutually repulsed. If, however, one or both has sufficient kinetic energy to overcome this repulsion (the coulomb barrier), then a reaction can occur. In Figure 3.12 we show a compound nucleus breaking up into ^{3}He and a neutron. Note that, according to Eq. 2.8, the neutron will have 75% of the Q-value as its kinetic energy, and the ^{3}He nucleus will have 25%. (This neglects any initial energy of the deuterium nuclei.)

The important point of Figure 3.12 is that the reacting nuclei must possess some minimum relative kinetic energy. Here is the crucial difference between fission and fusion. The important fission reactions are those that either require no minimum energy of the fission-inducing neutron or require energy well below that which the neutron possesses as it emerges from the fission reaction. In short, the energy requirement for the reactants in the fission reaction does not

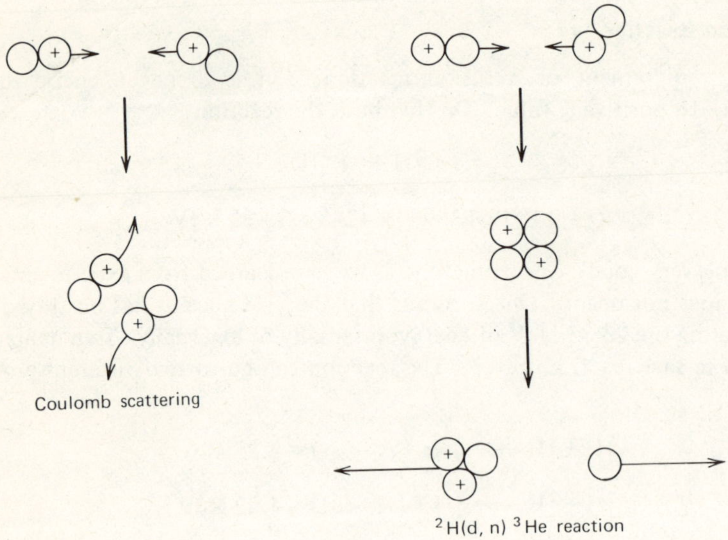

Figure 3.12 Two deuterium nuclei reacting.

present an obstacle to the conduct of the reaction. That is not the case with fusion.

How much energy is required to bring about a fusion reaction? Figure 3.13 addresses this question. It shows that for either the DD or the DT reaction, the reacting nuclei must have at least 10 keV relative kinetic energy. The chances of

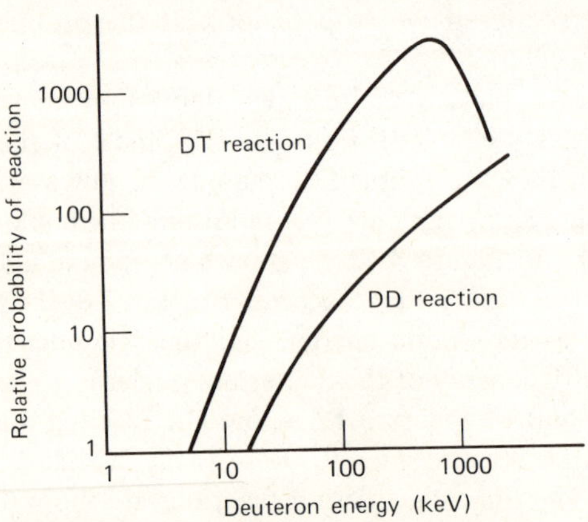

Figure 3.13 The effect of particle energy on fusion reaction probabilities.

reaction are much higher at 100 keV. The DT reaction probability peaks at 150 keV, while that of the DD reaction continues to rise. It must be remembered that even at these energies, a scattering reaction is 1000 to 10,000 times more probable than the fusion reaction.

The coulombic forces, which cause this energy requirement, increase as the product of the Z values of the two interacting nuclei. This is the reason that the leading-candidate fusion reactions involve hydrogen ($Z = 1$) nuclei. It is not too useful to look at heavier reactants, because they will require greater energy. Otherwise, for example

$$^{7}_{3}\text{Li} + ^{1}_{1}\text{H} \longrightarrow ^{8}_{4}\text{Be} \longrightarrow 2\,^{4}_{2}\text{He}; \qquad Q = 17.35$$

could be an interesting fusion reaction. In fact, it cannot be ruled out, but the DT and DD reactions appear easier to conduct. The DT reaction is receiving most attention in fusion development work at the present time.

Example 3.10
If all the deuterium (^{2}H) in one m^3 of water reacts in the DD reaction, how much energy is released?

Solution: Deuterium comprises 0.0148 a/o of natural hydrogen. The density of water is 1000 kg/m^3. The number of water molecules in a cubic meter is

$$1000 \times \frac{1000}{18} \times 6.023 \times 10^{23} = 3.346 \times 10^{28} \text{ molecules/m}^3$$

We can use the value of 18 (16 + 2) for the molecular weight of water, because it contains so predominantly the hydrogen-1 isotope. The number of deuterium atoms is

$$3.346 \times 10^{28} \times 2 \times 0.000148 = 9.905 \times 10^{24} \text{ atoms/m}^3$$

As noted in the text, the DD reaction goes two ways with about equal frequency. We can take the average Q-value.

$$Q = \frac{3.25 + 4.02}{2} = 3.64 \text{ MeV}$$

The energy to be obtained from reacting all the deuterium is

$$\frac{9.905 \times 10^{24}}{2} \times 3.64 \times 1.602 \times 10^{-19} = 2.884 \times 10^{6} \text{ MJ/m}^3$$

or 2884 MJ/kg-water. We divide by two because each reaction takes two nuclei of deuterium. A typical value for the combustion of coal is 29 MJ/kg.

The Fusion Reactor

The energy requirement for the DT reaction, say 10 keV, is very small compared to the 17.59 MeV released in the reaction. Moreover, there is no problem in bringing about a fusion reaction. There are different accelerators on the market that can impart an energy of 150 keV to a deuteron nucleus, which then impinges on a tritium target. The DT reaction occurs. These accelerators are a popular source of neutrons in many laboratories. The problem is that, due to the scattering reactions, it is necessary to accelerate many—more than a thousand—deuterons into the target to get one reaction. The others scatter and either lose sufficient energy to cause the reaction or leave the target. The result is that more energy is spent on the acceleration of the deuterons than is obtained from the DT reactions. The reaction is not self-sustaining.

What is needed is a method of holding the deuterium and tritium nuclei in a reacting zone long enough for the fusion reaction to have a chance to occur even after thousands of scattering reactions. This would be the case if we could have a reacting zone in which the nuclei possessed the necessary energy, say 10 keV, by virtue of their thermal motion. Such a reacting mixture would have a spectrum of velocities; since it would be a perfect gas, it would be a Maxwellian spectrum (see Distribution Functions, Chapter 4). The average energy would be equal to $\frac{3}{2}kT$, where k is the Boltzmann constant. For an average energy of 10 keV,

$$T = \frac{2\overline{E}}{3k} = \frac{2 \times 10 \times 10^3 \times 1.602 \times 10^{-19}}{3 \times 1.3805 \times 10^{-23}} = 77 \times 10^6 \ {}^\circ\mathrm{K}$$

Such a gas would indeed be hot. It would be so hot that all molecules would be atoms and all atoms would be ions. Note that in hydrogen, ions and nuclei are identical. An electrically neutral mixture of ions and electrons is called a plasma; this is the state envisioned for the reacting mixture in a fusion (often called a thermonuclear) reactor.

The requirement of any fusion reactor is to confine a reacting plasma—deuterium and tritium, for example—at a sufficiently high temperature and density and for sufficient time, permitting enough fusion reactions to take place to obtain a net energy gain over and above that energy required to create the reacting conditions in the first place. There can be some trade-off between the values of temperature, density, and time. For example, if the density of a reacting plasma could be doubled, the time that reacting nuclei would have to remain in the reaction zone could be cut in half.

There are two rather different approaches to attainment of a fusion reactor. One involves electromagnetic confinement; the other involves compression by laser beams or by beams of charged particles. These two approaches will be described briefly.

Example 3.11
An accelerator that imparts 150 keV kinetic energy to a beam of deuterons has a rated beam current of 1 mA. The beam strikes a tritium-containing target, resulting in the production of 10^{12} neutrons per second from the DT reaction. What percent of the beam deuterons undergoes the DT reaction?

Solution: The charge on a single deuterium nucleus or ion is 1.6021×10^{-19} coulomb. A current of 1 mA, therefore, represents

$$\frac{10^{-3}}{1.6021 \times 10^{-19}} = 6.24 \times 10^{15} \text{ deuterons/s}$$

The percent undergoing the DT reaction is

$$\frac{10^{12}}{6.24 \times 10^{15}} \times 100 = 0.016\%$$

Electromagnetic Fusion Reactors

Because a fusion reaction plasma must be at a very high temperature, it is not feasible to consider conventional confinement such as by the walls of a vessel. The plasma, however, does consist of electrons and positively charged ions or nuclei, a fact that offers the possibility of confinement in a magnetic field. Figure 3.14 shows the path of individual electrons and ions in a magnetic field. This behavior is at the heart of electromagnetic confinement reactors. Over a period of 25 years there have been many experimental electromagnetic confinement devices that differ in several features such as field strength, field geometrical configuration, magnetic flux lines closed or open, constant field or with changes on some pulsed or oscillatory mode, and so on. It has been discovered that the plasma state has characteristics leading to instabilities that could not be predicted

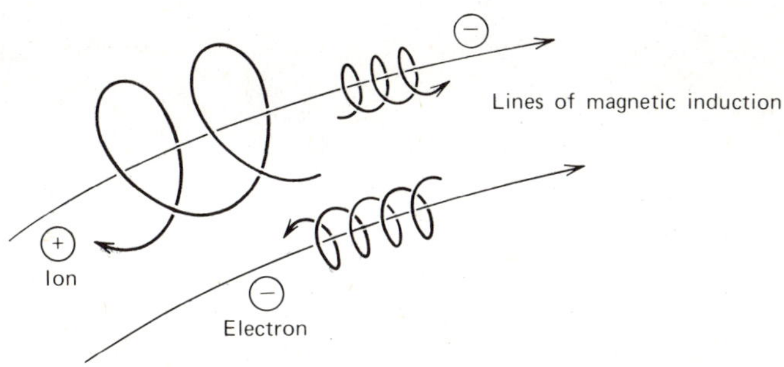

Figure 3.14 The influence of a magnetic field on the path of charged particles.

simply by summing the behavior of individual electrons and ions. The experimental programs, primarily in the Soviet Union, the United Kingdom, and the United States, have represented an extraordinary effort to realize the achievement of a fusion reactor. The task is proving very difficult. Although there have been steady gains in terms of high plasma densities and long reacting times, the achievement of the goal of a fusion reactor still lies in the indefinite future.

To date, the closest approach to the achievement of a fusion reactor has been the Russian design known as the Tokamak. Figure 3.15 is a simplified schematic of the Tokamak. The plasma is confined in a circular ring by the magnetic field created by the windings around the ring. There is further electromagnetic confinement, however. The plasma, being a conductor, is made to act as the secondary loop of a transformer, and a current is induced in the plasma by passing an alternating current through a primary winding, as shown in Figure 3.15. The current flowing in the plasma creates its own magnetic field, which serves to compress the plasma region and also to add thermal energy to the plasma by resistive heating.

Figure 3.16 gives a sketch of a controlled thermonuclear reactor based on the DT reaction. The particular characteristics of the reaction

$$^{2}H + {}^{3}H \longrightarrow {}^{4}He + n$$

greatly influence the design. In the first place, 80% of the energy of the reaction appears as the kinetic energy of the neutron (Eq. 2.8). These neutrons must be slowed down and their kinetic energy recovered as thermal energy at high temperature. A second characteristic is that the reaction requires tritium, an isotope

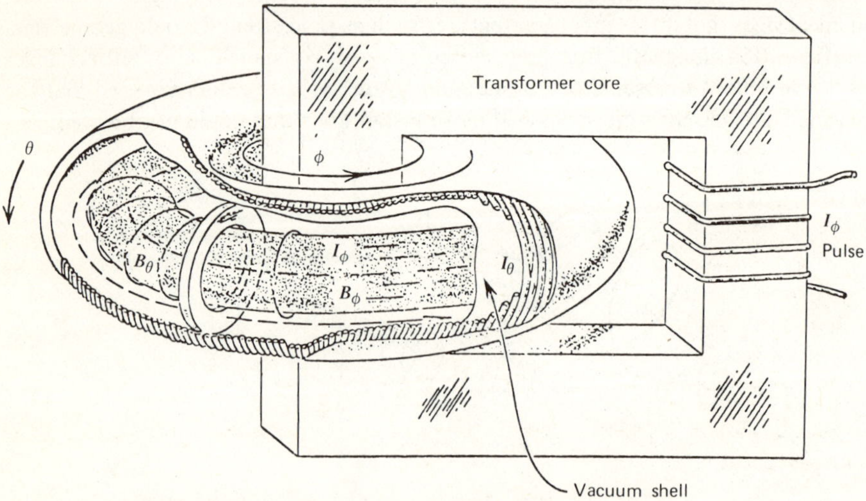

Figure 3.15 Schematic of the Tokamak fusion reactor principle.

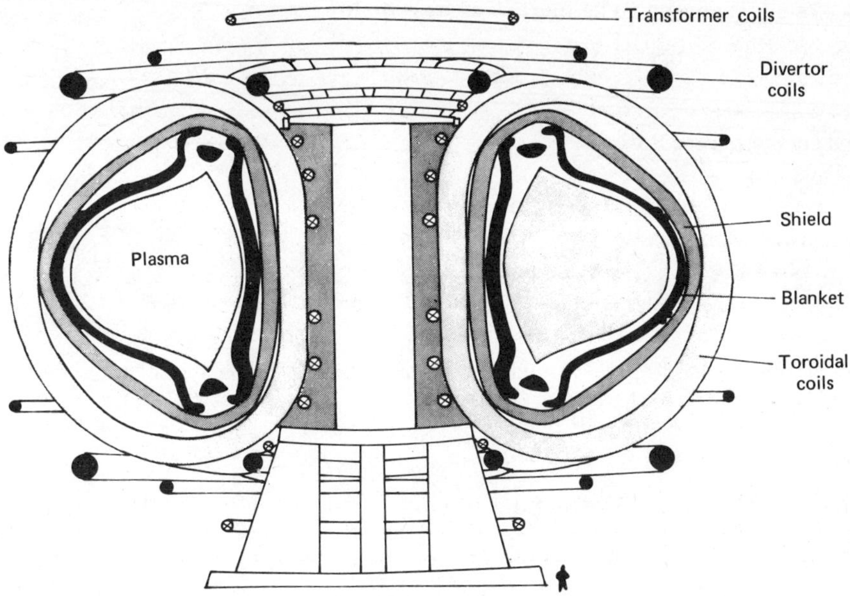

Figure 3.16 Fusion reactor concept based on Tokamak principle and the DT reaction. From G. L. Kolcinski, R. W. Conn et al., "VWMAK-I, The Wisconsin Tokamak Reactor Design," Report VWFDM-68, University of Wisconsin, March 1974.

which, unlike deuterium, is not found in nature. It must be produced. The reaction currently favored to produce it is

$$^{6}\text{Li} + n \longrightarrow {}^{4}\text{He} + {}^{3}\text{H}$$

As indicated in Figure 3.16, the plasma region of a DT fusion reactor is surrounded by a blanket region in which the neutron energy is recovered and tritium is produced. Outside of this blanket is located the magnet that produces the field that confines the plasma.

Example 3.12

Assume that all the tritium for a fusion electric nuclear station producing 6.57×10^{9} kWh(e) per year at a thermal efficiency of 40% is produced by the reaction

$$^{6}\text{Li} + n \longrightarrow {}^{4}\text{He} + {}^{3}\text{H}$$

How many kilograms of ^{6}Li are consumed annually?

Solution: We will assume that the thermal energy required is produced by the two reactions

$$^{2}\text{H} + {}^{3}\text{H} \longrightarrow {}^{4}\text{He} + n; \qquad Q = 17.59 \text{ MeV}$$

$$^{6}\text{Li} + n \longrightarrow {}^{4}\text{He} + {}^{3}\text{H}; \qquad Q = 4.78 \text{ MeV}$$

on a one-to-one ratio. The thermal energy required is

$$\frac{6.57 \times 10^9}{0.4} \times 3600 \times 1000 = 5.91 \times 10^{16} \text{ J/y}$$

The energy per reaction pair is

$$(17.59 + 4.78) \times 1.602 \times 10^{-13} = 3.584 \times 10^{-12} \text{ J}$$

Therefore, we need

$$\frac{5.913 \times 10^{16}}{3.584 \times 10^{-12}} = 1.650 \times 10^{28} \text{ reaction pairs per year}$$

There must be 1.650×10^{28} nuclei of ^{6}Li consumed. The mass of ^{6}Li is

$$\frac{1.650 \times 10^{28}}{6.023 \times 10^{23}} \times 6 \times \frac{1}{1000} = 164 \text{ kg/y}$$

Since ^{6}Li is only 7.5% of natural lithium, considerably more lithium must be supplied.

Inertial Confinement Fusion Reactor (Laser Fusion)

A radically different approach to a fusion reactor is to supply energy very rapidly to a small amount of reacting materials, causing some reactions to occur before the released energy disperses the material. This concept is at least imaginable, because if energy can be transferred to the surface of a tiny sphere (perhaps 1 mm in diameter) by means of a high-powered laser pulse (possibly 10^5 J lasting 10^{-9} s), the outer atoms will be ionized, and some of the resulting plasma will ablate, or leave the surface, at high velocity. There will thus be a momentum transfer to the sphere, which will create very high pressures (10^8 to 10^9 Pa). The inner portion of the sphere will reach fusion reaction temperatures. The theory is that before the sphere has a chance to disassemble, a number of fusion reactions will occur. There would be a small explosion with a release of nuclear energy. A reactor would be a device in which dozens of such explosions would occur every second. As in the electromagnetic confinement devices, the neutrons must be captured and the thermal energy converted to electrical energy.

Energy Source Potential of the Fusion Reaction

We can now test the fusion reaction against the criteria, previously discussed, for a large-scale energy source. The fusion reactions have a very large and

and positive Q-value. The specific energy release in the DT reaction is one billion megajoules per kilogram of deuterium. Fusion reactions pass this test easily. The major point of failure is the current fact that a suitable reactor for conducting the reaction on a self-sustaining basis and converting the released energy to electricity does not exist. On the other hand, no inherent reason why such a reactor cannot be developed has ever been found. So the work goes on.

The large incentive for developing the fusion reactor is to be found in the abundance of its most important raw material, deuterium. Since deuterium comprises 0.015% of all the hydrogen in nature, it is very abundant indeed. (See Example 3.10.) One way of expressing that abundance is that the fusion energy represented by the deuterium in the oceans is equal to 100 billion times the world's energy consumption in 1973. Furthermore, satisfactory methods for extracting deuterium from water exist. It is clear that should a fusion reactor be developed, deuterium availability will not limit the energy that can be produced. It would be limited by the availability of some other material, such as the lithium used to produce tritium, or the thermal loading of the atmosphere, and so on.

The environmental impact of fusion appears to be less than fission, although by no means negligible. A fusion reactor would contain large quantities of radioactive tritium, which must be contained. Furthermore, a large amount of radioactivity will be produced in the reactor components due to the (n, γ) and other neutron reactions induced by the neutrons produced in the fusion reactions.

PROBLEMS

3.1 How many grams of ^{210}Po would be required now to supply 100 watts of heat 1 year from now?

3.2 A radioisotope having a half-life of one hour is counted for a period of 15 minutes. By what factor must the average counting rate over this time interval be multiplied to obtain the counting rate at zero time?

3.3 If 1 g of freshly prepared radium is sealed in a platinum container of 1 cm^3 free volume, estimate how much the gas pressure will increase in a year as a result of the helium formed from radioactive disintegration.

3.4 A neutron is itself radioactive, having a half-life for decay to a proton and electron of about 12 minutes. If a neutron having a kinetic energy of 1 MeV is traveling in a perfect vacuum, what is the probability that it will travel 93 million miles (1.5×10^{11} m), the distance to the sun, without decaying?

3.5 It is estimated that by the year 2000 there will be 4.5×10^{12} kWh(t) per year of fission energy generated in the United States. Cesium-137 is a fission product that is formed in 6% of fissions. How many pounds and how many curies of ^{137}Cs would be formed in the United States in the year 2000?

3.6 Suppose that the annual production of ^{137}Cs in the above problem is assembled as pure metallic cesium (density = 1.87 g/cm^3). What would be the rate of heat generation (W and W/cm^3)?

3.7 Assume that when the earth was created, there was the same quantity of ^{235}U as ^{238}U. On the basis of that hypothesis, how old is the earth?

3.8 If a nuclear fission power plant generates 1000 MW(e) at a thermal efficiency of 33% and operates at a 80% capacity factor, how much mass is converted to energy each year? If a coal-fired plant has the identical ratings, how much mass is converted?

3.9 In recent testimony before the Joint (Congressional) Committee on Atomic Energy (JCAE), it was stated that a certain nuclear fission power reactor in California puts some 2400 Ci (curies) of tritium (^3H) into the atmosphere annually. The source of about one half of this tritium is a rarely occurring fission reaction. The reactor in question has a thermal power of 1400 MW(t) and operates 80% of the time. Calculate the apparent percentage of fissions that yield a tritium nucleus.

3.10 A hypothetical isotope, ^{100}X, has a decay scheme represented diagrammatically:

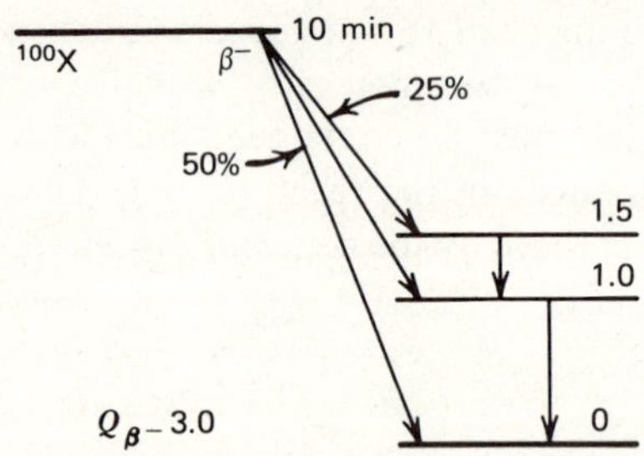

(a) How much beta energy is produced in units of MeV per average decay? MeV per second per gram?

(b) Give the energy and number per average decay of each gamma photon emitted.

3.11 A national average price for coal runs around 80 cents/million Btu. What price would uranium-235 have to have to give the same thermal energy cost? Use the most appropriate conversion factors you can recollect. Perhaps it will help to know that one watt-second is equivalent to 0.000948 Btu.

3.12 In the fuel cycle of a fission reactor, it is possible for ^{235}U to become contaminated with ^{232}U. What limit in concentration of ^{232}U in ^{235}U (ppm) must be specified to keep the decay rate of each the same? (There is nothing magic about having the decay rates equal; it is just that such a mixture would be only twice as radioactive as pure ^{235}U.)

3.13 Krypton-85 is a fission product that causes much concern. Being both a gas and also resistant to chemical reaction, it promises to be difficult to keep contained. It also has a rather long half-life. If it is contained in purified form, there may be a heating problem. Estimate the heat release in watts/g-^{85}Kr. If the fission yield of this isotope is 0.3%, how many grams will be formed in one year by a 1000-MW(e) nuclear electric plant?

3.14 In the case of isotopes that decay by spontaneous fission as well as by another mode (such as ^{240}Pu), one will often find a value of half-life given for each mode. A little thought will reveal some conceptual difficulty with a material having two half-lives. Explain what you think is meant by the two values. To make the discussion more quantitative, estimate the number of spontaneous fission neutrons given off by ^{240}Pu. Consider a quantity that at time zero amounts to 1 g ^{240}Pu. Estimate the spontaneous fission neutrons at $t = 0$ and $t = 25{,}000$ years.

3.15 If one could conduct the reaction

$$^{2}\mathrm{H} + {}^{2}\mathrm{H} \longrightarrow {}^{1}\mathrm{H} + {}^{3}\mathrm{H}$$

and then obtain the deuterium by extraction from a regular water supply, how much energy could be extracted from a cubic meter of water?

3.16 One "fusion" reaction considered to have a possibility as an eventual commercial energy source is

$$^{2}\mathrm{H} + {}^{2}\mathrm{H} \longrightarrow {}^{3}\mathrm{He} + n$$

(a) Calculate the energy released in this reaction in MeV per reaction and in Btu per pound of deuterium (^{2}H) reacting.
(b) In this reaction, compute the fraction of the nuclear energy released that appears initially as the kinetic energy of the neutron.
(c) If we can make an 80% recovery of the deuterium in natural water, how many gallons of water must be processed each day to supply a 4-million-kW(t) power plant with deuterium?
(d) Assuming that the deuterium costs $150 per pound and that the above (c) plant produces 1 million kW(e), what is the deuterium contribution to electric energy cost in mills/kWh? A mill is 10^{-3} $.

3.17 A radioisotope decays with emission of a negative beta. Two gamma photons of 1 and 2 MeV occur on each decay. The Q-value for the beta decay is 5 MeV. Sketch a decay scheme approximately to scale.

3.18 How long would it take a sample of ^{137}Cs (half-life, 30 years) to decay to 0.1% of some initial value?

3.19 A reaction on which one might base a power system is

$$^{13}_{7}\mathrm{N} + {}^{1}_{1}\mathrm{H} \longrightarrow {}^{12}_{6}\mathrm{C} + {}^{4}_{2}\mathrm{He}$$

How much energy would be obtained in terms of Btu per pound of naturally occurring nitrogen? Discuss the advantages and disadvantages of this reaction for a practicable power reactor. Compare with those reactions under more active consideration for a fusion reactor.

3.20 Cerium-144 is a fission product formed in 6.0% of fissions. Consider a fission power reactor operating at 3200 MW(t) steadily for one year. At that time, how many grams and curies of ^{144}Ce would be in the fuel?

3.21 Natural uranium is purchased from mining companies in the form of U_3O_8 (called "yellow cake"). A current market price is $20 per pound of yellow cake. If the utility can fission 75% of the ^{235}U contained in this material, what does this cost represent when expressed in cents/million Btu thermal?

3.22 A design objective for fuel in current-generation utility reactors is to achieve an energy yield of 33,000 MWd(t)/MTU (megawatt-days thermal per metric ton of uranium) charged to the reactor. The fuel is then discharged from the reactor for processing. What weight of fission products (kg per metric ton of original uranium) is in the discharged fuel? If the conversion efficiency of thermal to electric energy in the reactor is 32%, what unit costs of caring for the fission products ($/kg) would equal 1 mill/kWh of electricity produced?

3.23 A certain radioisotope is known to emit negative beta particles. The following information on its decay scheme is available:

β^-	γ
0.9 MeV max	0.158 MeV 20%[a]
0.47 MeV max	0.442 MeV 75%[a]
0.31 MeV max	0.601 MeV 20%[a]

Construct a plausible decay scheme sketch. Include in your sketch as much quantitative data of the kind presented in Figure 3.4 as the above information permits.

[a]Means % of decays in which a photon of that energy is observed.

Chapter 4

The interactions of radiation in matter

We have now touched on all classes of nuclear reactions that are of interest as large-scale energy sources. In order to gain a deeper understanding of energy systems based on these reactions, we must have a more quantitative understanding of the behavior of nuclear photons and particles (collectively called "radiation") that cause or are produced by these reactions. We need to know the kinds of reactions they may cause and the probability of each kind. We need to know how and where their kinetic energy is converted to thermal energy within the medium in which they are traveling. All of these subjects are generally grouped under the topic of the interaction of radiation with matter.

The radiation particles of interest to us are the neutral particles (neutrons and gamma photons) and the charged particles (electrons, protons, alpha particles, and so on). The neutral particles, by virtue of the greater variety of their interactions and the greater distance of their travel, will receive most attention.

NEUTRON REACTIONS

We wish to consider the progress of a neutron traveling in a homogeneous medium. By homogeneous, we mean simply that the probability of an interaction with the medium is not a function of the position or direction of the neutron. There is also the concept of the medium having discrete reaction sites separated by free space. A homely analogy (in two dimensions) is to liken it to a person walking through a rockstrewn field rather than through a bog or a swamp. The neutron flies along, its direction and speed undisturbed, until it encounters a discrete reaction site.

To begin with, we know nothing about the size or number of these reaction sites.

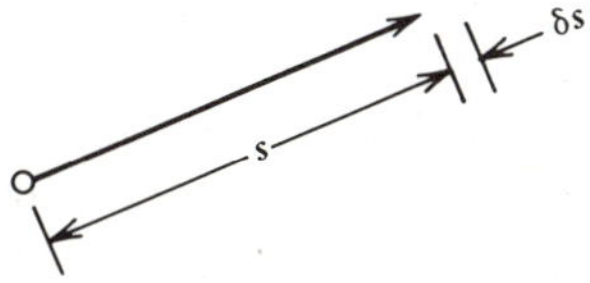

Given that a particle is at any arbitrary point, 0, we state that the probability of a future reaction is unrelated to its past history. The probability of a reaction in any distance, s, is only a function of s. We will express this probability as a MacLaurin expansion in s,

$$P_R(s) = \alpha + \beta s + \gamma s^2 + \cdots \tag{4.1}$$

Since a reaction cannot occur with zero travel, $\alpha = 0$. We can always find a value of s sufficiently small that

$$P_R(\delta s) = \beta \delta s = \Sigma \delta s \tag{4.2}$$

The change of β to Σ is simply to conform to conventional usage. The probability that the neutron will not have undergone a reaction in distance δs is

$$P_{NR}(\delta s) = 1 - \Sigma \delta s \tag{4.3}$$

Note that there are only two possibilities, a reaction or no reaction. Therefore we call this a binary situation. Because the probability of a future event is independent of past history, we can write the probability of no reaction in distance $s + \delta s$ as the product of the probabilities of no reaction in distances s and δs.

$$P_{NR}(s + \delta s) = P_{NR}(s) P_{NR}(\delta s) = P_{NR}(s) \; [1 - \Sigma \delta s] \tag{4.4}$$

Rearranging terms,

$$\frac{P_{NR}(s + \delta s) - P_{NR}(s)}{\delta s} = -\Sigma P_{NR}(s) \tag{4.5}$$

In the limit as $\delta s \to 0$

$$\frac{dP_{NR}}{ds} = -\Sigma P_{NR} \tag{4.6}$$

Integrating

$$P_{NR}(s) = e^{-\Sigma s} \tag{4.7}$$

and

$$P_R(s) = 1 - e^{-\Sigma s} \tag{4.8}$$

The constant, Σ, is called the macroscopic cross section. In the field of gamma-ray interactions it is called the linear attenuation coefficient and designated μ. The usual unit of these quantities is the reciprocal centimeter, cm^{-1}.

Example 4.1

In a medium in which Σ is 1.0 cm^{-1}, what is the probability that a neutron will travel at least 10 cm without an interaction? What is the probability that it will undergo a reaction at some point between 10 cm and 11 cm?

Solution: The probability of traveling 10 cm without an interaction is, by Eq. 4.7

$$P_{NR}(10) = e^{-10\Sigma} = e^{-10} = 4.54 \times 10^{-5}$$

We obtain the probability that it will have a reaction between 10 cm and 11 cm by computing first the probability of traveling 11 cm without a reaction.

$$P_{NR}(11) = e^{-11\Sigma} = e^{-11} = 1.67 \times 10^{-5}$$

Now, the probability of a reaction in the interval is just the difference in these two probabilities, or

$$4.45 \times 10^{-5} - 1.67 \times 10^{-5} = 2.78 \times 10^{-5}$$

The Mean Free Path

A concept that is useful in judging the nature of a shielding problem is the mean free path, or the average distance a particle will travel in a medium without reacting. In order to get this value, we must know the probability that a particle will react in the distance interval between s and $s + ds$. That is the product of the probability of *not* reacting in distance s and of reacting in distance ds. These we obtain from Eq. 4.2 and 4.7. The product is

$$e^{-\Sigma s}\, \Sigma ds$$

In order to get the average or mean distance traveled, we must multiply each value of s by the probability that it would be the distance. In a discrete value problem, for example, we would multiply the fraction of persons in a population having a given age *by* that age, repeating for each age and summing. For example:

$$\begin{array}{ll} 0.25 \times 20 = & 5.0 \\ 0.40 \times 21 = & 8.4 \\ 0.30 \times 22 = & 6.6 \\ \underline{0.05} \times 23 = & \underline{1.15} \\ 1.00 & 21.15 \text{ average age} \end{array}$$

We could represent this averaging process

$$\bar{a} = \sum_{0}^{\infty} af(a)$$

In the mean free path problem, the possible values of s are continuous, not discrete, and we write

$$\bar{s} \equiv \lambda = \int_0^{\infty} se^{-\Sigma s}\, \Sigma ds = \frac{1}{\Sigma} \qquad (4.9)$$

Just as the sum of the fractions had to be 1.0, so the integral of our distribution function

$$\int_0^\infty e^{-\Sigma s}\,\Sigma ds$$

must equal 1.0, which it does. One says that the function $e^{-\Sigma s}\,\Sigma$ is normalized. The mean free path, therefore, is the reciprocal of the macroscopic cross section. It is traditionally designated λ.

Example 4.2

If a neutron with a kinetic energy of 1 eV is moving in a medium with $\Sigma = 0.1\ \text{cm}^{-1}$, what is the expected value of the time to its first interaction?

Solution: The mean free path or the expected value of the distance to the first interaction is

$$\lambda = \frac{1}{\Sigma} = 10\ \text{cm} = 0.010\ \text{m}$$

The neutron is traveling with speed

$$v = \left(\frac{2E^1}{m}\right)^{1/2} = \left(\frac{2 \times 1 \times 1.602 \times 10^{-19}}{1.009 \times 1.6604 \times 10^{-27}}\right)^{1/2} = 13{,}830\ \text{m/s}$$

The expected value of time to first interaction is, therefore,

$$\frac{0.010}{13{,}830} = 7.23 \times 10^{-7}\ \text{s} = 0.72\ \mu\text{s}$$

Distribution Functions

The previous section deals with a class of phenomena occurring so often in the nuclear field that it deserves some special note. It often happens that we cannot assign a definite value to some variable (for example, the distance a neutron travels before reacting) but can only speak of its value in terms of probabilities. When the values that the variable can assume are continuous, we have a distribution function, $D(x)$, which is defined as a function such that $D(x)\,dx$ is the probability of the variable's assuming a value between x and $x + dx$. Note that the probability of the value being just x is zero. The probability of a reader being 20 years old is zero. There is no time interval expressed. Of course, a person

[1] Note that although in Chapter 2 we designated kinetic energy as T, we will now follow the more conventional designation, E. The use of T is particularly helpful when one wishes to distinguish the kinetic energy from the total energy (mc^2). However, in dealing with neutrons, the use of E for kinetic energy should not cause confusion.

will give an age of 20, but what is *really* meant is that her age is between 20 and 21 years or between 19.5 and 20.5. The interval is understood.

In the case of the distance a neutron will travel before reacting, the distribution function is

$$D(s) = \Sigma e^{-\Sigma s} \tag{4.10}$$

This is a function that has its maximum at $x = 0$, and declines monotonically. The *expected* value of s for a single neutron—or the *average* value, in the case of a population of neutrons—is given

$$\bar{s} = \int_0^\infty sD(s)\,ds \tag{4.11}$$

The integral

$$\int_0^\infty D(s)\,ds$$

should have the value of one, because it covers all possible values of s that the variable can assume.

When we deal with populations that show a distribution of some variable (such as energy or age), we sometimes refer to the distribution function as a spectrum. One distribution function often encountered is the Maxwellian distribution function. It is exhibited by the molecules or atoms of a perfect gas and also by neutrons that are in thermal equilibrium with the medium. We can write this function

$$n(E) = \frac{2\pi n}{(\pi kT)^{3/2}} E^{1/2}\, e^{-E/kT} \tag{4.12}$$

where $n(E)\,dE$ is the fraction of the population possessing energies between E and $E + dE$. The constants in Eq. 4.12 are adjusted so that

$$\frac{1}{n}\int_0^\infty n(E)\,dE = 1$$

The value n is the number of the entire population.

Example 4.3

Given the Maxwellian distribution as defined by Eq. 4.12, compute the most probable value and the average value for the energy of a population of particles having such a spectrum. What is the probability that a member of this population would have an energy greater than $5kT$?

Solution: The most probable value is, by definition, the value of E at which $n(E)$ is a maximum. We can obtain this value by setting $dn(E)/dE$ equal to zero.

$$\frac{dn(E)}{dE} = \frac{\pi n}{(\pi kT)^{3/2}} E^{-1/2} e^{-E/kT} - \frac{2n}{\pi^{1/2}(kT)^{5/2}} E^{1/2} e^{-E/kT} = 0$$

$$E^{-1/2} = \frac{2E^{1/2}}{kT}$$

$$E = \frac{kT}{2}$$

The most probable value of energy is one half kT.

The average value of energy is obtained by multiplying $n(E)\,dE$ times E and integrating as in Eq. 4.11.

$$\overline{E} = \frac{1}{n} \int_0^\infty E\, n(E)\, dE$$

$$= \frac{2\pi}{(\pi kT)^{3/2}} \int_0^\infty E^{3/2}\, e^{-E/kT}\, dE$$

The solution to the definite integral in this equation can be obtained by integrating by parts to obtain the normal probability integral. The solution is

$$\int_0^\infty E^{3/2}\, e^{-E/kT}\, dE = \frac{3\pi^{1/2}}{4} (kT)^{5/2}$$

and

$$\overline{E} = \frac{3}{2}\, kT$$

The probability of a member of this population having an energy greater than $5kT$ is simply formulated as

$$P(E > 5kT) = \frac{1}{n} \int_{5kT}^\infty n(E)\, dE$$

This integral can be solved by integrating by parts and reducing it to either the error function, or the normal probability integral, plus an algebraic term. The value $P(E > 5\,kT)$ is 0.017. Note that this value could just as well be described as the expected value of the fraction of the population that has energies above $5\,kT$.

Reaction Rate of a Neutron Population

We are often concerned, not with the reaction detail of a single neutron, but with the reaction rate of a large number, or population, of neutrons. The most general case is that in which we have a large number of neutrons traveling with some mean speed, $\bar{v}$, in random directions. At some point in the reacting medium of macroscopic cross section Σ, we envision a concentration of n neutrons per cm^3. Then

$n\bar{v}$ = total distance traveled by neutrons in one cubic centimeter every second

$\frac{1}{\Sigma}$ = average distance to a single interaction

$$\therefore \frac{n\bar{v}}{1/\Sigma} = \Sigma n\bar{v} = \text{reactions/cm}^3 \cdot \text{s} \tag{4.13}$$

The product, $n\bar{v}$, is a very important one in neutron reactor terminology. It is called the neutron flux (or simply flux) and is designated ϕ. We shall be using this quantity frequently.

The Microscopic Cross Section

Although the macroscopic cross section of a material is the property that is measured experimentally, the data so obtained can be greatly extended if the macroscopic cross section can be related to a property of the individual nuclei. This property is an apparent reaction area exhibited by each nucleus. It is called the microscopic cross section. If we know the microscopic cross sections of the various constituent nuclei, we can calculate the macroscopic cross section of any medium whether or not it has been measured.

Figure 4.1 gives a schematic basis for relating the macroscopic to the microscopic cross section. We envision a neutron traveling in any given direction at any point in a homogeneous medium. We express the probability of the neutron undergoing a reaction in the next differential distance of travel, δs, by Eq. 4.2.

$$P_R(\delta s) = \Sigma \delta s \tag{4.2}$$

We can also express this probability in terms of some reaction area presented by all of the nuclei in the volume $A\delta s$, where A is an arbitrary area in Figure 4.1 normal to the neutron path.

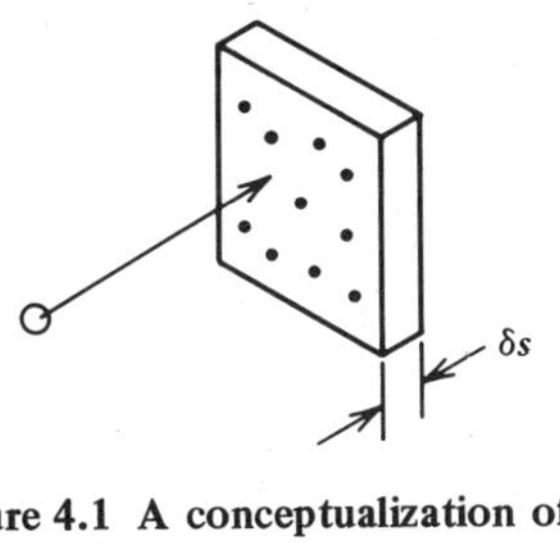

Figure 4.1 A conceptualization of the microscopic cross section.

$$\Sigma\delta s = \frac{\text{reaction area}}{\text{total area}}$$

If we ascribe a value of σ to the area of each nucleus and a value of zero to the area of the neutron, then the reaction area is the number of nuclei in volume $A\delta s$ times σ

$$\Sigma\delta s = \frac{N\sigma A\delta s}{A} = N\sigma\delta s \tag{4.14}$$

where N is the number of nuclei per unit volume. Note that it is important that δs be thin enough for the areas of the nuclei not to overlap. The relationship between macroscopic and microscopic cross section follows directly from Eq. 4.14.

$$\Sigma = N\sigma \tag{4.15}$$

In a material of a single nuclear species, the value of N can be obtained from the density, ρ, and the mass number, A.

$$N = \frac{6.023 \times 10^{23}\ \rho}{A} \tag{4.16}$$

In this case, ρ must be in units of g/cm^3, since ρ/A must be in moles/cm^3. The quantity, N, is often referred to as the number density.

In Eq. 4.15 the units of σ must be in cm^2. In data compilations the common unit is the barn, which is 10^{-24} cm^2. When one uses the values of σ in barns, it is common to express the number density in units of 10^{24} cm^{-3}. Combining Eq. 4.15 and 4.16, we write

$$\Sigma = \frac{0.6023\ \rho\sigma}{A} \tag{4.17}$$

It should be emphasized that the microscopic cross section, σ, is an *apparent* area. For some reactions, such as the bouncing or scattering of a neutron off a nucleus, the value of σ is very close to an actual area, πR^2. Values of σ for other kinds of neutron interactions can vary by many orders of magnitude.

Example 4.4

Manganese has a single isotope, ^{55}Mn, and a density of 7.43 g/cm^3. Compute a microscopic cross section based on the geometric area and the empirical formula $R = 1.4A \times 10^{-16}$ m. Compute the corresponding macroscopic cross section.

Solution: The microscopic cross section is calculated

$$R = 1.4 \times 55 \times 10^{-16} = 7.7 \times 10^{-15}\ \text{m}$$

$$\sigma = \pi R^2 = 1.86 \times 10^{-28}\ \text{m}^2 = 1.86\ \text{barns}$$

The number density of manganese can be calculated from Eq. 4.16.

$$N = \frac{6.023 \times 10^{23}\, \rho}{A}$$

$$= \frac{6.023 \times 10^{23} \times 7.43}{55} = 8.14 \times 10^{22} \text{ nuclei/cm}^3$$

$$\Sigma = N\sigma = 8.14 \times 10^{22} \times 1.86 \times 10^{-24} = 0.151 \text{ cm}^{-1}$$

Different Kinds of Neutron Reactions

Up to this point we have referred to neutron interactions with a nucleus without distinguishing between the different outcomes that may result. The outcomes do indeed differ, and the distinctions are important. Figure 4.2 gives a repre-

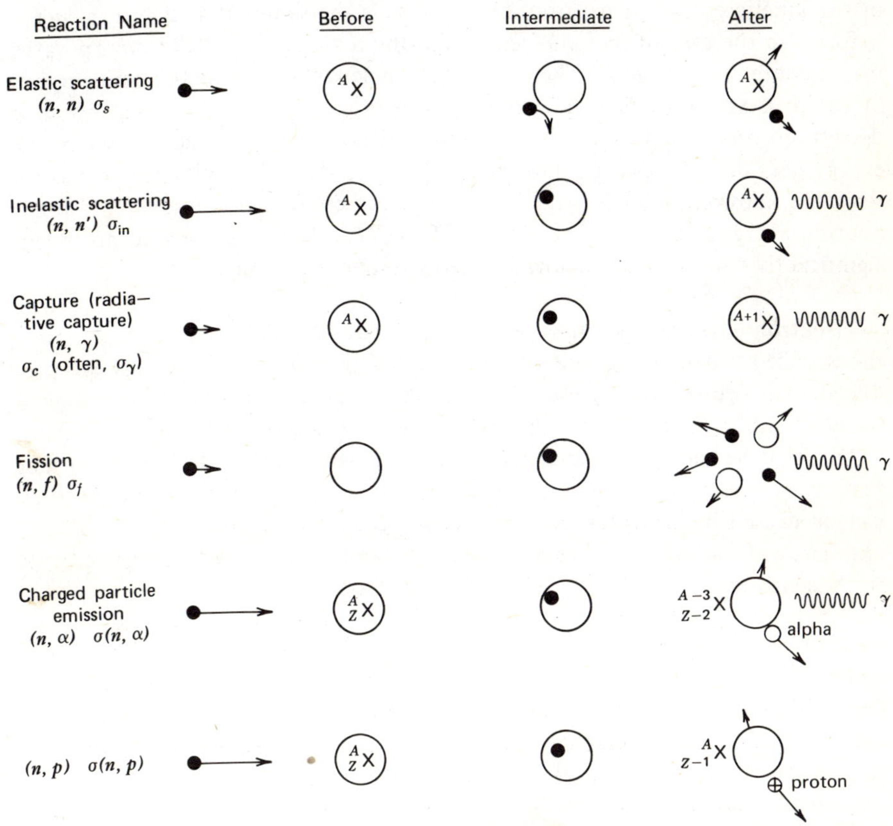

Figure 4.2 Different kinds of neutron reactions.

sentation of the different kinds of reactions with which we will be concerned. It is useful to think of these reactions in three stages: the conditions prior to the interaction, an intermediate stage in which often the neutron has been incorporated into the nucleus forming what is called the compound nucleus, and the situation following the reaction in which the product nucleus or nuclei and the various radiation forms appear. We shall discuss characteristics of the reactions shown in Figure 4.2, omitting the fission reaction, which was discussed in Chapter 3.

Scattering Reactions. All nuclei to some extent scatter neutrons. As indicated in Figure 4.2, there are two kinds of scattering. In *elastic* scattering (also called *potential* scattering), the neutron is repulsed by short-range nuclear forces as it comes very close to the nucleus. There is an exchange of momentum between the neutron and the nucleus, with the result that the neutron's direction and speed are changed. Kinetic energy is conserved, which means that the sum of the kinetic energy of neutron and nucleus is the same after the scattering as before. In the case of *inelastic* scattering, the neutron is actually incorporated into a compound nucleus. The compound nucleus then rejects the neutron, but in the process the original nucleus is placed in an excited state, from which it decays by gamma emission. Kinetic energy is not conserved, since some of the energy goes into the gamma photon(s). Inelastic scattering belongs to a class of reactions known as threshold reactions; these require some minimum (threshold) neutron energy in order to take place. Typically, these reactions do not occur significantly with neutrons below an energy of about one MeV.

Neutron Capture. Another very common neutron reaction is capture, sometimes called radiative capture because gamma radiation is almost always produced. The neutron simply joins the nucleus, forming the isotope of the original nucleus with a mass number higher by one. Of course, the product nucleus, ^{A+1}X, may be radioactive and decay to its daughter product. An important feature of most capture reactions is that they require no neutron energy; they can be caused by neutrons of any energy. Such reactions are almost always exothermic (that is, they have a positive Q-value) because the binding energy of ^{A+1}X is greater than that of ^{A}X plus a free neutron.

Charged Particle Emission. Another common neutron reaction leads to the emission of a charged particle, usually either an alpha particle (n, α) or a proton (n, p). If the Q-value of the reaction is positive, then no neutron energy is required. Such cases are relatively rare, however. The (n, α) reactions in lithium-6 and boron-10 and the (n, p) reaction in nitrogen-14 are notable exceptions. Otherwise, these reactions are threshold reactions, requiring that the neutron possess sufficient energy to make up the negative-Q deficit.

Cross Sections for Different Neutron Reactions

The designation of the microscopic cross section for each kind of reaction is shown in Figure 4.2. For example, the microscopic fission cross section is written σ_f. Likewise, the macroscopic cross section for fission is designated Σ_f. The relationship between microscopic and macroscopic cross section is that given by Eq. 4.15.

$$\Sigma_f = N\sigma_f \tag{4.18}$$

The values of cross sections (microscopic or macroscopic) express the relative probability of the different reactions. As an example, consider pure uranium-235. Table A.3 shows it to have cross sections:

$\sigma_c = 98.6$ barns

$\sigma_f = 582.2$ barns

$\sigma_s = 13.8$ barns

We presume that these three are the only reactions that occur. We can then say of a neutron traveling in pure uranium-235 that the probability of a capture reaction occurring *on the next interaction* is

$$\frac{98.6}{98.6 + 582.2 + 13.8} = 0.142$$

The probability of a capture reaction occurring eventually is

$$\frac{98.6}{98.6 + 582.2} = 0.145$$

because scattering does not terminate a neutron's career, whereas capture and fission do.

In the more general case of a material made up of several nuclides, we must use macroscopic cross sections to obtain the relative probabilities of the different kinds of reactions. If we consider that the small piece of matter represented in Figure 4.1 contains several nuclides, then the macroscopic cross section is simply the sum of their contributions.

$$\Sigma = N_1\sigma_1 + N_2\sigma_2 + N_3\sigma_3 + \cdots$$

or (4.19)

$$\Sigma = \sum_i N_i\sigma_i$$

The σ_i values can be for any kind of reaction designated. For example, the macroscopic scattering cross section of the material is

$$\Sigma_s = N_1\sigma_{s,1} + N_2\sigma_{s,2} + N_3\sigma_{s,3} + \cdots$$

or (4.20)

$$\Sigma_s = \sum_i N_i \sigma_{s,i}$$

The number densities, N_i, are the number of nuclei of each isotopic constituent per cubic centimeter of the material. We shall discuss calculation of these number densities shortly.

The Absorption Cross Section and Total Cross Section

For many calculations it is convenient to place in one category all reactions that terminate a neutron's career or life. These are called absorption reactions. Expressed most simply, they are all nonscattering reactions. The absorption cross section, therefore, is

$$\sigma_a = \sigma_c + \sigma_f + \sigma(n, \alpha) + \sigma(n, p) + \cdots \tag{4.21}$$

The reader may wonder why fission is included in this group, since fission produces neutrons. It is an arbitrary matter, but for many purposes it is convenient to treat the neutron causing a fission differently from those produced in fission, because the energy in the two cases can differ so widely.

In a previous section we computed a mean free path

$$\lambda = \frac{1}{\Sigma} \tag{4.9}$$

We did not describe the kind of reaction represented by Σ. Note, however, that in the derivation leading to Eq. 4.7 and 4.8, we used binary probabilities. There is either a reaction or there is not. No provision is there for different kinds of reactions. In Eq. 4.7, 4.8, and 4.9, therefore, Σ *must* be for all reactions.

$$\sigma = \sigma_s + \sigma_{\text{in}} + \sigma_c + \sigma_f + \cdots$$

$$\Sigma = \Sigma_s + \Sigma_{\text{in}} + \Sigma_c + \Sigma_f + \cdots \tag{4.22}$$

This cross section (microscopic or macroscopic) is called the total cross section. If it is necessary for clarity, it is written σ_t and Σ_t.

Because of the binary situation required for its derivation, Eq. 4.9 cannot usefully be applied to any cross section, Σ_f, for example. However, we can obtain a useful result substituting Σ_a into Eq. 4.9, if we understand the value obtained, λ_a, to be the mean or expected total distance traveled by the neutron before disappearing. Thus

$$\lambda_a = \frac{1}{\Sigma_a} \tag{4.23}$$

is the mean path to absorption and is the sum of all the straight-line flights. The mean free path, λ, is a straight-line path. In the application to absorption we have a binary situation, since we only recognize either no reaction or else absorption. Scattering in this context is a nonevent.

Example 4.5

A neutron with a kinetic energy of 0.025 eV is traveling in a medium in which the macroscopic absorption cross section, Σ_a, is 0.1 cm^{-1}, and the scattering cross section, Σ_s, is 0.2 cm^{-1}. These two categories cover all the neutron interactions in this material. (a) What is the expected value of the total distance the neutron will travel before disappearing? (b) How long will that take? (c) What is the expected distance the neutron will travel in a single straight-line flight? (d) What is the probability that the next interaction will be an absorption? (e) What is the probability that the neutron will undergo just three scatters and then be absorbed?

Solution: The expected distance to disappearance or absorption is

(a)
$$\lambda_a = \frac{1}{\Sigma_a} = 10 \text{ cm}$$

The neutron speed is

$$v = \left(\frac{2E}{m}\right)^{1/2} = \left(\frac{2 \times 0.025 \times 1.602 \times 10^{-19}}{1.009 \times 1.6604 \times 10^{-27}}\right)^{1/2} = 2187 \text{ m/s}$$

The time to disappearance, or the neutron lifetime, is

(b)
$$l = \frac{\lambda_a}{v} = \frac{10 \times 10^{-2}}{2187} = 4.57 \times 10^{-5} \text{ s}$$

The average value of straight-line paths is the mean free path, or

(c)
$$\lambda = \frac{1}{\Sigma} = \frac{1}{0.1 + 0.2} = 3.33 \text{ cm}$$

The probability of the next reaction being a scattering reaction is the fraction that the scattering macroscopic cross section contributes to the total macroscopic cross section, or

(d)
$$\frac{\Sigma_s}{\Sigma_a + \Sigma_s} = \frac{0.2}{0.1 + 0.2} = 0.667$$

The probability of any particular chain of events is the product of the probabilities of each of the events. Therefore, the probability of just three scatterings and then an absorption is

(e)
$$(0.667)^3 (0.333) = 0.0988$$

Such a sequence should occur about 10% of the time.

Calculation of Number Densities in Various Materials

The composition of a homogeneous material is generally specified in terms of the weight percent or fraction of different constituents. These constituents will be either chemical compounds, elements, or isotopes. We are interested in reducing such information to a set, N_i, of number densities of the isotopic constituents, where N_i is the number of atoms or nuclei of isotope i per cubic centimeter of the material in question. Given a set, N_i, we are halfway to the computation of macroscopic cross sections (Eq. 4.19).

From the weight fraction information we can calculate

$$N_k = 6.023 \times 10^{23} \frac{w_k \rho}{A_k} \tag{4.24}$$

w_k = weight fraction of constituent k
A_k = molecular weight, atomic weight, or isotope weight, depending on the nature of k
ρ = density of the material (g/cm^3)

If the material is a single isotope, then Eq. 4.24 is the same as Eq. 4.16. If k represents an element, then N_i is calculated

$$N_i = a_i N_k \tag{4.25}$$

where a_i is the atomic fraction of element k that is isotope i. If the element is just as it occurs in nature, then the a_i values are the abundance values obtained from the *Chart of the Nuclides*, LHP, or Table A.3. Otherwise, the isotopic composition of the element must be obtained from other information.

If constituent k in Eq. 4.24 is a chemical compound, then we can obtain the number density of the elements in the compound from its formula. For example, if the formula of compound k is $X_m Y_n$,

$$N_X = mN_k$$
$$N_Y = nN_k \tag{4.26}$$

The number densities of the isotopes of elements X and Y are calculated from Eq. 4.25. The application of these recipes to a variety of materials is illustrated in the following examples.

Example 4.6
Elemental boron has a density of 2.45 g/cm^3. From the data in the *Chart of the Nuclides* or Table A.3 on isotopic abundance, calculate the number densities of the naturally occurring isotopes.

Solution: From Table A.3, there are two naturally occurring isotopes, ^{10}B and ^{11}B, with abundances of 19.8 a/o and 80.2 a/o, respectively. The atomic weight of boron is, therefore

$$A = 0.198 \times 10 + 0.802 \times 11 = 10.80$$

This value can also be obtained from data tables. The number densities of each isotope is calculated by combining Eq. 4.24 and 4.25. The weight fraction of boron in this case is 1.0.

$$^{10}\text{B}: N_{10} = \frac{6.023 \times 10^{23} \times 0.198 \times 2.45}{10.80} = 0.02701 \times 10^{24} \text{ nuclei/cm}^3$$

$$^{11}\text{B}: N_{11} = \frac{6.023 \times 10^{23} \times 0.802 \times 2.45}{10.80} = 0.110 \times 10^{24} \text{ nuclei/cm}^3$$

It is convenient to express these values $\times 10^{24}$, because σ values are in units of 10^{-24} cm^2.

Example 4.7
The compound boron carbide, B_4C, has a density of 2.52 g/cm^3. Compute the isotopic number densities.

Solution: The molecular weight is computed from the atomic weights (*Chart of the Nuclides*).

$$M = 4 \times 10.81 + 12.01 = 55.25$$

The atomic number densities are obtained by combining Eq. 4.24 and 4.26.

$$\text{Boron:} \quad N_\text{B} = \frac{6.023 \times 10^{23} \times 4 \times 2.52}{55.25} = 0.1099 \times 10^{24} \text{ atoms/cm}^3$$

$$\text{Carbon:} \quad N_\text{C} = \frac{1}{4} N_\text{B} = 0.0275 \times 10^{24} \text{ atoms/cm}^3$$

The isotopic number densities are obtained from natural abundances and Eq. 4.25.

$$^{10}\text{B}: \quad N_{10} = 0.198\, N_\text{B} = 0.0218 \times 10^{24} \text{ atoms/cm}^3$$

$$^{11}\text{B}: \quad N_{11} = 0.802\, N_\text{B} = 0.0881 \times 10^{24} \text{ atoms/cm}^3$$

$$^{12}\text{C}: \quad N_{12} = 0.9889\, N_\text{C} = 0.0272 \times 10^{24} \text{ atoms/cm}^3$$

$$^{13}\text{C}: \quad N_{13} = 0.0111\, N_\text{C} = 0.0003 \times 10^{24} \text{ atoms/cm}^3$$

Example 4.8
An aqueous solution that is 10 weight percent (w/o) H_2SO_4 has a density of 1.08 g/cm^3. Calculate the number densities of the isotopic constituents of this solution.

Solution: First, calculate the number density of the two compounds, H_2O and H_2SO_4. If H_2SO_4 is 10 w/o, then H_2O must be 90 w/o. From Eq. 4.24

$$H_2O: \quad N = 6.023 \times 10^{23} \times \frac{0.9 \times 1.08}{18} = 3.252 \times 10^{22} \text{ molecules/cm}^3$$

$$H_2SO_4: \quad N = 6.023 \times 10^{23} \times \frac{0.1 \times 1.08}{98} = 6.638 \times 10^{20} \text{ molecules/cm}^3$$

We can obtain the number densities of the elements from Eq. 4.26. Note that hydrogen and oxygen are in both compounds.

$$\text{Hydrogen:} \quad N_H = 2 \times 3.252 \times 10^{22} + 2 \times 6.638 \times 10^{20} = 6.638 \times 10^{22}$$

$$\text{Oxygen:} \quad N_O = 3.252 \times 10^{22} + 4 \times 6.638 \times 10^{20} = 3.518 \times 10^{22}$$

$$\text{Sulfur:} \quad N_S = 6.638 \times 10^{20}$$

Substituting natural abundances in Eq. 4.25, we obtain isotopic number densities.

$$\text{Hydrogen-1:} \quad N_1 = 0.99985 \times 6.638 \times 10^{22} = 6.637 \times 10^{22}$$

$$\text{Hydrogen-2:} \quad N_2 = 0.00015 \times 6.638 \times 10^{22} = 9.956 \times 10^{18}$$

$$\text{Oxygen-16:} \quad N_{16} = 0.99756 \times 3.518 \times 10^{22} = 3.509 \times 10^{22}$$

$$\text{Oxygen-17:} \quad N_{17} = 0.00039 \times 3.518 \times 10^{22} = 1.372 \times 10^{19}$$

$$\text{Sulfur-32:} \quad N_{32} = 0.95 \times 6.638 \times 10^{20} = 6.306 \times 10^{20}$$

$$\text{Sulfur-33:} \quad N_{33} = 0.0075 \times 6.638 \times 10^{20} = 4.978 \times 10^{18}$$

$$\text{Sulfur-34:} \quad N_{34} = 0.042 \times 6.638 \times 10^{20} = 2.788 \times 10^{19}$$

$$\text{Sulfur-36:} \quad N_{36} = 0.00015 \times 6.638 \times 10^{20} = 9.956 \times 10^{16}$$

In many calculations it is not necessary to calculate all such numbers, because several are completely negligible.

It sometimes happens that microscopic cross sections are given for a nucleus of an element. Such a cross section is clearly an average of the separate isotopes weighted by the abundances. The *Chart of the Nuclides* gives some microscopic cross sections for elements. In converting to macroscopic cross sections, these values must be multiplied by the element number density.

Number Densities in Heterogeneous Media

It frequently happens that a neutron reacting medium is heterogeneous, that is, composed of two or more phases. A mixture of water and steel plates would be an example. Sometimes these mixtures can be treated as a single homogeneous region for the purpose of calculating macroscopic cross sections. In this case, the approach is to compute $N_{i,j}$ values for each homogeneous region, j, just as described in previous sections. Then the set, N_i, is obtained for the heterogeneous region.

$$N_i = \frac{\sum_j V_j N_{i,j}}{\sum_j V_j} = \sum_j v_j N_{i,j} \tag{4.27}$$

where V_j is the volume of each phase or homogeneous region and v_j is the volume fraction $(=V_j/\Sigma V_j)$.

Example 4.9

A fission reactor fuel rod is made up of a long rod of UO_2 (density 10 g/cm^3) 7 mm in diameter enclosed in a stainless steel tube with 0.5 mm wall thickness. For simplicity, assume the stainless steel is iron of density 7.8 g/cm^3. Compute the elemental number densities, treating the clad fuel rod as a single homogeneous region.

Solution: The individual region number densities are computed.

UO_2: $N = 6.023 \times 10^{23} \times \frac{10}{(238 + 32)} = 2.23 \times 10^{22}$ molecules/cm^3-UO_2

U: $N = 2.23 \times 10^{22}$ atoms/cm^3-UO_2

O: $N = 2 \times 2.23 \times 10^{22} = 4.46 \times 10^{22}$ atoms/cm^3-UO_2

Fe: $N = 6.023 \times 10^{23} \times \frac{7.8}{55.85} = 8.41 \times 10^{22}$ atoms/cm^3-iron

The volume of each region is calculated.

UO_2: $V_1 = \frac{\pi}{4}(0.7)^2 = 0.3849$ cm^2

Fe: $V_1 = \frac{\pi}{4}(0.8^2 - 0.7^2) = 0.1178$ cm^2

Since the length dimension is identical for each region, there is no need to include it. The homogeneous number densities are (Eq. 4.27)

U: $N = \frac{0.3849 \times 2.23 \times 10^{22}}{(0.3849 + 0.1178)} = 1.71 \times 10^{22}$ atoms/cm^3

O: $N = 2 \times 1.71 \times 10^{22} = 3.42 \times 10^{22}$ atoms/cm^3

Fe: $N = \frac{0.1178 \times 8.41 \times 10^{22}}{(0.1178 + 0.3849)} = 1.97 \times 10^{22}$ atoms/cm^3

Microscopic Cross Section Values

Unfortunately for the simplicity of calculations of reaction rates in a neutron field, cross sections are often a sensitive function of neutron energy. The functions are frequently rather badly behaved ones, not lending themselves to simple algebraic expressions. Figure 4.3 attempts to make the point that many neutron reaction cross sections are strongly dependent on neutron energy. Moreover,

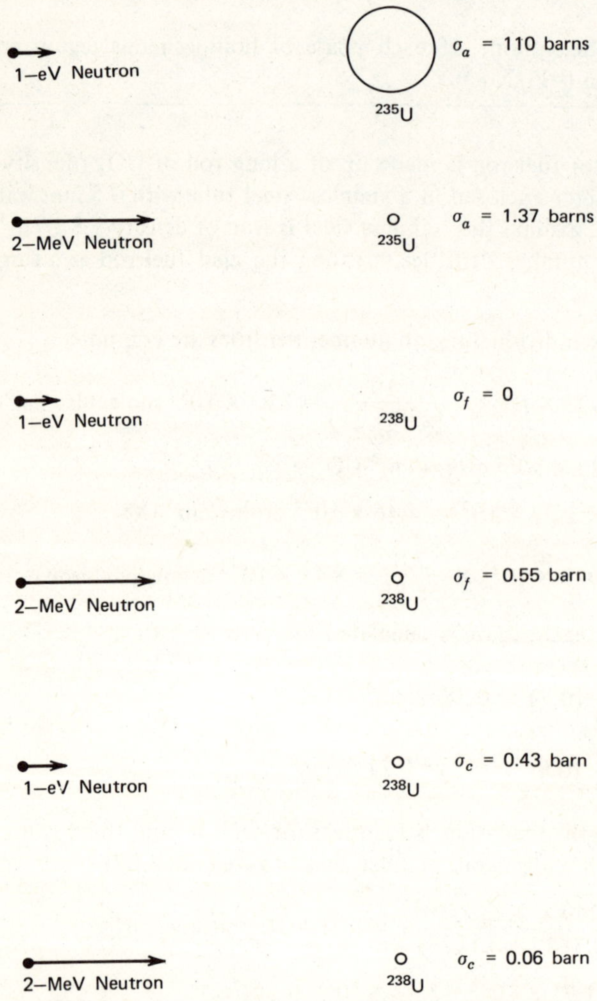

Figure 4.3 Examples of the dependence of cross sections on neutron energy.

the nature of the dependence varies from isotope to isotope and from one kind of reaction to another. Reactions such as capture and fission in uranium-235 and capture in uranium-238 are no-threshold reactions. These reactions can have large cross sections for low-energy neutrons. A threshold reaction, such as fission in uranium-238, has a zero cross section for low-energy neutrons. Although the threshold varies in each case, for ^{238}U the fission cross section becomes significant only above 1 MeV.

Figure 4.4 presents an overview of cross section energy dependence for three different classes of reactions: elastic scattering, threshold reactions, and no-threshold reactions. The energy range of interest varies from the energy of source neutrons down to very low energies. The upper range is determined by the characteristics of the neutron source. The sources of interest to us are primarily fission neutrons (which have a spectrum around a mean value of 2 MeV) and the neutrons produced in the DD and DT fusion reactions (2 MeV and 14 MeV, respectively). These source neutrons have very high speeds relative to the nuclei of the medium. They therefore lose energy in scattering collisions and slow down. If there is a sufficiently high scattering cross section, Σ_s, the neutrons come into thermal equilibrium with the various nuclear constituents of the medium. These nuclei have thermal motion, and there is a give-and-take exchange of momentum as the neutrons scatter off them. The neutrons then exhibit Maxwellian behavior and have a spectrum around an average value of $3/2\ kT$, where T is the absolute temperature of the medium. The neutron population in any material, therefore, has a spectrum composed of a complicated mixture of source neutrons and thermal neutrons at the two ends of the spectrum, and intermediate-energy or slowing-down neutrons in between. The particular mix of these population sectors depends very strongly on the material. The spectrum can vary all the way from one only slightly degraded in energy from the source spectrum (in the case of materials having a high ratio of absorption to scattering, Σ_a/Σ_s) to one with predominantly thermal neutrons (in a material with a high Σ_s/Σ_a ratio).

The discussion of the character of neutron spectra is relevant here because we are only interested in cross sections in the neutron energy range encountered in reacting systems. Moreover–and this is just a coincidence–the three spectral regions listed above correspond roughly to three regions of different characteristic energy-dependent behavior of cross sections. Figure 4.4 makes this point.

Consider first the energy-dependent behavior of the no-threshold class of reactions. At low energies, corresponding to the thermal neutron portion of the spectrum, the cross section typically is a smooth function of energy. The relationship is

$$\sigma_c \propto \frac{1}{\sqrt{E}} \text{ or } \frac{1}{v} \tag{4.28}$$

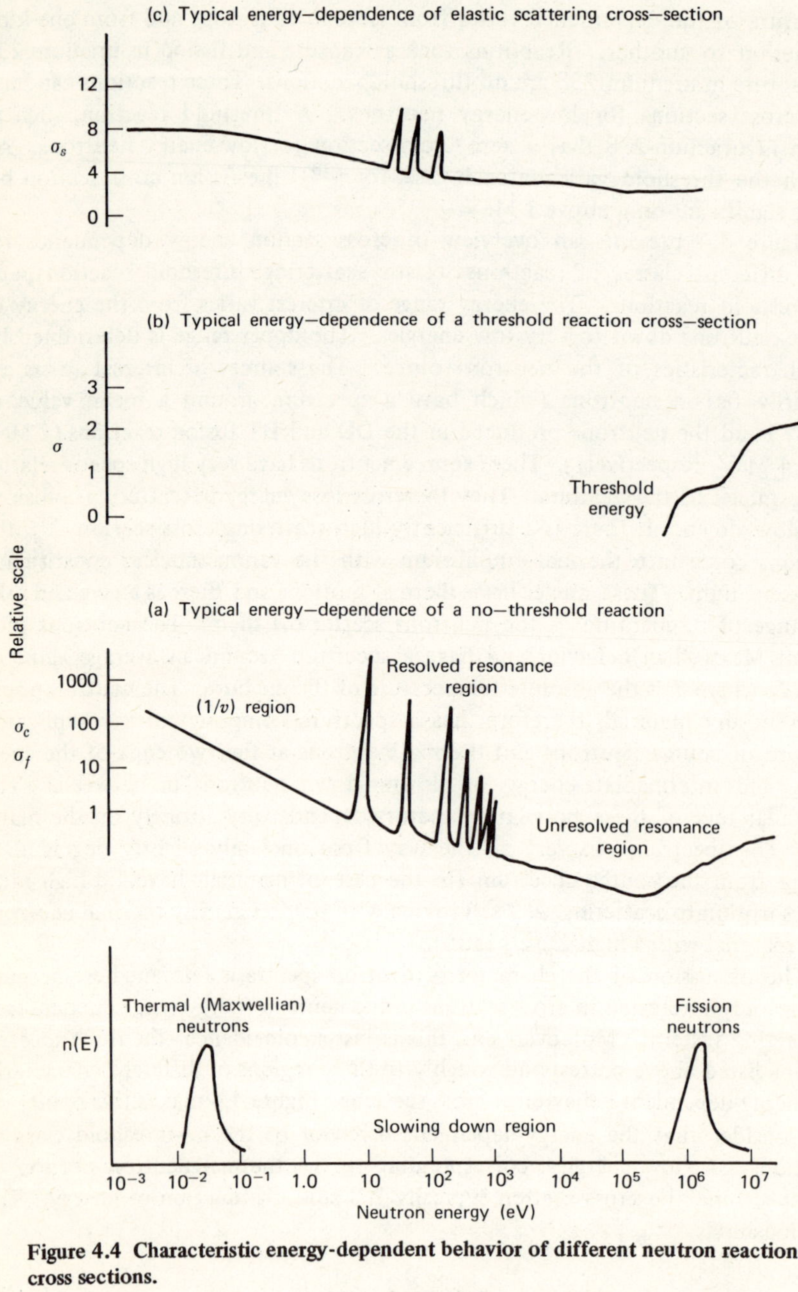

Figure 4.4 Characteristic energy-dependent behavior of different neutron reaction cross sections.

In the parlance of the trade, these are called one-over-vee (or $1/v$) cross sections. There are some significant deviations from $1/v$ behavior exhibited by a few nuclides, but these need not concern us here.

Data sources for cross sections conventionally quote capture and fission cross sections for low-energy neutrons at a speed of 2200 m/s, which corresponds to a kinetic energy of 0.0253 eV and a temperature ($E = \frac{3}{2} kT$) of about 293.2°K (20°C). As long as $1/v$ behavior holds, the cross section at any energy can be calculated

$$\sigma(E) = \sigma(E_0)\sqrt{\frac{E_0}{E}} \tag{4.29}$$

or

$$\sigma(v) = \sigma(2200\ \text{m/s})\,\frac{2200}{v} \tag{4.30}$$

These equations will usually apply up to an energy of a few electron volts.

As the neutron energy increases, the cross section begins to show sharp peaks (Figure 4.4). These peaks can go as high as thousands of barns. They are called resonances. They correspond to discrete energy levels of the compound nucleus. Neutrons possessing energy in these narrow bands are much more "in tune" with the nucleus, and the chance of a reaction rises sharply. Obviously, in this energy region it is not useful for data presentation purposes to quote a value or even many values of the cross section, because these values are changing so rapidly with energy. The practice is to quote the value of an integral cross section. The integral is

$$\text{Resonance integral} = \int_{\check{E}}^{\hat{E}} \sigma(E)\,\frac{dE}{E} = \int_{\check{E}}^{\hat{E}} \sigma(E)\,d\ln E \tag{4.31}$$

It could also be expressed (but seldom is) as an average cross section.

$$\bar{\sigma} = \frac{\displaystyle\int_{\check{E}}^{\hat{E}} \sigma(E)\,\frac{dE}{E}}{\ln \hat{E}/\check{E}} \tag{4.32}$$

The resonance integral can be measured directly in experiments, or it can be calculated from experimental data on the individual resonances. The reason for integrating over $\ln E$ rather than E is not apparent; it relates to the nature of the neutron spectrum in this energy region.

The lower and upper energy limits ($\check{E}$ and $\hat{E}$, respectively) of the resonance integral are not of great significance to us. The lower limit is about 0.5 eV, a value that is determined by the experimental method for measuring the resonance integral. The upper limit is arbitrary, but 0.1 MeV is a reasonable value. The integral is very insensitive to the upper limit.

Referring again to Figure 4.4*a*: as the energy increases, the resonances draw closer and closer together. At some point it is no longer possible for an experimentalist to resolve (separate) these peaks, and we have what appears to be a relatively smooth curve, although in theory there are still resonances. For this reason, the two regions are often referred to as the resolved and unresolved resonance regions.

The highest energy region, beginning in the 0.1–1 MeV region, is characterized by a smooth, though ill-defined, function. The fission cross sections tend to decline. Cross sections tend to be small, rarely rising above 5 barns. It is in this region that threshold reactions of interest (Figure 4.4*b*) become significant. These can be fission reactions, (n, α) reactions, (n, p) reactions, (n, n') reactions (inelastic scattering), and so on. There is little that can be generalized about cross section values in this region.

Figure 4.4*c* shows the typical behavior of the elastic scattering cross section, σ_s. Fortunately, it does not exhibit as complex an energy dependence as the other reaction cross sections. At low energies it is energy-independent. There are scattering resonances, but these are not of great import. At the higher energies the elastic scattering cross section declines somewhat. There also is relatively little variation in scattering cross section from one isotope to another. Almost all elastic scattering cross sections fall in the range of one to ten barns.

Cross Section Data Sources

In this text, microscopic cross section data are given in Table A.3 for most isotopes of interest. A single value of scattering cross section is given. Capture and non-zero fission cross sections at 2200 m/s neutron speed are given. The resonance fission and capture integrals (Eq. 4.31) are given for several isotopes. Finally, capture and fission cross sections at 2 MeV are given for a number of isotopes. This limited set of data will not equip the reader to do detailed reactor or shielding-design calculations, but it does provide a basis for performing many significant calculations and for understanding the important aspects of nuclear energy systems that are related to cross section values.

The principal source of cross section data in the United States is the National Neutron Cross Section Center at Brookhaven National Laboratory. The best known publications are the various editions and supplements of the report, BNL-325. There is also an international consortium of organizations that produces cross section data compilations known as the Evaluated Neutron Data File. This information, however, is highly formalized and is of use primarily to professionals in the field.

The Rate of Various Kinds of Neutron Reactions

The total rate of neutron reactions in any medium is given by Eq. 4.13

$$\text{Reaction rate} = \Sigma\phi = \Sigma_t \phi$$

where $\phi = n\bar{v}$, and Σ_t is the total cross section. Since the fraction of any one kind of reaction, say fission, is Σ_f/Σ_t, it follows that

$$\text{Fission rate} = \Sigma_f \phi \qquad (4.33)$$

and so on for all the various kinds of reactions.

Example 4.10
A piece of manganese weighing one gram is irradiated in a thermal (low-energy) neutron flux of 10^{12} $n/\text{cm}^2 \cdot \text{s}$. What level of radioactivity (curies) will be reached in five hours?

Solution: This is a problem similar to Example 3.5. The radioactivation equation, Eq. 3.8, is

$$\lambda N(t) = \dot{P}(1 - e^{-\lambda t})$$

where $\lambda N(t)$ is the activity produced by the (n, γ) reaction in ^{55}Mn after time t at a reaction rate, $\dot{P}$. We are now in a position to calculate $\dot{P}$. It is $\Sigma_c\phi$ nuclei/cm$^3 \cdot$ s or $\Sigma_c\phi/\rho$ nuclei/g $\cdot$ s. Since manganese is 100% ^{55}Mn, from Eq. 4.17

$$\Sigma_c/\rho = \frac{N\sigma_c}{\rho} = \frac{0.6023\ \rho}{\rho A}\ \sigma_c = \frac{0.6023\ \sigma_c}{A}$$

From Table A.3, σ_c for thermal neutrons is 13.3 barns.

$$\Sigma_c/\rho = \frac{0.6023 \times 13.3}{55} = 0.1457\ \text{cm}^2/\text{g}$$

$$\dot{P} = 0.1457 \times 10^{12}\ \text{reactions/g} \cdot \text{s}$$

The (n, γ) reaction produces ^{56}Mn with a half-life of 2.58 h. From Eq. 3.8

$$\lambda N(t) = 0.1457 \times 10^{12}(1 - e^{-\lambda t})$$

$$\lambda = \frac{\ln 2}{2.58} = 0.2687\ \text{h}^{-1}$$

$$\lambda N(5\text{h}) = 0.1457 \times 10^{12}[1 - \exp - (0.2687 \times 5)]$$

$$= 1.076 \times 10^{11}\ \text{dps/g}$$

$$= \frac{1.076 \times 10^{11}}{3.7 \times 10^{10}} = 2.91\ \text{Ci/g}$$

Example 4.11

One concern in the design of fission and fusion reactors is a loss of structural integrity in key metallic components due to accumulation of helium gas, which results from (n, α) reactions. To take a highly simplified case, compute the volume of helium, at 1000°K and 10^5 Pa pressure, generated in a cubic centimeter of iron exposed to a neutron flux of 10^{14} neutrons/cm$^2 \cdot$ s for one year. The effective microscopic cross section, $\sigma(n, \alpha)$, in this flux is 0.1 barn.

Solution: An atom of helium is produced for every (n, α) reaction that occurs. Iron has a density of 7.8 g/cm^3 and an atomic weight of 55.85. The rate of (n, α) reactions is $\Sigma_\alpha \phi$.

$$\Sigma_\alpha \phi = N\bar{\sigma}_\alpha \phi = \frac{.6023 \times 7.8}{55.85} \times 0.1 \times 10^{14}$$

$$= 8.41 \times 10^{11} \text{ reactions/cm}^3 \cdot \text{s}$$

In one year the number of helium atoms formed is $8.41 \times 10^{11} \times 3.156 \times 10^7$ or 2.65×10^{19} or 4.41×10^{-5} moles. The volume occupied by this quantity can be calculated according to the perfect gas law.

$$V = 4.41 \times 10^{-5} \frac{RT}{P}$$

$$R = 8.314 \text{ J/mole} \cdot K \text{ (or N} \cdot \text{m/mole} \cdot K)$$

$$V = 4.41 \times 10^{-5} \times \quad 8.314 \quad \times 1000 \times \frac{1}{10^5} = 3.66 \times 10^{-6}$$

$$\text{mole} \qquad \frac{\text{N} \cdot \text{m}}{\text{mole} \cdot K} \qquad K \qquad \frac{\text{m}^2}{\text{N}} \qquad \text{m}^3$$

The volume of helium generated in one cm^3 of iron would be 3.66 cm^3.

Although the derivation of $\Sigma_x \phi$ as the rate of reaction of kind x is straightforward enough, it probably is not very satisfying at this point, since we have not yet discussed how one is to calculate a neutron flux value. Perhaps a few words on this subject here will not be too distracting. Remember that $\phi = n\bar{v}$ and n, the neutron density or neutrons/cm^3, is the important quantity. Neutrons can exist only if there is a source of neutrons. That is simple enough. Neutrons also disappear at a rate $\Sigma_a \phi$. It follows that if we are to have a steady population of neutrons (steady state),

$$S = \Sigma_a \phi \tag{4.34}$$

where S must be the source of neutrons per cubic centimeter per second.

$$\phi = \frac{\Sigma_a}{S}$$

Actually, it is more complex than this, because S and Σ_a are functions of position—different functions of position—so that Eq. 4.34 should be more complex to represent the general case. However, it will do for now. There are a number of neutron sources, but the principal ones are the fission reaction and the fusion reactions, DD and DT. In the case of fission, the neutron source term is $\bar{\nu}\Sigma_f\phi$, where $\bar{\nu}$ is the average number of neutrons per fission. If we substitute $\bar{\nu}\Sigma_f\phi$ into Eq. 4.34, however, the result

$$\bar{\nu}\Sigma_f\phi = \Sigma_a\phi \tag{4.35}$$

is obviously not useful in evaluating ϕ. We can see, however, that if $\bar{\nu}\Sigma_f\phi > \Sigma_a\phi$ then the neutron population would grow. This, in fact, is how fission reactors are controlled. The value of $\bar{\nu}\Sigma_f\phi$ is made greater than $\Sigma_a\phi$ until a desired value of ϕ or $\Sigma_f\phi$ (depending on what sensing instrumentation is being used) is reached. The values are made equal when the reactor is operating at steady state. In order to shut down a reactor, $\Sigma_a\phi$ is made larger than $\bar{\nu}\Sigma_f\phi$.

Example 4.12

A fission reactor fuel is in the form of UO_2 (density, 9.0 g/cm^3) in which the uranium is 3 a/o ^{235}U. If the neutron flux in the reactor is 10^{13} neutrons/cm^2 · s and the effective energy is 0.05 eV, what is the thermal power density generated in the fuel (W/cm^3)?

Solution: This problem requires the synthesis of information from preceding sections and chapters. We must obtain the fission rate, $\Sigma_f\phi$, and so must calculate Σ_f of UO_2. From Eq. 4.20

$$\Sigma_f = \sum_i N_i \sigma_{f,i}$$

Of the constituents of UO_2, only ^{235}U has a non-zero fission cross section for low-energy neutrons. From Table A.3, the fission cross section of ^{235}U at 0.0253 eV is 582 barns. From Eq. 4.29, we can obtain the fission cross section at 0.05 eV.

$$\sigma_f(0.05\text{ eV}) = 582\left(\frac{0.0253}{0.05}\right)^{1/2} = 414 \text{ barns}$$

The number density of uranium-235 is calculated by combining Eq. 4.24 and 4.25.

$$^{235}\text{U:}\quad N = 0.03 \times 0.6023 \times \frac{9.0}{(238+32)} = 6.023 \times 10^{-4}\ (10^{24} \text{ nuclei/cm}^3)$$

$$\Sigma_f = 6.023 \times 10^{-4} \times 414 = 0.2494 \text{ cm}^{-1}$$

$$\Sigma_f\phi = 0.2494 \times 10^{13} \text{ fissions/cm}^3 \cdot \text{s}$$

In Chapter 3, we assigned a value of 200 MeV to the energy released per fission. Therefore, the thermal power density in the UO_2 (q''') is

$$q''' = 0.2494 \times 10^{13} \times 200 \times 1.602 \times 10^{-13} = 79.9 \text{ W/cm}^3$$

This thermal energy must be transferred to the reactor coolant.

Neutron Reaction Rates with Energy-Dependent Cross Sections

We have made the point in Figures 4.3 and 4.4 that reaction cross sections are energy dependent. Further, a neutron population usually has a spectrum of energies. These two facts mean that the expression for the rate of any reaction, given by Eq. 4.33 as $\Sigma_x \phi$, is of limited use because it leaves open the question of the correct value of the energy-dependent cross section to use. In order to solve this problem we must introduce an energy distribution function for the neutron flux. (See section on distribution functions.) This function is $\phi(E)$. It is defined so that $\phi(E)\,dE$ is the flux of neutrons having energies between E and $E + dE$. It can also be written

$$\phi(E) = vn(E) \tag{4.36}$$

where v is $(2E/m)^{1/2}$ and $n(E)$ is a distribution function in terms of the neutron density. The neutron flux is

$$\phi = \int_0^\infty \phi(E)\,dE \tag{4.37}$$

Note that $\phi(E)$ has units of neutrons/cm$^2 \cdot$ s $\cdot$ eV. Now Eq. 4.33 for reaction rate can be applied to those neutrons with energy between E and $E + dE$, so that the value of the cross section is unambiguously the value at E. The absorption rate, for example, is $\Sigma_a(E)\,\phi(E)\,dE$ for neutrons in the interval dE. The rate for all neutrons is obtained by integrating over all neutron energies.

$$\text{Absorption Rate} = \int_0^\infty \Sigma_a(E)\,\phi(E)\,dE \tag{4.38}$$

In this text, when we express an absorption rate as $\Sigma_a \phi$, or a fission rate as $\Sigma_f \phi$, it will be understood that the integration required by Eq. 4.38 is intended. In other words

$$\Sigma_a \phi \equiv \int_0^\infty \Sigma_a(E)\,\phi(E)\,dE \tag{4.39}$$

The evaluation of the integral in Eq. 4.39 for realistic situations is beyond the scope of this text. However, it should be noted that the usual approach to evaluation of the integral is to write it

$$\int_0^\infty \Sigma_a(E)\,\phi(E)\,dE = \sum_g \int_{\check{E}_g}^{\hat{E}_g} \Sigma_a(E)\,\phi(E)\,dE = \sum_g \Sigma_{a,g}\phi_g \qquad (4.40)$$

The integral is broken into a number of discrete groups, g, with energy limits $\check{E}_g$ and $\hat{E}_g$. This approach is called multigroup theory. It has the advantage that it is a simpler matter to average $\Sigma_a(E)$ over the relatively narrow group energy limits than over all energies.

Example 4.13
Assume that a neutron population is well described by the Maxwellian function (Eq. 4.12).

$$n(E) = \frac{2\pi n}{(\pi kT)^{3/2}} E^{1/2} e^{-E/kT}$$

A certain nuclide has a $1/v$ absorption cross section. Show that the correct average cross section for this population of neutrons is $(\pi/4)^{1/2}\,\Sigma_a(kT)$.

Solution: The distribution function, $\phi(E)$, is written

$$\phi(E) = vn(E)$$

$$v = \left(\frac{2}{m}\right)^{1/2} E^{1/2}$$

$$\phi(E) = \frac{2\pi n}{(\pi kT)^{3/2}} \left(\frac{2}{m}\right)^{1/2} Ee^{-E/kT}$$

From Eq. 4.29

$$\Sigma_a(E) = \left(\frac{kT}{E}\right)^{1/2} \Sigma_a(kT)$$

We define the correct average cross section by Eq. 4.39

$$\Sigma_a = \frac{\int_0^\infty \Sigma_a(E)\,\phi(E)\,dE}{\int_0^\infty \phi(E)\,dE}$$

$$\Sigma_a = \frac{\Sigma_a(kT)\int_0^\infty \left(\frac{kT}{E}\right)^{1/2} Ee^{-E/kT}\,dE}{\int_0^\infty Ee^{-E/kT}\,dE}$$

Evaluating both integrals gives

$$\Sigma_a = \left(\frac{\pi}{4}\right)^{1/2} \Sigma_a(kT)$$

To illustrate with a specific example: Table A.3 indicates that at $v = 2200$ m/s, $E = 0.0253$ eV, σ_a of ^{235}U is 680.8 barns. Assuming this cross section follows a $1/v$ rule

$$\overline{\sigma_a}(kT = 0.0253) = \left(\frac{\pi}{4}\right)^{1/2} \times 680.8 = 603.3 \text{ barns}$$

If the temperature of the population is 600°K

$$\sigma_a(600°\text{K}) = \sigma_a(293.2°\text{K}) \left(\frac{293.2}{600}\right)^{1/2}$$

$$= 476 \text{ barns}$$

and the correct average cross section is

$$\overline{\sigma_a}(600°\text{K}) = \left(\frac{\pi}{4}\right)^{1/2} \times 476 = 422 \text{ barns}$$

Neutron Slowing Down by Elastic Scattering

We have already described elastic scattering as an event in which the neutron bounces off a nucleus, imparting some of its kinetic energy to the nucleus. We wish to be more quantitative about this energy-sharing process. We shall consider only those cases in which the kinetic energy of the nucleus prior to the collision can be neglected. The case is similar to that leading to Eq. 2.11, except that the reaction is modified.

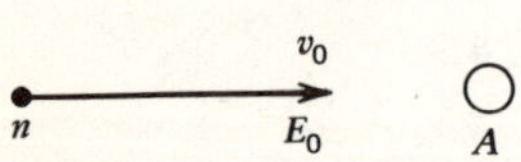

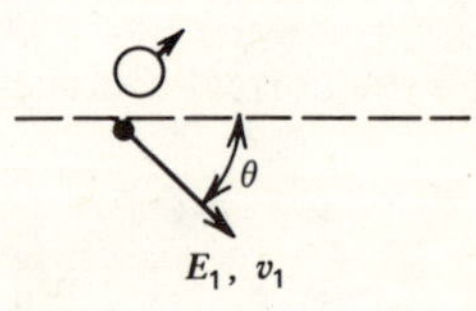

In a scattering reaction, Q is zero. Equation 2.11 becomes

$$E_1\left(1 + \frac{1}{A}\right) - E_0\left(1 - \frac{1}{A}\right) - \frac{2}{A}(E_0E_1)^{1/2} \cos\theta = 0 \qquad (4.41)$$

This result is obtained by solving the mechanics of the scattering in the laboratory system, that is, one in which the observer is fixed relative to the reacting particles. It proves to lead to more useful solutions if we solve it in the so-called center of mass system. This is a system that is moving in the laboratory system with a speed, v_c, calculated from the momentum balance.

$$v_c = \frac{v_o}{1 + A} \tag{4.42}$$

In the center of mass system, there is zero momentum, so the neutron and the nucleus must scatter at 180° to each other.

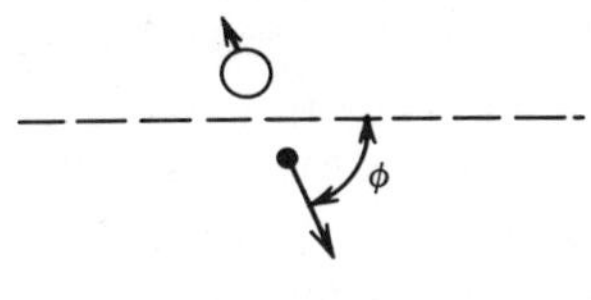

The angle ϕ is the angle at which the neutron scatters in the center of mass system relative to its original direction. Note also that since the momenta must sum to zero and the kinetic energy remain unchanged, the speeds of neutron and nucleus are independent of the scattering angle ϕ in the *center of mass system*. The speed of the neutron in the center of mass system is

$$v_o - v_c = \frac{A}{A + 1} v_o \tag{4.43}$$

To obtain the final neutron velocity, v_1, in the laboratory system it is necessary to add $(v_o - v_c)$ and v_c vectorially.

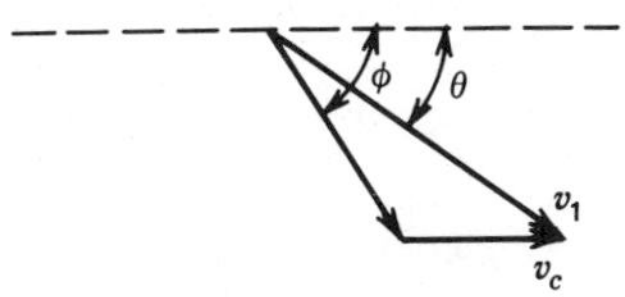

$$v_1^2 = \frac{A^2}{(A + 1)^2} v_0^2 + \frac{v_0^2}{(A + 1)^2} - \frac{2A}{(A + 1)^2} v_0^2 \cos(\pi - \phi)$$

or

$$\frac{E_1}{E_0} = \frac{v_1^2}{v_0^2} = \frac{A^2 + 2A \cos \phi + 1}{(A + 1)^2} \tag{4.44}$$

We obtain one useful conclusion immediately from this equation. E_1/E_0 has a minimum when $\phi = \pi$, that is, the neutron is bounced straight back.

$$\alpha \equiv \left(\frac{E_1}{E_0}\right)_{\min} = \left(\frac{A - 1}{A + 1}\right)^2 \tag{4.45}$$

Since A is the mass number of the scattering nucleus, we can see that in scattering off a hydrogen nucleus, a neutron could lose all of its kinetic energy. The reader may have seen the experiment in a science museum in which a sphere on a fine wire is allowed to hit another identical stationary sphere that is similarly suspended. The first comes to a stop, while the second acquires the kinetic energy of the first. Such a demonstration models the case of a neutron scattering off a proton at $\phi = \pi$.

The minimum value of E_1/E_0 is certainly of interest, but of much more interest is the *average* value. The problem is how to calculate it. What can be said about the values that ϕ will assume is that ϕ is a variable that can have any value from 0 to π and that we can only treat it with a distribution function (see the section on that subject earlier in this chapter). We obtain a distribution function, $p(\phi)$, on the basis of the well-established physical observation that scattering *in the center of mass system* is *isotropic*. This means that if you circumscribe a sphere with the neutron-nucleus system as its center, the probability of the scattered neutron flying out through any given area of the sphere's surface *per unit area* is everywhere the same. Without going into the derivation, we claim that

$$p(\phi) = \frac{\sin \phi}{2} \tag{4.46}$$

where $p(\phi)\,d\phi$ is the probability of the scattering angle being between ϕ and $\phi + d\phi$. The average value of E_1/E_0 can now be calculated

$$\overline{E_1/E_0} = \frac{1}{2}\int_0^{\pi} \frac{E_1}{E_0} \sin \phi \, d\phi \tag{4.47}$$

Substituting for E_1/E_0 from Eq. 4.44 and integrating, we obtain the solution

$$\overline{E_1/E_0} = \frac{A^2 + 1}{(A + 1)^2} \tag{4.48}$$

Substituting Eq. 4.45

$$\overline{E_1/E_0} = \frac{\alpha + 1}{2} \tag{4.49}$$

This means that *on the average* a neutron loses one half of the maximum energy it can lose. We note that the average value of E_1/E_0 is independent of neutron energy. We can also approximate an expected value or average value for the number of collisions it will take to decrease a neutron's energy by any given fraction. After n collisions

$$E_{\text{final}}/E_{\text{initial}} = (\overline{E_1/E_0})^n$$

or

$$n = \frac{\ln (E_{\text{final}}/E_{\text{initial}})}{\ln \overline{E_1/E_0}}$$

Table 4.1. Slowing Down Characteristics of Some Nuclei

NUCLEUS	A	α	ξ	AVERAGE COLLISIONS TO REDUCE E BY $10^6/1$
Hydrogen-1	1	0	1.0	14
Deuterium	2	0.1111	0.726	19
Lithium-7	7	0.5625	0.260	53
Beryllium	9	0.6400	0.207	68
Carbon	12	0.7160	0.158	87
Oxygen	16	0.7786	0.120	115
Sodium	23	0.8403	0.0845	164
Iron	56	0.9311	0.0353	391
Uranium	238	0.9833	0.0084	1645

It is usual practice to use the mean logarithm of E_1/E_0 rather than the logarithm of the mean of $\overline{E_1/E_0}$. The mean logarithm is obtained by the integration of Eq. 4.47, but with the substitution of ln (E_1/E_0) for (E_1/E_0). If this operation is performed we obtain

$$\xi \equiv -\overline{\ln E_1/E_0} = 1 + \frac{(A-1)^2}{4A} \ln \frac{(A-1)^2}{(A+1)^2} = 1 + \frac{\alpha}{1-\alpha} \ln \alpha \qquad (4.50)$$

Table 4.1 gives values of the neutron-slowing-down characteristics of several nuclides over a wide range of mass number. Values of α and ξ are computed from Eq. 4.45 and Eq. 4.50, respectively. It is apparent that the light nuclides are most effective in slowing down neutrons. When a material is used in a reactor for the purpose of slowing down neutrons, the material is called a moderator. The moderator is an important component in the design of thermal neutron (slow neutron) reactors. These equations only hold, of course, when the speed of the neutron is so much greater than that of the nucleus that the latter can be treated as zero. When a neutron's kinetic energy is reduced to the region of 0.1 to 1 eV, then the speeds of the nuclei become significant and the neutrons begin to pick up kinetic energy in some collisions. Such neutrons are called thermal neutrons because they approach thermal equilibrium with the medium.

Example 4.14

A neutron having a kinetic energy of 0.1 MeV is scattered by a nucleus of lithium-7. The scattered neutron has an angle of 45° in the center of mass system. What is its energy? What fraction of its energy did it lose? What is the scattering angle in the laboratory system?

Solution: From Eq. 4.44

$$\frac{E_1}{E_0} = \frac{A^2 + 2A\cos\phi + 1}{(A+1)^2} = \frac{49 + 7\sqrt{2} + 1}{64} = 0.936$$

The scattered neutron's energy is 93,600 eV. The fraction energy loss is

$$\frac{E_0 - E_1}{E_0} = 0.064 \ (6.4\%)$$

From the diagram showing the scattering angles in both systems

$$v_1 \sin\theta = \frac{A}{A+1} v_0 \sin\phi$$

$$\sin\theta = \frac{A}{(A+1)\sqrt{E_1/E_0}} \sin\phi$$

$$= \frac{7}{8(0.967)} \frac{\sqrt{2}}{2} = 0.640$$

$$\theta = 39.8°$$

Example 4.15
A neutron slows down in pure heavy water, D_2O. On the average, how many collisions will it take to lose 90% of its kinetic energy?

Solution: The first thing to realize here is that neutrons do not bump into molecules; they only interact with nuclei. Therefore, we *cannot* treat D_2O as a single scattering entity. Sometimes a neutron will scatter off of a deuteron and lose, on the average, about 50% of its energy (see Table 4.1). Other times it will scatter off of an oxygen nucleus and lose 11% of its energy. What we need to know is the relative number of scatters in each nuclide. We obtain this from the scattering cross sections (Table A.3).

deuterium: $\sigma_s = 4.9$ barns

oxygen: $\sigma_s = 3.76$ barns

The fraction of scatters in deuterium is

$$\frac{2 \times 4.9}{2 \times 4.9 + 3.76} = 0.72$$

The other 28% occur in oxygen. Now, we can take the weighted average of ξ

$$\xi_{D_2O} = 0.72\ \xi_D + 0.28\ \xi_O$$

$$= 0.72 \times 0.726 + 0.28 \times 0.120$$

$$= 0.556$$

The average number of collisions, n, which it takes to absorb 90% of a neutron's energy in this medium is

$$n = \frac{\ln 10}{0.556} = 4.1$$

Neutron Beams

One physical situation that merits special attention is the neutron beam and its interaction in matter. By a beam we mean a stream of neutrons all having the same direction, that is, monodirectional. It is also called a collimated beam.

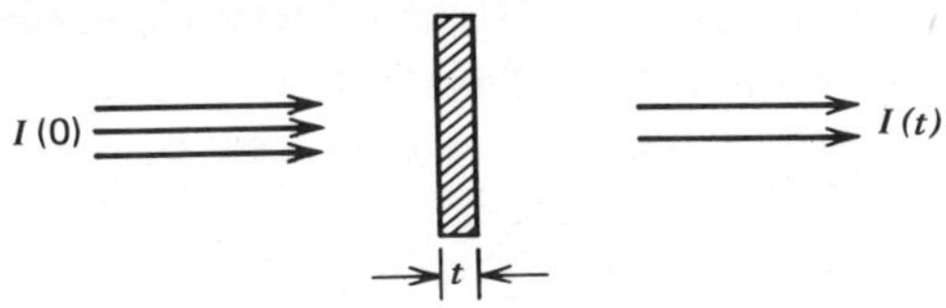

We characterize the strength of a beam with an intensity I neutrons/cm$^2 \cdot$ s, where the area is taken normal to the beam flow direction. The neutrons are usually visualized as flowing in a vacuum or a weakly attenuating medium such as air. We have already demonstrated that the probability of a neutron traveling a distance, s, without an interaction is $e^{-\Sigma s}$. If we visualize the beam of neutrons entering a medium of thickness, t, and total cross section, Σ

$$I(t) = I(0)e^{-\Sigma t} \tag{4.51}$$

This assumes that the number of neutrons in the beam is sufficiently great that we can equate the probability of a reaction to a frequency. In this simple case, $I(0) - I(t) = I(0)(1 - e^{-\Sigma t})$ represents the number of reactions occurring. One note of caution: if the reactions taking place include scattering, then the scattered neutrons are not included in $I(t)$. We sometimes say that what is calculated here are the uncollided neutrons. Those that have been scattered one or more times are referred to as secondary radiation.

Example 4.16

Consider a beam of thermal (0.0253 eV) neutrons with an intensity 10^{12} neutrons/cm$^2 \cdot$ s. A sheet of iron one centimeter thick is placed normal to this beam. How many of the beam neutrons get through?

Solution: From Table A.3, we find that iron has a density of 7.87 g/cm^3, an absorption cross section for low-energy neutrons of 2.55 barns and a scattering cross section of 10.9 barns. We calculate the total cross section

$$\Sigma_t = N\sigma_t = N(\sigma_a + \sigma_s)$$

$$\Sigma_t = \frac{7.87}{55.85} \times 0.6023 \times (2.55 + 10.9)$$

$$\Sigma_t = 1.142 \text{ cm}^{-1}$$

From Eq. 4.51

$$I(t) = 10^{12} e^{-1.142} = 3.19 \times 10^{11} \text{ neutrons/cm}^2 \cdot \text{s}$$

This number of beam neutrons goes through the iron plate unscathed. We can say more, however. Of the 6.81×10^{11} neutrons interacting

$$\frac{2.55}{2.55 + 10.9} \times 6.81 \times 10^{11} = 1.29 \times 10^{11} \text{ neutrons/cm}^2 \cdot \text{s}$$

are absorbed. The rest are scattered. Some of these will be absorbed on second and subsequent flights in the iron. Others will make it out. These latter will be going in all directions, so they will not be a significant part of the beam.

GAMMA PHOTONS

Like neutrons, gamma photons are neutral and travel in undisturbed flight for considerable distances. Precisely the same reasoning applies to their interactions in matter. Thus we can go through the derivation of Eq. 4.1 and 4.2 to obtain the probability of no reaction in distance s

$$P_{NR}(s) = e^{-\mu s} \tag{4.52}$$

and the probability of a reaction

$$P_R(s) = 1 - e^{-\mu s} \tag{4.53}$$

By historical tradition, the nomenclature for gamma interactions differs from that of neutron interactions. The symbol μ, which replaces Σ, is called the linear attenuation coefficient. The mean free path of a gamma photon is $1/\mu$. A beam of collimated photons of intensity $I(0)$ photons/cm$^2 \cdot$ s will be reduced to $I(0)\, e^{-\mu t}$ in traveling through a medium of thickness t.

Gamma interactions differ from neutron reactions in one very important respect. In the energy range of interest to us, gamma photons interact with electrons, while neutrons interact with nuclei. The variation of gamma attenuation coefficients from one atomic species to another is relatively small and shows a regular variation with atomic number, Z. This behavior is in marked contrast to neutron reaction cross sections, which show great variations from one isotope to another.

Example 4.17

The linear attenuation coefficient for 1-MeV gamma photons in iron of density 7.8 g/cm^3 is 0.467 cm^{-1}. If we assign an average microscopic cross section in barns to each orbital electron of an iron atom, what is the value?

Solution: The linear attenuation coefficient is a macroscopic cross section that we can write

$$\mu = N_e \bar{\sigma}$$

N_e is the number density of electrons.

$$N_e = 6.023 \times 10^{23} \frac{Z\rho}{A}$$

$$= 6.023 \times 10^{23} \times \frac{26 \times 7.8}{55.85} = 2.187 \times 10^{24} \text{ electrons/cm}^3$$

$$\bar{\sigma} = \frac{0.467}{2.187 \times 10^{24}} = 2.135 \times 10^{-25} \text{ cm}^2 = 0.2135 \text{ barns}$$

Values of the Linear Attenuation Coefficient

Although the linear attenuation coefficient for gamma photons corresponds directly to the macroscopic cross section for neutrons, the gamma interaction data are not usually presented as microscopic cross sections. They could be, as indicated in Example 4.17; it simply is not the tradition. The usual data presenta-

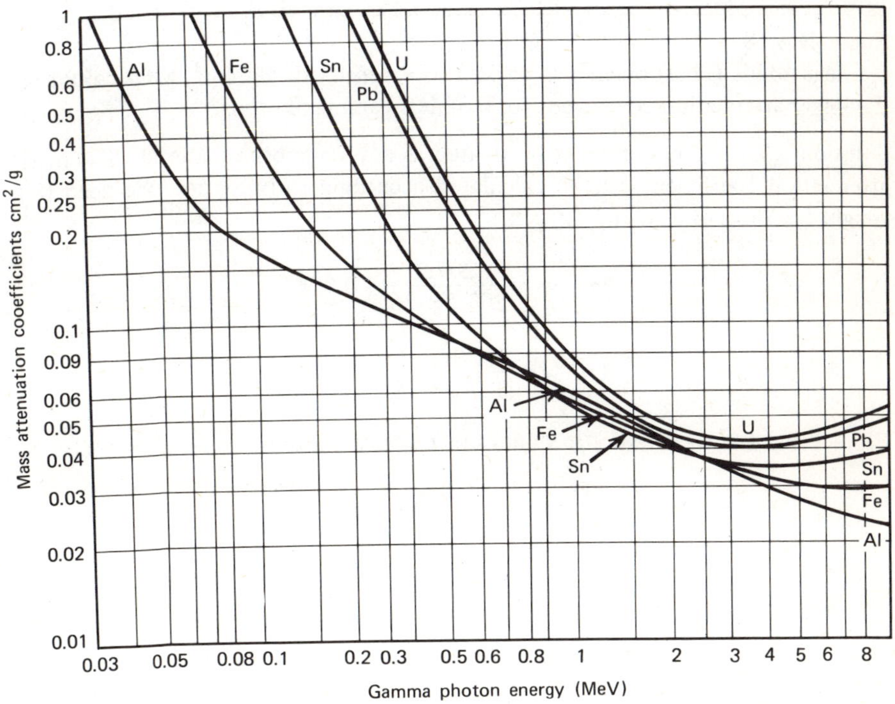

Figure 4.5 Energy dependence of some mass attenuation coefficients.

tion is in the form of μ/ρ values, where ρ is the density of the material and μ/ρ is called the mass attenuation coefficient. This quantity varies much less from one material to another than the linear attenuation coefficient. Values of mass attenuation coefficients are given in Table 4.2. In order to show the trends with gamma energy and with atomic number, some values are plotted in Figure 4.5. It can be seen that at low gamma energies, the high-Z materials (Pb, U) have noticeably higher mass attenuation coefficients; but at values from one to five MeV, the variations are not great. Interpolation by Z value for some material for which data are lacking will be reasonably accurate.

For a composite material the mass attenuation coefficient can be calculated from the formula:

$$\frac{\mu}{\rho} = w_1 \left(\frac{\mu}{\rho}\right)_1 + w_2 \left(\frac{\mu}{\rho}\right)_2 + \cdots \qquad (4.54)$$

where w_i is the weight fraction of element or material i in the composite material, and $(\mu/\rho)_i$ is the mass attenuation coefficient of material i. Although Eq. 4.54 does not look similar to Eq. 4.19, which applies to neutron cross sections in mixtures, the two are in fact quite similar.

Example 4.18
If a lead oxide (PbO) powder is packed to a density of 5 g/cm^3, what would be the linear attenuation coefficient for 0.5-MeV gamma photons?

Solution: This compound can be treated as any composite material. The mass attenuation coefficient can be calculated according to Eq. 4.54. We need the weight fractions of lead and oxygen.

$$\text{Pb:}\quad w = \frac{207.2}{207.2 + 16} = 0.928$$

$$\text{O:}\quad w = \frac{16}{207.2 + 16} = 0.072$$

From Table 4.2, we obtain μ/ρ values for 0.5-MeV photons.

$$\text{Pb:}\quad \mu/\rho = 0.161 \text{ cm}^2/\text{g}$$

$$\text{O:}\quad \mu/\rho = 0.0873 \text{ cm}^2/\text{g}$$

Substituting in Eq. 4.54

$$\text{PbO:}\quad \mu/\rho = 0.928 \times 0.161 + 0.072 \times 0.0873 = 0.156 \text{ cm}^2/\text{g}$$

$$\mu = 0.156 \times 5 = 0.778 \text{ cm}^{-1}$$

Kinds of Gamma Reactions

Gamma photons interact with orbital electrons in different ways. Some knowledge of the possibilities is necessary to an understanding of the behavior of populations of gamma photons in matter. There are three distinct kinds of interactions. These are depicted in Figure 4.6 and are called the photoelectric effect, Compton scattering, and pair production. In the photoelectric interaction, a photon displaces an orbital electron. The photon disappears; its energy goes into

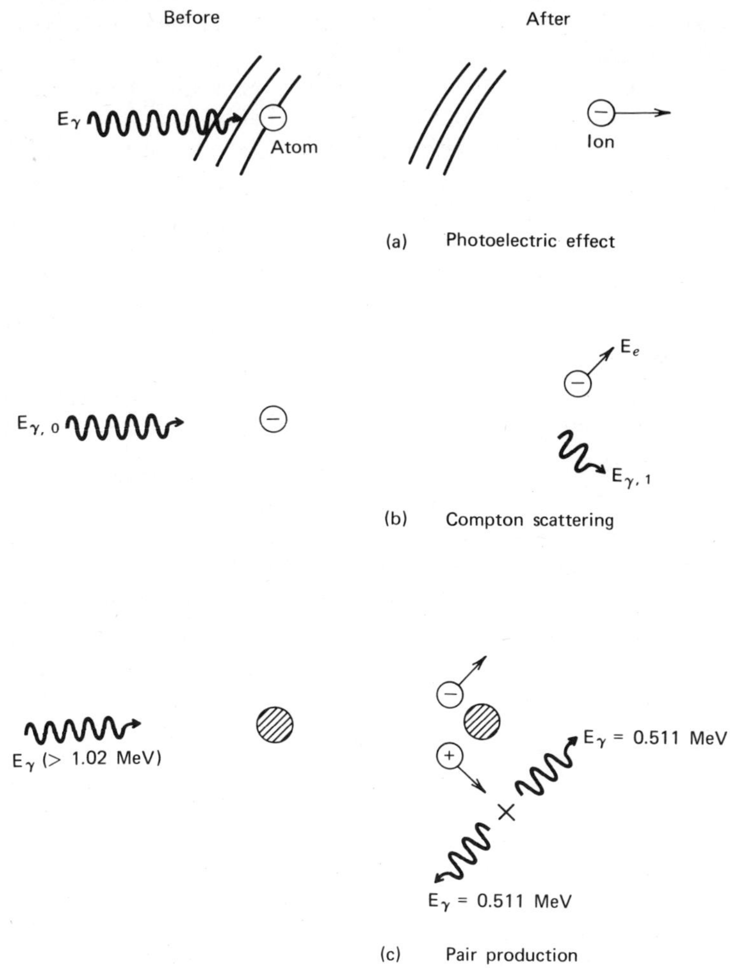

Figure 4.6 Kinds of gamma-electron interactions. *(a)* Photoelectric effect. *(b)* Compton scattering. *(c)* Pair production.

the binding energy of the dislodged electron plus the kinetic energy of the electron. In Compton scattering, as its name implies, the gamma photon is not absorbed but has its direction changed and its energy decreased. We shall deal with this exchange quantitatively in the next section. The third phenomenon, pair production, is much more difficult to visualize. A gamma photon, on encountering some particle such as a nucleus, is converted into an electron and a positron. The nucleus involved plays a necessary role in the reaction mechanics but is not itself greatly affected. The energy of the gamma photons is divided between the rest mass energy of the electron and positron. The rest mass of an electron, positive or negative, is (Table A.2) 5.486×10^{-4} u, or 0.511 MeV. Consequently the gamma photon must have at least 1.022 MeV energy. Any amount over this goes into the kinetic energy of the pair. As indicated in Figure 4.6*c*, the positron gives rise to two photons, each of 0.511 MeV. What happens is that the positron quickly slows down and then interacts with an electron, in the reverse of pair production, to form the two photons. These two photons are called, appropriately enough, annihilation radiation.

Each of the three kinds of reactions just described has its own cross section. Just as we wrote the total neutron cross section as the sum of the cross sections for the individual reactions, we can do likewise with attenuation coefficients (mass or linear).

$$\mu = \underset{\text{photoelectric}}{\mu_e} + \underset{\text{Compton}}{\mu_c} + \underset{\text{pair production}}{\mu_p} \tag{4.55}$$

Figure 4.7 shows a typical breakdown of the mass attenuation coefficient into these three components. The breakdown is strongly energy dependent. The photoelectric effect is the dominant component for gamma energies below a few tenths of an MeV. (In light elements the photoelectric effect is much lower.) Compton scattering tends to predominate in the range from a few tenths MeV to a few MeV. Pair production dominates the other two reactions above a few MeV.

Secondary Radiation

In the case of Compton scattering and pair production (Figure 4.6), it is obvious that the incoming photon has given rise to another photon or photons, although of lower energy. Radiation that is the product of the interactions of radiation is often classified as secondary radiation. Of course, the incoming photons in Figure 4.6 may themselves be secondary radiation. "Primary radiation" usually refers to that emitted by a source, and the source is not evident in Figure 4.6. In practical situations the distinction is not difficult to make.

There are many other sources of secondary electromagnetic radiation, however. The electrons that receive some of the photon's energy can give off electromagnetic radiation in decelerating. This is called bremsstrahlung, or braking

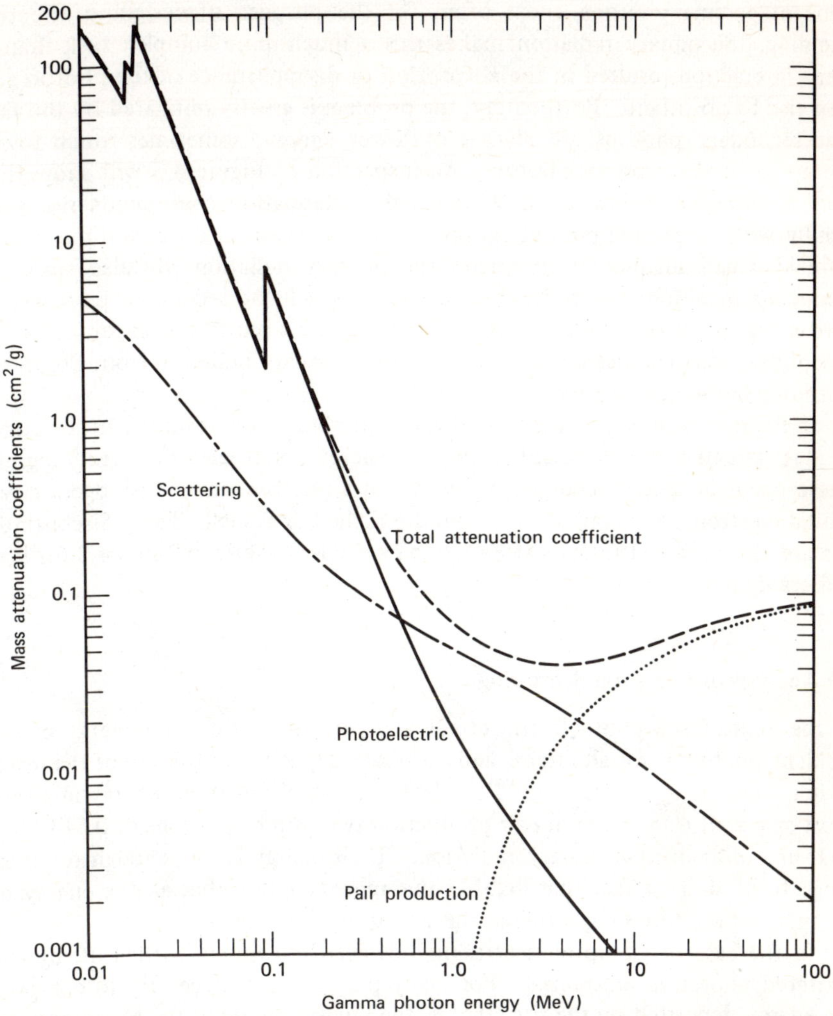

Figure 4.7 Characteristic relative values of components of the attenuation coefficient. Adapted from the *Radiological Health Handbook*, U.S. Department of Health, Education, and Welfare, January 1970.

radiation. The ion that is produced in the photoelectric reaction will quickly claim another electron to return to its normal atomic state. In the process, X rays will be emitted. Remember that from the point of view of their behavior, X rays, gamma photons, or bremsstrahlung are all the same form of radiation; their reactions depend only on their energy. The different names tell us about their origin.

From an engineering standpoint, we are interested in the behavior of populations of gamma photons, most often for the purpose of designing effective shielding. Secondary radiation makes this a much more complex task than if every interaction resulted in the absorption or disappearance of the photon and gave rise to no others. Fortunately, the problem is greatly mitigated by the fact that secondary photons are always of lower energy, sometimes much lower energy, than the primary photon. An inspection of Figure 4.5 will show that once a photon is below an MeV or so, the attenuation coefficients rise very rapidly with decreasing photon energy. In many cases, this fact will mean that a shield which adequately attenuates the primary radiation will also stop the secondary radiation. Compton scattering, in which the secondary photon can have a high fraction of the energy of the incoming photon, is an unfortunate exception. We shall discuss the treatment of secondary radiation from Compton scattering in the next section.

In Figure 4.7, it will be noted that there are some discontinuities in the curve of mass attenuation coefficient in the low-energy (photoelectric effect) region. These occur at specific energies below which a photon cannot dislodge a given orbital electron, for example, one from the K shell or L shell. These discontinuities are also reflected in mass attenuation coefficient values in Table 4.2 for lead and uranium.

An Analysis of Compton Scattering

In the case of the photoelectric effect and pair production, the energy of the incident photon is, for all intents and purposes, deposited at the site of the interaction. The energy is transferred to electrons, which lose their energy in a very short space. It is true that in pair production two gamma photons of 0.511 MeV each are formed as secondary radiation. Their energy is low enough for their range to be short, a fact that justifies the approximation that all the energy of the incident photon is deposited at the site of the reaction.

In the case of Compton scattering, however, the energy carried off by the scattered photon is substantial. For many purposes it is necessary to compute the energy deposited on the site (that is, the kinetic energy of the scattered electron), or the energy of the scattered (secondary) photon, or both. We can see this in an adaptation of Figure 4.6*b*. We can relate the energy of the secondary

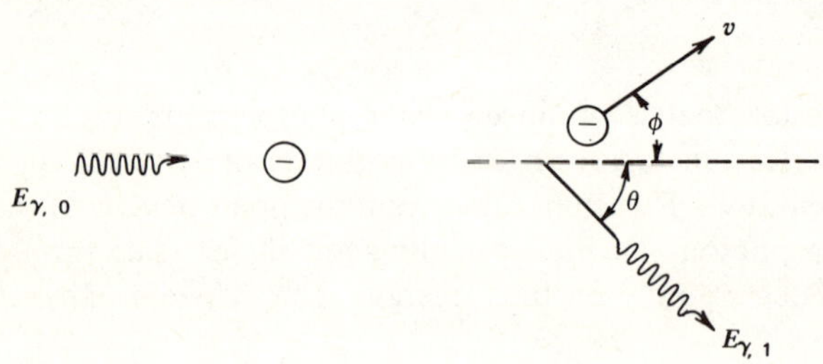

Table 4.2. Mass Attenuation Coefficients[a] (cm^2/g)

PHOTON ENERGY (MeV)	CHEMICAL ELEMENT OR SUBSTANCE											
	H	C	O	Al	Fe	Pb	U	H_2O	AIR	CONCRETE	SiO_2	MUSCLE
0.015	0.376	0.787	1.74	7.93	56.4	115	63.9	1.58	1.55	8.24	5.73	1.63
0.02	0.369	0.429	0.826	3.41	25.5	85.7	71.0	0.775	0.752	3.59	2.49	0.793
0.05	0.335	0.187	0.213	0.369	1.94	7.81	11.1	0.227	0.208	0.392	0.318	0.227
0.10	0.294	0.152	0.156	0.171	0.370	5.40	1.91	0.171	0.154	0.179	0.169	0.170
0.15	0.265	0.135	0.136	0.138	0.196	1.97	2.56	0.151	0.136	0.144	0.140	0.149
0.20	0.243	0.123	0.124	0.122	0.146	0.991	1.28	0.137	0.123	0.127	0.126	0.136
0.50	0.173	0.0872	0.0873	0.0844	0.0840	0.161	0.193	0.0968	0.0870	0.0877	0.0874	0.0960
1.00	0.126	0.0637	0.0637	0.0613	0.0599	0.0708	0.0776	0.0707	0.0636	0.0637	0.0636	0.0700
1.50	0.103	0.0519	0.0518	0.0500	0.0488	0.0517	0.0548	0.0575	0.0518	0.0519	0.0518	0.0570
2.00	0.0875	0.0445	0.0446	0.0432	0.0425	0.0455	0.0475	0.0494	0.0445	0.0448	0.0447	0.0489
5.00	0.0505	0.0271	0.0278	0.0284	0.0314	0.0424	0.0445	0.0303	0.0275	0.0290	0.0287	0.0300
10.00	0.0325	0.0196	0.0209	0.0231	0.0298	0.0484	0.0506	0.0222	0.0204	0.0231	0.0226	0.0219

[a]Values taken from *Radiological Health Handbook*, U.S. Department of Health, Education, and Welfare, January 1970.

photon, $E_{\gamma,1}$, to the scattering angle, θ, using the methods of Chapter 2. We simply balance energy and momentum. However, the gamma photon must be treated relativistically and, in the general case, so must the electron. The energy balance gives

$$E_{\gamma,0} = E_{\gamma,1} + \frac{m_0 c^2}{\sqrt{1 - v^2/c^2}} - m_0 c^2 \tag{4.56}$$

and a momentum balance gives

$$\frac{E_{\gamma,0}}{c} = \frac{E_{\gamma,1}}{c} \cos\theta + \frac{m_0 v}{\sqrt{1 - v^2/c^2}} \cos\phi \tag{4.57}$$

$$0 = \frac{E_{\gamma,1}}{c} \sin\theta - \frac{m_0 v}{\sqrt{1 - v^2/c^2}} \sin\phi \tag{4.58}$$

where m_0 is the rest mass of the electron. One form of the solution to these three simultaneous equations is

$$\frac{E_{\gamma,1}}{E_{\gamma,0}} = \frac{1}{1 + \dfrac{E_{\gamma,0}}{m_0 c^2}(1 - \cos\theta)} \tag{4.59}$$

Since the angle θ can assume any value from 0° to 180°, the energy of the secondary photon, $E_{\gamma,1}$, is variable from one scattering to the next. That means we have a spectrum and a distribution function for θ or $E_{\gamma,1}$. We shall not pursue this matter further except to note that the distribution functions are known. Table 4.3 expresses the results in terms of average values of $E_{\gamma,1}/E_{\gamma,0}$ for different values of $E_{\gamma,0}$.

Table 4.3 The Average Value of Energy Retained by the Secondary Photon in Compton Scattering

$E_{\gamma,0}$ (MeV)	$\overline{E_{\gamma,1}/E_{\gamma,0}}$	$E_{\gamma,0}$ (MeV)	$\overline{E_{\gamma,1}/E_{\gamma,0}}$
0.1	0.861	1.0	0.560
0.15	0.817	1.5	0.505
0.2	0.782	2.0	0.469
0.3	0.729	3.0	0.423
0.4	0.690	4.0	0.394
0.5	0.659	5.0	0.373
0.6	0.632	6.0	0.357
0.8	0.592	8.0	0.333

Example 4.19
Table 4.3 indicates that when a 1-MeV photon is Compton scattered, on the average the scattered photon has 56% (0.56 MeV) of the energy. To what scattering angle does a photon of this energy correspond?

Solution: It is a question of solving Eq. 4.59 for θ. Rearranging that equation

$$\cos\theta = 1 - \frac{m_0c^2}{E_{\gamma,1}} + \frac{m_0c^2}{E_{\gamma,0}}$$

$$\cos\theta = 1 - \frac{0.511}{0.560} + \frac{0.511}{1.0} = 0.598$$

$$\theta = 53.2^\circ$$

The Energy Absorption Coefficient

For at least two important reasons it is desirable to be able to compute the thermal energy release caused by gamma-ray interactions with matter. One is that the damage that radiation causes in animate and inanimate objects is often correlated with the release of such energy. A second is that in high radiation fields the thermal energy released will actually raise the temperature of materials to the point where they are damaged. The design of a cooling system adequate to dissipate this heat requires that the thermal energy generation rate be known.

In the previous sections on neutron reaction rates it was shown that the reaction rate is $\Sigma\phi$. The same expression applies to gamma interactions: the rate is $\mu\phi$ interactions/$\text{cm}^3 \cdot \text{s}$. Here we visualize the flux, ϕ, as the impingement rate of photons, photons/$\text{cm}^2 \cdot \text{s}$, from all directions on a sphere of indefinitely small size located at some point. The cm^2 term in the units of ϕ refer to the area, πr^2, that the small sphere projects in all directions. As we use it, the beam intensity, I photons/cm^2 . s, implies collimation in some direction (that is, a vector). We can then visualize flux as a scalar formed by integrating I over all directions

$$\phi(E, \mathbf{r}) = \int_{4\pi \text{ steradians}} I(\mathbf{\Omega}, E, \mathbf{r})\, d\mathbf{\Omega} \tag{4.60}$$

In the case of a collimated beam, $\phi = I$.

If all of the energy of the reacting photon were deposited at the site of the interaction, the thermal energy release rate would be

$$q'''_{\text{max}} = \mu\phi E_{\gamma,0} \text{ MeV/cm}^3 \cdot \text{s} \tag{4.61}$$

However, we have seen that in Compton scattering a considerable portion of this energy is carried off in the secondary photon. For that reason, Eq. 4.61 gives an

Table 4.4. Mass Energy Absorption Coefficients[a] (cm^2/g)

PHOTON ENERGY (MeV)	CHEMICAL ELEMENT OR SUBSTANCE									
	H	C	O	Al	Fe	Pb	U	AIR	H_2O	MUSCLE
0.015	0.110	0.536	1.44	7.48	55.5	112.0	59.6	1.27	1.28	1.36
0.02	0.0135	0.208	0.575	3.06	24.9	83.3	55.5	0.512	0.512	0.544
0.05	0.0271	0.0233	0.0437	0.184	1.69	7.05	9.30	0.0406	0.0418	0.0409
0.10	0.0406	0.0215	0.0237	0.0386	0.223	5.51	1.54	0.0234	0.0256	0.0252
0.15	0.0481	0.0235	0.0251	0.0285	0.0820	1.90	1.12	0.0250	0.0277	0.0276
0.20	0.0525	0.0265	0.0268	0.0276	0.0498	0.893	0.712	0.0268	0.0297	0.0297
0.50	0.0593	0.0297	0.0297	0.0287	0.0295	0.108	0.122	0.0297	0.0330	0.0327
1.00	0.0555	0.0279	0.0280	0.0270	0.0261	0.0407	0.0458	0.0280	0.0311	0.0308
1.50	0.0507	0.0256	0.0256	0.0248	0.0243	0.0303	0.0317	0.0256	0.0284	0.0281
2.00	0.0465	0.0237	0.0238	0.0232	0.0233	0.0286	0.0278	0.0237	0.0263	0.0257
5.00	0.0320	0.0178	0.0185	0.0194	0.0228	0.0350	0.0324	0.0182	0.0200	0.0188
10.00	0.0229	0.0147	0.0160	0.0185	0.0254	0.0457	0.0432	0.0156	0.0168	0.0154

[a]Values taken from R. G. Jaeger (editor), *Engineering Compendium on Radiation Shielding*, Springer-Verlag, New York, 1968. See Tables 4.1–4.9.

upper bound to the energy release rate termed q'''_{max}. In order to calculate the thermal energy release correctly, the attenuation coefficient, μ, is adjusted to reflect the Compton phenomenon. The result is an energy absorption coefficient, μ_E, often called simply absorption coefficient.

$$q''' = \mu_E \phi E_{\gamma,0} \text{ MeV/cm}^3 \cdot s \tag{4.62}$$

There is also the mass absorption coefficient, μ_E/ρ. Values of this quantity are tabulated in Table 4.4.

Example 4.20
A beam of 1-MeV photons with an intensity of 10^{12} photons/cm^2 · s impinges on a steel plate one centimeter thick. Estimate the rate at which thermal energy must be removed from this plate (W/cm^2) in order to hold its temperature steady.

Solution: Although steel contains other alloying metals, their absorption coefficients are all close enough to iron to permit us to assume the steel is all iron. From Table 4.4

$$\mu_E/\rho = 0.0261 \text{ cm}^2/\text{g}$$

For this case, the rate of energy release is, by Eq. 4.62

$$q''' = \mu_E \phi E = \mu_E I E$$

We know the intensity of the impinging beam is $I(0) = 10^{12}$ photons/cm^2 · s. However, the beam may be significantly attenuated through the plate. Emerging from the plate, the beam strength is

$$I(t) = I(0)\, e^{-\mu t}$$

where μ is the attenuation coefficient. From Table 4.2, this value is

$$\mu/\rho = 0.0599 \text{ cm}^2/\text{g}$$

Assume a density of 7.8 g/cm^3.

$$\mu = 0.0599 \times 7.8 = 0.467 \text{ cm}^{-1}$$

$$t = 1 \text{ cm}$$

$$I(t) = 10^{12} \times e^{-0.467} = 6.27 \times 10^{11}$$

The beam is indeed significantly attenuated. The value $\mu_E IE$ should be integrated over the thickness of the plate

$$q''' = \mu_E E \int_0^t I(x)\, dx$$

$$q''' = \mu_E EI(0) \int_0^t e^{-\mu x}\, dx$$

$$q''' = \frac{\mu_E}{\mu} EI(0)[1 - e^{-\mu t}]$$

$$= \frac{0.0261}{0.0599} \times 1 \times 10^{12} \times 0.373 \times 1.602 \times 10^{-13}$$

$$= 0.0260 \text{ W/cm}^2$$

This value is in fact a minimum value, since it considers only the interactions of photons in the incident beam—the uncollided photons. The secondary photons, which also contribute to the flux, ϕ, are not included. Their calculation is too complex for treatment in this text. However, we can compute an upper bound, q'''_{max}, from Eq. 4.61, integrating as before in this example. Then

$$q'''_{max} = EI(0)[1 - e^{-\mu t}]$$

$$= 0.0598 \text{ W/cm}^2$$

A thermal power between 0.026 and 0.060 W/cm^2 must be transferred from the plate in order to hold it at a steady state.

CHARGED PARTICLES

We have dealt with the passage, interaction, and energy loss of neutral particles (neutrons and photons) in matter. It has been feasible to treat the interactions as discrete countable events, because the number of these that a neutron or photon undergoes in its lifetime is a manageable one. Charged particles (electrons, protons, alpha particles, ions), on the other hand, undergo coulombic interactions with electrons of atoms. These events are so numerous in the life of a charged particle that it is only feasible to treat the passage as one of continuous slowing down. To use the analogy on page 91, it is more like a person walking into a bog than like a person walking through a rockstrewn field.

A few generalizations regarding charged particles are easily understood and will help to place the matter in perspective. The distance traveled by charged particles of a given energy in a given material is much shorter than that traveled by a neutron or gamma photon. A 10-MeV proton, for example, will travel about 1 mm in water. One frequently encounters semiquantitative statements to the effect that a sheet of paper will stop alpha particles and a sheet of glass will stop beta particles. (It is necessary to know that these statements refer to typical particles from radioactive decay, such as 5-MeV alpha particles and 1-MeV betas.) For many purposes such generalizations are sufficient. We wish to be a bit more quantitative here.

Before presenting our treatment, we should say something about *why* we are interested in charged-particle slowing down. For one thing, the energy loss and life-termination reactions of neutrons and gamma photons almost always produce charged particles. In order to understand several aspects of these terminal processes—for example, over what distance the energy is converted to thermal energy, and the location and extent of resulting cell damage in biological systems—we must know the range and energy-loss rate of charged particles. In some cases, such as shielding of manned space vehicles from protons in the Van Allen belt (where protons have energies of 1–100 MeV), the charged particle is relatively penetrating. We do not encounter charged particles of such penetrability in radioactive decay, fission, or fusion, however.

Interaction of Electrons

It is convenient to categorize charged particles as light (electrons) and heavy (protons and heavier), although their interactions or energy loss in matter involve the same mechanisms. For low-energy electrons ($E \lesssim 2m_0c^2$ or $T \lesssim m_0c^2$), the mechanisms are ionization and excitation of the atoms in the medium. Table 4.5 gives the relevant quantitative information on the interaction of electrons in various materials. The data also apply to positrons. For a series of values of electron energy, the electron behavior is expressed as "stopping power," range, and radiation yield. The stopping power is the instantaneous rate of kinetic energy loss, expressed $-dT/\rho ds$ in units of MeV · cm^2/g. The range, as its name implies, expresses the average distance the electron of the given energy will travel in the medium. The range is given as ρR in units of g/cm^2. As is common practice in shielding work, the data are given in terms that are independent of the density of the material in question. In order to convert the range values to linear distance units, the range value in g/cm^2 must be divided by the material density. The range is calculated from the stopping power by integrating

$$R = \int_{T_0}^{0} \frac{dT}{dT/ds} \tag{4.63}$$

The radiation yield values given in Table 4.5 refer to the fact that electrons, in decelerating, lose some of their energy by emission of electromagnetic radiation. This radiation is called bremsstrahlung. The fraction of the electron's energy that appears in this form is proportional to TZ, where Z is the atomic number of the medium. An inspection of Table 4.5 will show that the radiation yield does indeed increase with T and Z. Bremsstrahlung radiation is worthy of

Table 4.5 Energy Loss Rate and Range of Electrons in Various Materials[a]

T_0 (MeV)	STOPPING POWER $-dT/\rho ds$ (MeV · cm²/g)	RANGE (g/cm²)	RADIATION (BREMSSTRAHLUNG) YIELD (%)	STOPPING POWER $-dT/\rho ds$ (MeV · cm²/g)	RANGE (g/cm²)	RADIATION (BREMSSTRAHLUNG) YIELD (%)
	Hydrogen			H_2O		
0.1	8.768	0.0066	0.01	4.202	0.0140	0.07
0.25	5.293	0.0301	0.03	2.580	0.0625	0.15
0.5	4.209	0.0845	0.05	2.071	0.1735	0.26
0.75	3.926	0.1464	0.07	1.939	0.2990	0.38
1.0	3.832	0.2110	0.09	1.893	0.3771	0.44
2.0	3.846	0.4730	0.18	1.889	0.9613	0.89
5.0	4.152	1.222	0.46	2.014	2.499	2.08
10.0	4.490	2.377	0.97	2.183	4.880	4.15
	Tissue			Air		
0.1	4.154	0.0142	0.07	3.632	0.0163	0.08
0.25	2.550	0.0632	0.14	2.240	0.0722	0.17
0.5	2.052	0.1754	0.26	1.810	0.1996	0.30
0.75	1.927	0.3019	0.37	1.705	0.3428	0.43
1.0	1.890	0.4332	0.48	1.676	0.4910	0.55
2.0	1.922	0.9604	0.87	1.714	1.084	1.08
5.0	2.125	2.442	1.98	1.913	2.736	2.23
10.0	2.374	4.661	3.84	2.159	5.188	4.28

	Aluminum			Iron		
0.1	3.20	0.0186	0.16	2.857	0.0211	0.37
0.25	1.998	0.0816	0.33	1.808	0.0911	0.77
0.5	1.621	0.2243	0.59	1.485	0.2475	1.34
0.75	1.529	0.3840	0.84	1.411	0.4211	1.84
1.0	1.502	0.5493	1.06	1.394	0.5997	2.31
2.0	1.528	1.212	1.88	1.447	1.307	3.94
5.0	1.680	3.227	8.20	1.692	3.076	4.08
10.0	1.924	5.841	7.72	2.033	5.926	14.40
	Lead					
0.1	2.024	0.0310	1.67			
0.25	1.350	0.1267	3.34			
0.5	1.167	0.3303	5.48			
0.75	1.150	0.5471	7.19			
1.0	1.168	0.7631	8.67			
2.0	1.316	1.571	13.30			
5.0	1.779	3.519	22.92			
10.0	2.493	5.879	33.63			

[a]Values taken from M. J. Berger and S. M. Seltzer, "Tables of Energy Losses and Ranges of Electrons and Positrons," in document *Studies in Penetration of Charged Particles in Matter*, NAS-NRC Publication 1133, 1964.

note, because it means that one can do a perfectly good job of stopping or shielding electrons, only to find that this form of secondary radiation causes problems. If bremsstrahlung is to be avoided, one prefers a low-Z material such as aluminum for shielding electrons rather than a high-Z material such as lead.

If it is desired to extend the information of Table 4.5 to materials not given there, a useful approximation is that the stopping power varies directly, and the range inversely, as Z/A of the medium. For materials that contain a mixture of elements, Z/A is calculated by summing $w_i Z_i/A_i$, where w_i is the weight fraction of the element having atomic and mass numbers Z_i and A_i, respectively.

Example 4.21
What distance in iron ($\rho = 7.8$ g/cm^3) would an electron travel in slowing down from 10 MeV to 1.0 MeV?

Solution: From Table 4.5, the range of a 10-MeV electron in iron is 5.926 g/cm^2.

$$R = \frac{5.926}{7.8} = 0.760 \text{ cm}$$

For a 1-MeV electron

$$R = \frac{0.5997}{7.8} = 0.077 \text{ cm}$$

The electron travels 0.760 - 0.077 = 0.683 and deposits 9 MeV of kinetic energy (converted to thermal energy) in that distance. The average dT/ds is 9/0.683 = 13.2 MeV/cm, or $dT/\rho ds$ is 1.69 MeV · cm^2/g.

Example 4.22
One μCi of a beta-emitting radioisotope is embedded in tissue. If it emits a single beta with E_{max} of 1 MeV, what is the range of the beta electrons and what is the beta energy deposition (J/kg · h) in a sphere of tissue having a radius equal to the range?

Solution: Assume the density of tissue is 1.0 g/cm^3. Table 4.5 indicates that the range is 0.4332 g/cm^2.

$$R = \frac{0.4332}{\rho} = 0.4332 \text{ cm}$$

The volume of tissue with radius R is

$$V = \tfrac{4}{3}\pi R^3 = 0.3405 \text{ cm}^3$$

The rate of energy deposition, remembering that the average beta energy is only one third of the maximum, is

$$\dot{q} = 3.7 \times 10^4 \times 1 \times \tfrac{1}{3} \times 1.602 \times 10^{-13} = 1.976 \times 10^{-9} \text{ J/s}$$

The energy is deposited over 0.3405×10^{-3} kg of tissue.

$$\frac{\dot{q}}{m} = \frac{1.976 \times 10^{-9} \times 3600}{0.3405 \times 10^{-3}} = 0.0209 \text{ J/kg} \cdot \text{h}$$

In the next chapter we will learn that this is a local absorbed dose rate of about 2 rad/h.

Interaction of Heavy Charged Particles

The deposition of kinetic energy in material is more complicated for charged particles such as protons, alpha particles, and so on, than for electrons, because two additional variables are introduced; namely, the charge and mass of the particle in question. Table 4.6 gives the stopping power for protons, ^1H, in a variety of materials. These stopping-power values may be extended to other charged particles of the same kinetic energy per unit mass (MeV/u) by multiplying by Z^2, the square of the particle's charge. For example, an 8-MeV alpha particle has an energy of 2 MeV/u. Its rate of energy loss in aluminum is 2^2 times 112 MeV · cm² /g, or 448 MeV · cm² /g.

The range of protons in various materials is given in Table 4.7. These values may be extended to other charged particles by the formula

$$R = R_H \frac{M_p}{Z^2} \tag{4.64}$$

where M_p is the mass of the particle in atomic mass units and Z is the particle's charge. The value R_H must be read for a particle energy per atomic mass unit, T/M_p (MeV/u), corresponding to the first column of Table 4.7.

Table 4.6 Rate of Energy Loss (Stopping Power) of Protons in Various Materials[a]

PROTON ENERGY T/M_p (MeV/u)	$-dT/\rho ds$ (MeV · cm² /g) H	H_2O	MUSCLE	AIR	Al	Fe	Pb
1.0	673	274	271	224	170	132	63
2.0	386	170	168	142	112	90	46
4.0	220	100	99	84	69	56	31
6.0	158	71	70	60	50	41	24
8.0	124	56	55	47	39	33	20
10.0	102	47	46	40	34	29	17

[a]Values taken from W. H. Barker and M. J. Berger, "Tables of Energy Losses and Ranges of Heavy Charged Particles," in document *Studies in Penetration of Charged Particles in Matter,* NAS-NRC Publication 1133, 1964.

Table 4.7 Range of Protons in Various Materials[a]

PROTON ENERGY T/M_p (MeV/u)	H	H_2O	MUSCLE	R_H (g/cm^2) AIR	Al	Fe	Pb
1.0	0.0008	0.0023	0.0024	0.0029	0.0041	0.0055	0.0122
2.0	0.0029	0.0071	0.0072	0.0087	0.0115	0.0149	0.0309
4.0	0.0101	0.0230	0.0234	0.0277	0.0350	0.0438	0.0844
6.0	0.0209	0.0471	0.0477	0.0562	0.0697	0.0858	0.1577
8.0	0.0353	0.0791	0.0801	0.0941	0.1157	0.1406	0.2496
10.0	0.0532	0.1183	0.1198	0.1402	0.1703	0.2047	0.3561

[a]Values taken from W. H. Barker and M. J. Berger, "Tables of Energy Losses and Ranges of Heavy Charged Particles," in document *Studies in Penetration of Charged Particles in Matter*, NAS-NRC Publication 1133, 1964.

Example 4.23

What linear distance would you expect a 10-MeV proton to travel in air at standard conditions of temperature and pressure? Apply the same question to a 4-MeV alpha particle.

Solution: The density of air at 10^5 Pa and 20°C is 1.22 kg/m^3 or 1.22×10^{-3} g/cm^3. The range of a 10-MeV proton is read from Table 4.7 to be 0.1402 g/cm^2. Therefore the linear range is

$$\frac{0.1402}{1.22 \times 10^{-3}} = 115 \text{ cm}$$

For a 4-MeV alpha particle, $T/M = 1.0$ MeV/u. The proton range in air, R_H, at this value is 0.0029 g/cm^2. The range of an alpha particle, by Eq. 4.64, is by Eq. 4.64, is

$$R = 0.0029 \times \tfrac{4}{4} = 0.0029 \text{ g/cm}^2$$

In linear distance, the range is

$$\frac{0.0029}{1.22 \times 10^{-3}} = 2.4 \text{ cm}$$

REFERENCES

1. "Radiological Health Handbook," U. S. Department of Health, Education, and Welfare, January, 1970.
2. R. G. Jaeger (editor), *Engineering Compendium on Radiation Shielding*, Springer-Verlag, New York, 1968.

3. M. J. Berger, S. M. Seltzer, "Tables of Energy Losses and Ranges of Electrons and Positrons," in document, "Studies in Penetration of Charged Particles in Matter," NAS-NRC Publication 1133, 1964.
4. W. H. Barkus, M. J. Berger, "Tables of Energy Losses and Ranges of Heavy Charged Particles," in document, "Studies in Penetration of Charged Particles in Matter," NAS-NRC Publication 1133, 1964.

PROBLEMS

4.1 A collimated neutron beam has an intensity of 10^7 neutrons/cm$^2 \cdot$ s, an area of 10 cm^2, and the neutrons have a velocity of 2200 m/s. The half-life of neutrons for decay to a proton and electron is 12 min. How many neutron disintegrations will occur in a section of the beam 1 m long?

4.2 A large sheet of lead (density 11.35 g/cm^3) one half inch thick is placed normal to the beam of the previous problem. What fraction of the neutrons would you expect to get through this plate without absorption?

4.3 One gram of manganese is placed in a neutron flux of 10^7 neutrons/cm$^2 \cdot$ s for one hour and then removed. What is the activity (curies) of ^{56}Mn at the time of removal? What is the activity one hour later?

4.4 Dry air has a density of 0.0013 g/cm^3 and is roughly 79 wt % nitrogen and 21 wt % oxygen.

(a) What is the macroscopic cross section for scattering and for absorption of thermal (2200 m/s) neutrons in air?

(b) What percentage of neutrons absorbed in air are absorbed in oxygen?

(c) What is the probability of a neutron traveling 100 m in air without reacting?

4.5 A gold foil weighing 0.2 g is irradiated for 5 days at constant thermal neutron flux. Three days after the termination of this irradiation it is counted and found to have an activity of 50 disintegrations per second.

(a) Estimate the neutron flux, assuming all neutrons have a speed of 2200 m/s.

(b) If this foil is allowed to decay for a long time, how many grams of mercury-198 will be in the foil?

4.6 When a 0.0025 cm thick foil of material of mass number 10 and density 2 g/cm^3 is bombarded with a neutron beam perpendicular to the foil plane, it is observed that only 50% of the neutrons pass through. Assuming that scattering is negligible, calculate the microscopic cross section for neutron absorption of this material.

4.7 The following information is available on Type 347 stainless steel:

density = 7.98 g/cm^3

COMPONENT	WT %
Iron	72
Chromium	18
Nickel	10

Calculate the macroscopic cross section of this steel for capture of thermal neutrons.

4.8 A neutron population has a speed distribution function

$$n(v) = Cv^{-3} \frac{\text{neutrons}}{\text{cm}^3 \cdot \text{cm/s}} \qquad \text{for } v > 10^6 \text{ cm/s}$$

What is the average speed of neutrons in this population?

4.9 Consider pure H_2O, in which the thermal neutron flux is 10^{13} neutrons/cm^2 · s. Suppose you were able to identify one particular hydrogen nucleus. What is the probability that it would absorb a neutron during a one-hour period?

4.10 In a certain region in a reactor, the uranium-235 density is 1 g/cm^3 and the neutron flux is 10^{13} neutrons/cm^2 · s. The microscopic fission cross section of uranium-235 is 580 barns. Calculate the heat generation in this region in watts/cm^3.

4.11 A fuel material for a reactor has the composition:

1 a/o ^{235}U
1 a/o ^{239}Pu
10 a/o molybdenum
88 a/o ^{238}U

How many fission neutrons are produced per absorption of a thermal neutron in this fuel? The $\bar{\nu}$ values (Chap. 3) of ^{235}U and ^{239}Pu are 2.5 and 3.0, respectively.

4.12 A collimated beam of neutrons having an intensity of 10^5 neutrons/s cm^2 · s impinges on a 0.0625-inch plate of gold (^{197}Au). The plate is placed across the beam, making an angle of 45° with it. The neutrons all have speeds close to 2200 m/s. Estimate the activity of 1 cm^2 of the plate after 5 days' irradiation.

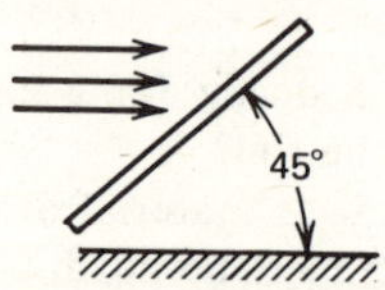

4.13 The spectrum or distribution function of neutrons above 1 eV is

$$n(E) = \frac{C}{E^2} \frac{\text{neutrons}}{\text{cm}^3 \cdot \text{eV}}$$

(a) What fraction of this population has an energy below 10 eV?
(b) What is the average energy of neutrons in the 1–10 eV range?

4.14 At a certain location in a reactor, the neutron flux is 10^{12} neutrons/cm^2 · s. One gram of an isotope ^{60}X, which has a microscopic absorption cross section of 10 barns, is placed in this location. At what rate are neutrons absorbed in this sample?

4.15 A certain medium has the properties

$$\Sigma_s = 0.1 \text{ cm}^{-1}$$
$$\Sigma_a = 0.01 \text{ cm}^{-1}$$

(a) What is the probability of a neutron traveling 1 cm in this medium without interacting with a nucleus?

(b) If the neutron speed is 2×10^5 cm/s, what is the average time between scattering collisions?

4.16 A sample of material is placed in a fission reactor. A certain radioisotope having a half-life of 2 days is being produced in this sample by an (n, γ) reaction at the rate of 10^{12} reactions/s. What will be the activity (disintegrations/s) of this isotope at the end of 3 days in the reactor?

4.17 A homogeneous mixture of UO_2 in BeO has a density of 3.20 g/cm^3. The UO_2 forms 5 wt % of the mixture and the uranium is 90% enriched. At a temperature of 1000°C, the thermal-neutron flux is 10^{14} neutrons/cm^2 · s. What is the power density (watts/cm^3) in this material?

4.18

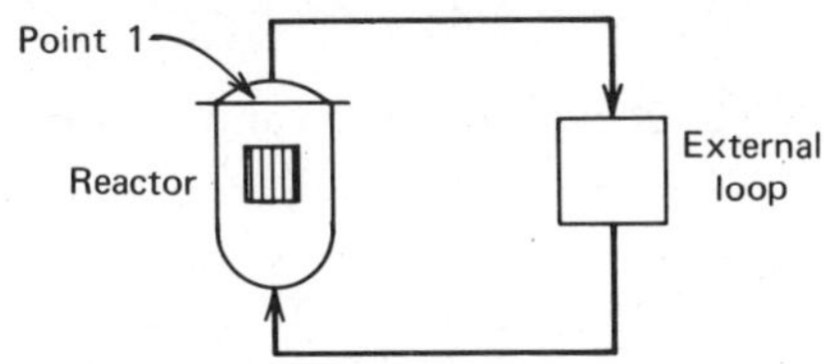

A radioisotope is formed by activation in the circulating coolant of a reactor system. On each pass, the coolant spends t_i seconds in the reactor and t_c seconds in the external loop. The radioisotope (decay constant λ) is produced at a constant rate, P nuclei/cm^3 . s, while the coolant is in the reactor and, of course, simply decays while the coolant is in the rest of the loop. Assuming that the activation process reaches a steady state, what will be the activity (dps/cm^3) in the coolant at Point 1 (at the exit of the reactor) at equilibrium? Show how the relationship can be simplified for the cases when the isotope half-life is very short and very long relative to t_i, t_c.

4.19 Consider a 14.3-MeV neutron slowing down in graphite. Assume only elastic scattering occurs and that the cross section is constant at 4 barns.

(a) How many scatters, on the average, will it take to reduce the neutron to 0.1 eV?

(b) What is the gross distance the neutron travels in slowing down?

(c) What is the expected or average value of the neutron after its first collision?

4.20 Consider a slab made up of a laminate of alternating layers of aluminum and iron, each 1 mm thick. A 0.025-eV neutron enters normal to this slab. The outer layer is aluminum.

(a) What is the probability that the neutron will make its first interaction in the third iron layer?

(b) In which layer is the neutron's first interaction most likely to occur?

(c) If one treats this laminate as a homogeneous mixture, what is the macroscopic cross section for neutron scattering and capture? What is the linear attenuation coefficient (μ) for 1-MeV gamma photons?

4.21 A stable nuclide, ^{100}X, has a cross section of 2 barns for neutron scatter and of 1 barn for neutron capture. Its density in pure form is 10 g/cm^3.

(a) If this material is in a flux of 10^{10} neutrons/cm^2 · s, what is the neutron capture rate in captures/g-X · s?

(b) If a neutron is in a large volume of X (so that it will not escape), what is the probability that it will undergo just two scatters and then be captured?

(c) On the average, how much energy would a 1-MeV neutron lose in a collision with a nucleus of X?

4.22 If a neutron enters normal to a 0.25-inch-thick sheet of iron, what is the probability of its undergoing a reaction (scattering or capture) if (a) it is a slow (0.025 eV) neutron, and if (b) it is a fast (~2 MeV) neutron? What is the relative probability of these two neutrons being captured (n, γ) in the sheet?

4.23 Following are slow (2200 m/s) neutron capture cross sections (in barns) for the principal isotopes found in 304 stainless steel:

Fe		Cr		Ni	
A	σ_c	A	σ_c	A	σ_c
54	2.9	50	17	58	4.4
56	2.7	52	0.8	60	2.6
57	2.5	53	18	61	2
58	1.1	54	0.38	62	15
				64	1.5

From the handbook properties of 304 stainless, calculate the macroscopic capture cross section.

4.24 A typical concrete has a density of 2.7 g/cm^3 and has the following approximate composition.

COMPONENT	w/o
H_2O	10
CaO	15
SiO_2	75

The mass absorption coefficients of 3-MeV gamma photons are

ELEMENT	cm^2/g
H	0.0693
O	0.0359
Si	0.0367
Ca	0.0378

What minimum thickness of concrete is required to ensure that an impinging 3-MeV photon has only one chance in a million of getting through without any interaction? Assume that concrete is poured without holes.

4.25 If a 1-MeV neutron scatters elastically off a ^{6}Li nucleus, what is the maximum energy it can lose? On the average, what energy would it lose? What is the expected number of scatters in ^{6}Li that would decrease the energy of a neutron from 1000 eV to 1 eV?

4.26 Uranium dioxide, UO_2, is by far the favored chemical form for fuel in current fission reactors. Its density is about 10 g/cm^3. In an LWR it is typically enriched to 3 w/o ^{235}U. Use the 0.025-eV and 2-MeV cross sections in Table A.3.

(a) Compute the macroscopic cross section for fission, absorption, and scattering for 0.025-eV and 1-MeV neutrons.

(b) In each neutron energy case, what is the probability that a neutron entering normal to a one-half-inch-thick slab of UO_2 would be absorbed on the first flight (that is, without first scattering)?

(c) If this fuel is in a power reactor in which the thermal neutron flux is 10^{13} neutrons/cm$^2\cdot$s, what will be the thermal power density (watts/cm^3) in the fuel? Consider only the thermal neutrons.

4.27 Helium has a microscopic neutron scattering cross section of 0.8 barns. This cross section is essentially energy independent. Its absorption cross section is essentially zero. What is the gross distance (not the crow-flight distance) that a neutron would travel in slowing down from 1 MeV to 1 eV in helium at 700 psia and 1000°F?

4.28 Consider a population of neutrons, all of 0.0253 eV energy (2200 m/s speed) in an infinite medium that has the following macroscopic cross sections:

$$\Sigma_s = 1.0 \text{ cm}^{-1}$$

$$\Sigma_f = 0.01 \text{ cm}^{-1}$$

$$\Sigma_c = 0.03 \text{ cm}^{-1} \quad \text{(for } n,\gamma \text{ reaction)}$$

$$\nu = 3 \text{ neutrons per fission}$$

(a) On the average, how many scatters occur in the life of these neutrons in this medium?

(b) What is the average lifetime of these neutrons? (How much time to disappearance?)

(c) For every 100 source (0.0253-eV neutrons, how many fission neutrons are produced?

(d) Imagine that the source of 0.0253-eV neutrons is shut off instantaneously and that at that instant there are one million source neutrons in the medium. Estimate how long afterward one might still expect to find neutrons in this medium. Remember that those in existence at the time of the shutoff will go on to form some fission neutrons, which in turn will slow down and cause more fissions, and so on. Assume that the slowing-down process takes no time.

4.29 In an experiment used to measure total neutron cross sections, neutrons impinge normal to a flat-plate target. In one experiment the target material is pure manganese of density 7.2 g/cm^3 and thickness 1 mm. The fraction transmission of the neutrons is measured to be 0.7. What value of microscopic total cross section is implied for ^{55}Mn for neutrons of the beam energy?

4.30 It is the same situation as above, but the material is manganese dioxide (MnO_2) of density 4 g/cm^3 and thickness 1 mm. The transmission is 0.85. The microscopic total cross section of oxygen is known to be 4 barns. What is the implied total cross section for ^{55}Mn?

4.31 A long, cylindrical rod of pure ^{235}U, having a density of 1.88 g/cm^3 and diameter of 0.5 cm, is placed in a collimated neutron beam with its axis perpendicular to the beam direction. The beam has a uniform current of 10^{13} neutrons/cm^2 · s impinging on the surface of the cylinder. How many watts of fission energy is generated in the rod per foot (30.48 cm) of length? You should allow for diminution of beam strength through the cylinder. Note that the neutron path length varies according to where the neutron impinges on the surface of the cylinder.

4.32 A collimated neutron beam of 10^{12} neutrons/cm^2 · s impinges on a flat plate of uranium metal normal to the plate. The plate is two centimeters thick, has a density of 18.8 g/cm^3, and is 50 wt % ^{235}U and 50 wt % ^{238}U. Assume scattering is negligible. What is the capture rate of thermal neutrons in the ^{238}U, in reactions/s · g-uranium?

4.33 The product of the ^{238}U (n, γ) reaction is, of course, ^{239}U, which has a half-life of 23.5 min. If the uranium plate of Problem 4.32 is placed in the beam for 1 hour, what activity of ^{239}U, in Ci/g-uranium, will be present at that time?

4.34 If a gram of sodium iodide (NaI), obtained from a chemistry stockroom, is placed in a slow (0.025 eV) neutron flux of 10^9 neutrons/cm^2 · s for 15 minutes, how much of what kind of radioactivity would be produced?

Chapter 5

Health effects of nuclear radiation

Nuclear radiation has effects on all matter on which it impinges. By nuclear radiation we mean electrons, protons, alpha particles, gamma photons, neutrons, and any other charged or uncharged particle or photon emitted as a result of nuclear reactions. In Chapter 4, we examined the interactions of this radiation in terms of distances traveled, kinds of reactions, and so on. In short, we studied the phenomena in terms of the physics of radiation effects at the microscopic level. Such effects (for example, the production of an ion in the photoelectric reaction of a gamma photon—see Figure 4.6) produce changes in animate and inanimate material. These changes can affect the macroscopic properties or, in the case of biological systems, the functioning of the materials.

By far the greatest concern in the field of radiation effects is directed to the human as the radiation-receiving object. We shall not attempt to trace a series of microscopic events, starting with one or many ionizations in the tissue of an individual and terminating with the diagnosis of cancer or other defect in that individual. For one thing, the details of such a mechanism are not known at the present time. For another, the information that is known is well beyond the scope of this text. The student wishing to study this subject in greater depth might consider either of the following:

K. Z. Morgan and J. E. Turner, *Principles of Radiation Protection*, Wiley, New York, 1967.

M. Eisenbud, *Environmental Radioactivity*, Second Edition, Academic Press, New York and London, 1973.

We shall use the quantitative causal relationships that various scientific bodies have found to be consistent with the state of knowledge and to be prudent from industrial health and public health standpoints.

QUANTIFICATION OF RADIATION RECEIVED

Sources of Radiation

In previous chapters we have discussed most of the sources of radiation of concern. They are summarized in Table 5.1.

Table 5.1 Sources of Radiation

RADIATION	PRINCIPAL SOURCES
Electrons	Beta-emitting radioisotopes Accelerators
Protons	$^1H(n, n)$–fast neutron scattering in hydrogen $^{14}N(n, p)$ reaction Cosmic rays
Alpha particles	Alpha-emitting radioisotopes
Gamma rays (and X rays)	Radioisotopes (n, γ) reactions X-ray devices Fission reactors
Cosmic rays	Outer space
Neutrons	Fission reactors Fusion reactors Accelerators

Units of Radiation Received

In order to be quantitative about the effects of radiation, we must have a measure of the amount of radiation that a person, or some organ of a person, receives. We need well defined quantities and units. There has been an evolution of the definition of these units over the years. The principal ones currently in use are described briefly.

The Roentgen (R), a Unit of Exposure. Exposure refers *only* to gamma rays or X rays (electromagnetic radiation). The unit, a Roentgen, is defined in terms of the number of ion pairs, or the resulting electric charge, formed by radiation in a specified mass of dry air.

$$1 \text{ R} = 2.58 \times 10^{-4} \text{ C/kg-air} \qquad (5.1)$$

Exposure is primarily of interest in instrument calibration, because electric charge is a quantity that instruments can readily measure.

The Rad, a Unit of Absorbed Dose. A quantity that relates more directly to the effect of radiation in any material is the radiation energy absorbed (that is, converted to thermal energy) in the material. This quantity is called the absorbed dose (D_{abs}). The unit of the absorbed dose is

$$1 \text{ rad} = 100 \text{ ergs/g} = 0.01 \text{ J/kg} \qquad (5.2)$$

The unit mass here is of the material that absorbs the radiant energy. Another common unit is the millirad, or mrad. The absorbed dose can refer to any kind of radiation. The unit of absorbed dose is not chosen completely independently of the exposure. For example, in a gamma field of 1 R exposure, the absorbed dose in air is close to one rad (0.88 rad).

The Rem, a Unit of Dose Equivalent. Although the absorbed dose is clearly related to the effects the radiation produces, it does not prove to correlate the biological damage from all kinds of radiation. For a given absorbed dose, greater damage is sustained when the radiation deposits its energy in a shorter lineal distance. This means that heavier particles—alphas and protons, for example—have a greater effect than electrons. A 1-MeV electron deposits energy at the rate of 1.9 MeV/cm in water; the value for a 1-MeV proton is 270 MeV/cm. A quantity with the curious name, quality factor, designated QF, is introduced to express the factor by which the absorbed dose should be increased to express the biological damage caused. The dose equivalent, D, is related to absorbed dose, D_{abs}

$$D = \mathrm{QF}\, D_{abs} \tag{5.3}$$

The unit of the dose equivalent (we shall simply call it dose) is the rem. Table 5.2 gives quality factors for common radiation forms.

Table 5.2 Quality Factors for Common Types of Radiation

RADIATION	QF
Gamma rays	1
X rays	1
Electrons	1
Positrons	1
Alpha particles	10
Protons	10
Recoil nuclei	20
Neutrons	2–10

The quality factor for electromagnetic radiation is, by definition, one. Therefore, when discussing electromagnetic radiation or electrons, absorbed dose (rads) and dose equivalent (rems) have the same value. If the absorbed dose of alpha particles is one rad, however, the dose is 10 rem. The underlying concept is that a dose of a given number of rems to a given organ causes the same biological damage regardless of the kind of radiation being absorbed.

Dosimetry, or the Estimation of Dose

Although we have a definition of dose and its unit, the rem, we lack a way of estimating the dose a person or organ receives. In principle it is straightforward: one measures the energy deposited in a given organ, multiplies by the quality factor for the radiation in question, and obtains the dose in rem. In practice, however, it is not that simple. For example, to measure the energy being deposited in a person's liver is simply too difficult to make this approach practicable. We must resort to indirect measurements and calculations. One technique, for example, is to construct a model of a person, a "phantom," and to place delicate radiation detectors within this phantom in order to obtain the dose received at some given organ under given conditions of irradiation. A complementary approach is to calculate the degree to which radiation coming from outside a person penetrates to various parts of the body.

As the reader may perceive, there are large numbers of painstaking experiments and calculations involved in dosimetry. The details are well beyond the scope of this text. However, the principles involved are relatively simple and require only the synthesis of principles we have treated in previous chapters. We shall attempt to demonstrate this fact by analyzing the dose from internal and external radiation.

Dose from Internal Radiation. Here we are referring to radiation that is received from radioisotopes within the body. The calculation of dose rate, if we know how many radioisotopes there are and where, is very direct, as illustrated by Example 5.1.

Example 5.1
If, at a given time, a person's lungs contain 1 μg of plutonium-239, what is the dose rate (rem/y) to the lung?

Solution: Plutonium-239 decays by alpha decay. We can assign the entire Q-value to the alpha energy. From Table A.3, the Q-value is 5.243 MeV. The rate of energy deposition in the lung is

$$\dot{q} = \lambda N Q_\alpha$$

$$t_{1/2} = 24{,}400 \text{ y}$$

$$\lambda = 9.007 \times 10^{-13} \text{ s}^{-1}$$

$$\dot{q} = 9.007 \times 10^{-13} \times \frac{1 \times 10^{-6}}{239} \times 6.023 \times 10^{23} \times 5.243$$

$$= 11{,}900 \text{ MeV/s} = 11{,}900 \times 1.602 \times 10^{-13} = 1.907 \times 10^{-9} \text{ J/s}$$

A common value for the mass of the lungs is 570 g, excluding blood. We can calculate the absorbed dose rate

$$\dot{D}_{abs} = \frac{1.907 \times 10^{-9}}{0.57 \times 0.01} = 3.34 \times 10^{-7} \text{ rad/s}$$

We must multiply by the quality factor in order to obtain the dose rate. The quality factor for alpha particles is 10.

$$D = 3.34 \times 10^{-7} \times 10 \times 3.154 \times 10^{7} = 105 \text{ rem/y}$$

This calculation is based on a "homogeneous" model. It does not provide for concentration of plutonium in any particular portion of the lungs.

In the case of Example 5.1, the fact that ^{239}Pu is an alpha emitter simplifies the calculation, because the alpha particles deposit their energy in the lung or in whatever organ the plutonium may be. The same would be true of a beta emitter–strontium-90, for example. In the case of a gamma emitter, however, the gamma energy may be deposited over a much wider range, depending on the photon energy. In estimating internal doses received from ingested or inhaled radioisotopes, by far the greatest uncertainties arise in the area of the temporal and spatial history of the isotope in the body. How much is where and for how long? At what rate is the isotope excreted? There is a considerable body of health physics literature that addresses these questions and provides approximate answers.

Dose from External Radiation. Here we refer to the situation in which the source of radiation is outside a person's body. There is a wide variety of sources: X-ray machines, radioisotopes, cosmic rays, and so on. Regardless of the source, the analysis usually starts from an important relationship.

$$\phi_u(r) = \frac{\dot{S}}{4\pi r^2} \tag{5.4}$$

This is the point-source equation. It tells us the flux of particles (or photons) originating in a source of strength, $\dot{S}$ particles/s, at a distance r from the source, assuming none have been absorbed or scattered in that distance. The designation, ϕ_u, is for the uncollided flux. The equation assumes the source is isotropic, that is, it emits without a preferred direction. Given all its restrictions, one might well ask whether Eq. 5.4 is not too limited. In practice it proves to be more useful than one would think. For example, if a source is fairly sizable but the distance, r, is a few times the source's characteristic dimensions, the source can be treated as a point. Further, even when a source is distributed over considerable space, the analysis consists basically of breaking it up into a series of point sources and summing or integrating to get the flux at some fixed point.

What does the calculation of the flux have to do with dose? Simply that the flux is proportional to the thermal energy deposition rate (see Eq. 4.62). Table 5.3 and Table 5.4 give values of dose rate (mrem/h) corresponding to the photon

Table 5.3 Dose Rate as a Function of Gamma and X-Ray Photon Flux[a]

PHOTON ENERGY, E (MeV)	$\phi/\dot{D}$ $\left(\frac{\text{photons/cm}^2\cdot\text{s}}{\text{mrem/h}}\right)$	$E\phi/\dot{D}$ $\left(\frac{\text{MeV/cm}^2\cdot\text{s}}{\text{mrem/h}}\right)$	PHOTON ENERGY, E (MeV)	$\phi/\dot{D}$ $\left(\frac{\text{photons/cm}^2\cdot\text{s}}{\text{mrem/h}}\right)$	$E\phi/\dot{D}$ $\left(\frac{\text{MeV/cm}^2\cdot\text{s}}{\text{mrem/h}}\right)$
0.05	8296	415	1.00	561	561
0.06	9031	542	1.50	410	615
0.08	8272	662	2.00	333	666
			3.00	255	765
0.10	6773	677	4.00	210	840
0.15	4173	626	5.00	182	910
0.20	2919	584	6.00	161	966
0.30	1812	544	8.00	131	1048
0.40	1322	529	10.00	110	1100
0.50	1051	526			
0.60	878	527			
0.80	675	540			

[a]R. H. Thomas, "The Conversion of Particle Fluence to Dose Equivalence," Health Physics and Occupational Health, Stanford University, 1969.

flux and to the neutron flux, respectively. First, we will check Table 5.3 by seeing if we can compute the photon flux of 1-MeV photons that will result in a dose of 1 mrem/h to tissue.

Since the quality factor is one, the absorbed dose rate equivalent to 1 mrem/h is 1 mrad/h, or 0.1 ergs/g · h, or 10^{-8} J/g · h (see definition of one rad, Eq. 5.2). From Eq. 4.62, the rate of energy deposited is

$$q''' = \mu_E \phi E \text{ MeV/cm}^3 \cdot \text{s}$$

In order to express the deposition rate on a mass basis, we need the density of tissue, which we will take to be 1.0 g/cm^3.

$$\dot{D}_{\text{abs}} = \dot{q} = \mu_E \phi E \rho = 10^{-8} \text{ J/g} \cdot \text{h}$$

From Table 4.4, for tissue (muscle)

$$\mu_E/\rho = 0.0308 \text{ cm}^2/\text{g}$$

$$\mu_E = 0.0308 \text{ cm}^{-1}$$

$$\phi = \frac{10^{-8}}{0.0308 \times 1 \times 1.602 \times 10^{-3} \times 3600} = 563 \text{ photons/cm}^2 \cdot \text{s}$$

Table 5.4 Dose Rate as a Function of Neutron Flux[a]

NEUTRON ENERGY	$\phi/\dot{D}$ $\left(\frac{\text{neuts/cm}^2\cdot\text{s}}{\text{mrem/h}}\right)$	NEUTRON ENERGY	$\phi/\dot{D}$ $\left(\frac{\text{neuts/cm}^2\cdot\text{s}}{\text{mrem/h}}\right)$	NEUTRON ENERGY	$\phi/\dot{D}$ $\left(\frac{\text{neuts/cm}^2\cdot\text{s}}{\text{mrem/h}}\right)$
		0.01 keV	210	0.01 MeV	160
0.025 eV	270	0.02	210	0.02	110
0.05	260	0.05	200	0.05	57
0.10	250	0.10	200	0.10	34
0.20	240	0.20	200	0.20	21
0.50	230	0.50	220	0.50	12
1.0	230	1.0	220	1.0	9.7
2.0	220	2.0	240	2.0	8.2
5.0	210	5.0	230	5.0	7.2
				10.0	6.8

[a]R. H. Thomas, "The Conversion of Particle Fluence to Dose Equivalence," Health Physics and Occupational Health, Stanford University, 1969.

The value in Table 5.3 is 561 photons/cm^2 · s. In Table 5.3, the value of $E\phi/\dot{D}$ is also given, solely because this quantity varies only slowly with energy, making interpolation much simpler.

Example 5.2

If a 100-Ci source of cesium-137 is exposed in a laboratory and a person stands 5 meters away, what dose rate is this person receiving?

Solution: This is an opportunity to review some material from earlier chapters. First, we must know the decay scheme of ^{137}Cs. LHP is our best source. ^{137}Cs emits a single gamma, 0.662 MeV, on 85% of decays. The source strength is, therefore

$$\dot{S} = 100 \times 3.7 \times 10^{10} \times 0.85 = 3.145 \times 10^{12} \text{ photons/s}$$

At 5m, assuming no attenuation in air, the flux is

$$\phi = \frac{3.145 \times 10^{12}}{4 \times \pi \times (5 \times 100)^2} = 1.00 \times 10^6 \text{ photons/cm}^2 \cdot \text{s}$$

The energy flux is 0.662×10^6 MeV/cm^2 · s. In order to convert to dose, we must interpolate in Table 5.3. For 0.662 MeV photons we interpolate

$$E\phi/\dot{D} = 530 \frac{\text{MeV/cm}^2 \cdot \text{s}}{\text{mrem/h}}$$

Therefore,

$$\dot{D} = \frac{0.662 \times 10^6}{530} = 1250 \text{ mrem/h}$$

The person would actually receive a somewhat higher dose, because photons leaving the source in other directions would be Compton scattered off the walls and would thereby add to the dose rate we have calculated here.

Table 5.4 gives $\phi/\dot{D}$ values for neutrons. The computation of these values is more complex for neutrons than for photons. In the case of thermal neutrons, the dose comes primarily from the (n, γ) reaction in hydrogen of tissue and from the (n, p) reaction in nitrogen.

Example 5.3
Given that tissue is 3.05 w/o nitrogen and has a density of 1.0 g/cm^3, what flux of thermal neutrons would it take to produce a dose rate of 1 mrem/h from the (n, p) reactions in nitrogen? Compare with Table 5.4.

Solution: From Table A.3, the microscopic (n, p) cross section for low-energy neutrons in nitrogen is 1.82 barns. From Eq. 4.15 and Eq. 4.24, the macroscopic cross section is

$$\Sigma_a = 0.6023 \times \frac{0.0305}{14} \times 1.82 = 2.39 \times 10^{-3} \text{ cm}^{-1}$$

The proton that is formed deposits its energy at the site of the reaction. We need to know the energy of that proton. It can be calculated from the Q-value of the reaction

$$n + {}^{14}\text{N} \longrightarrow {}^{1}\text{H} + {}^{14}\text{C}$$

From Eq. 2.4 and Table A.3

$$\begin{aligned} Q &= 8.0714 + 2.8637 - 7.2890 - 3.0198 \\ &= 0.6263 \text{ MeV} \end{aligned}$$

The energy is divided between the proton and the carbon-14 nucleus, with the proton getting fourteen fifteenths of it, according to Eq. 2.8. However, the recoil carbon is ionizing also, so all the energy can be used. The absorbed dose should be calculated

$$\dot{q} = \Sigma_a \phi E \rho \text{ MeV/g} \cdot \text{s}$$

However, a proton has a quality factor of 10, which means that this energy should be multiplied by 10 for the dose to be in rem units. One mrad/h is 0.1 erg/g · h or 10^{-8} J/g · h.

$$\dot{D} = 10^{-8} = 2.39 \times 10^{-3} \times \phi \times 10 \times 0.6263 \times 1.602 \times 10^{-13} \times 3600$$

$$\phi = 1158 \text{ neutrons/cm}^2 \cdot \text{s}$$

This compares with 270 neutrons/cm^2 · s. Presumably, the dose from the (n, γ) in hydrogen causes the lower value.

THE RELATIONSHIP OF DOSE TO DAMAGE OR INJURY

We have treated the subject of how radiation received may be measured or computed in a manner that correlates with damage to living organisms. A logical next question is: "What is the damage to a human on receiving some given dose equivalent?" Putting it more prosaically, "If I received one rem this morning, how bad off am I?" This question can be answered with the fewest qualifications in the case of relatively high doses received in a short period of time.

High-Dose or Acute Effects

A few effects of relatively high radiation doses received in a short time interval are given in Table 5.5.

Table 5.5 Typical Quoted Values of Whole-body Doses Corresponding to Acute Symtoms

DOSE	CLINICAL EFFECTS
5–25 rem	Threshold of detectable effects
25–100 rem	Detectable temporary blood changes
100–200 rem	Nausea, fatigue, blood changes. Expect 100% recovery from these symptoms
450 ± 100 rem	LD_{30}^{50} (LD-50-30). Wide range of symptoms. Expect that 50% of persons receiving such a dose would succumb in 30 days. LD stands for lethal dose

The values cited in Table 5.5 are for orienting purposes only. Obviously, there are not large numbers of opportunities to gather data. In the case of a dose in the lethal range, the subject's survival can depend strongly on the quality and timing of therapy measures. There are many other variables involved. The reader is justified in concluding from Table 5.5 that he or she would not *feel* any effect of receiving a dose of one rem and probably would not survive the effects of 1000 rem. To place it in a more concrete context, the 1.25 rem/h computed in Example 5.2 from a 100-curie cesium-137 source at five meters would not produce any observable effect in a worker so exposed for a few hours. On the other hand, should a worker be only one meter away from such a source for a period of eight hours, the dose received would be $1.25 \times 25 \times 8 = 250$ rem, and some severe symptoms would be experienced. If the worker should place the source in a pocket for a matter of minutes, he or she would be lethally exposed.

The Effects of Low-Dose Exposures: The Dose-Response Curve

Table 5.5 gives the short-term or acute effects of relatively high doses. Such data as do exist in this dose range derive from direct observations. For example, several persons who have received doses in excess of 1000 rem have succumbed in a matter of days or weeks. There is no doubt about the cause and the effect. In the low-dose range this is not the case. The reason is well expressed in the so-called BEIR report:

> *With few exceptions the somatic effects we have considered manifest themselves only after an interval of years or decades following irradiation and are indistinguishable from lesions that occur naturally in nonirradiated populations; thus, their relationship to radiation is detectable only in a statistical sense.*[1]

The word "somatic" is used in radiation-effects work to mean the class of effects that manifest themselves in the person actually receiving the radiation. The other class is genetic effects, which are manifest in the progeny of the person receiving the radiation. As noted in the above quote from the BEIR report, any genetic effects and low-dose somatic effects will occur a considerable time after the radiation is received. These effects are therefore often called delayed or late effects.

The somantic effect of exclusive concern in low-dose exposures is the induction of cancer of one form or another. The important question is the quantitative relationship between dose (rems) and cases of cancer. The approach to an answer takes the form of establishing a dose-response curve from the experimental and epidemiological data available. The form of the dose-response curve is very important in this case. Figure 5.1 shows some of the possible forms. Figure 5.1*a* is a form typical of many chemical materials. In such cases there is some level of exposure or dose below which no harm is caused. This level is the threshold. On Figure 5.1*a* are drawn three different curves; one is straight,

[1]National Academy of Sciences, National Research Council, "The Effects on Populations of Exposure to Low Levels of Ionizing Radiation," November 1972. This is one of the more concise and authoritative treatments of the subject and is recommended to the reader wishing deeper insight into the subject.

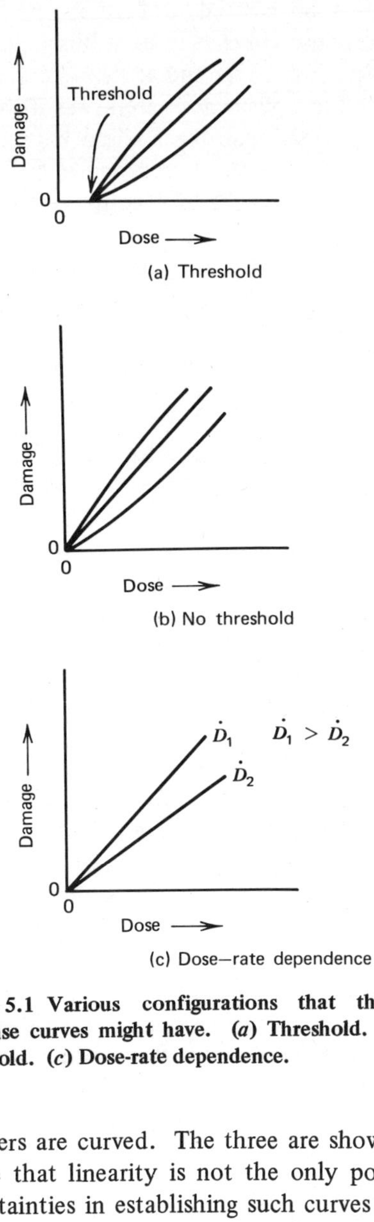

Figure 5.1 Various configurations that the dose-response curves might have. (*a*) Threshold. (*b*) No threshold. (*c*) Dose-rate dependence.

or linear, and the others are curved. The three are shown for the sake of completeness and to note that linearity is not the only possibility. On the other hand, given the uncertainties in establishing such curves for low-level radiation, the straight line is all that is empirically justifiable and is probably conservative. The more important question concerns the existence or nonexistence of the threshold. Figure 5.1*b* shows the no-threshold dose-response curve; it goes to the origin.

Still another variable is introduced in Figure 5.1*c*. The two curves there suggest that the dose-response curve may be a function of the rate at which the dose is received. At high dose levels and acute effects, there is no question that this is true. The effects for a given dose are lower if the dose is received over a longer period of time, because repair mechanisms have a chance to operate. For low-dose exposures, the evidence of a rate dependency is an unresolved question at present. Current policy guides for radioactive emission standards assume no rate dependency.

The student may well ask why the proper curve is not established experimentally for radiation doses, so that the matter is resolved once and for all. The answer is that it is practicable to establish a positive cause-effect relationship only when high doses are used (for example, see Example 5.4). The curve near the origin is determined by extrapolation from higher doses. The extrapolation does not have the accuracy necessary to determine with high confidence the existence or nonexistence of a threshold. Given the uncertainty, and while admitting that the question is not scientifically resolved, the BEIR report states:

> *At this time, then, the linear hypothesis, which allows the mean tissue dose to be used as the appropriate measure of radiation exposure, provides the only workable approach to numerical estimation of the risk in a population. Further, since there is no means at present of determining the value of the dose-effect slope in the low-dose region of interest, use of the linear extrapolation from data obtained at high doses and dose rates may be justified on pragmatic grounds as a basis for risk estimation.*

In short, as a matter of prudence, the BEIR report recommends a no-threshold, linear, dose-rate-independent, dose-response curve for cancer risk estimates. This recommendation is followed by the regulatory agencies, such as the Environmental Protection Agency and the Nuclear Regulatory Commission, in setting radioactivity emission standards.

The slope of the dose-response curve (the linear curve of Figure 5.1*b*) that the BEIR report regards as most probable is 1.8×10^{-4} fatal cancers per person-rem whole-body dose. This has one kind of interpretation for an individual and another for populations. Consider first the random individual. In 1970, the deaths in the U.S. attributed to cancers was one for every 617 people in the population. The random probability of death from cancer is, therefore, 0.00162 per year. If our random individual receives one rem of whole-body radiation, his chances of getting a fatal cancer over the next 30 years or so have increased by 1.8×10^{-4}. The random probability is 30×0.00162, or 0.0486. The value 1.8×10^{-4} added to this probability increases it by 0.4%. Naturally, this calculation will differ for a 10-year old and a 75-year old. No individual is average.

When the BEIR value of 1.8×10^{-4} fatal cancers/person-rem is applied to populations, the calculations are a little different. If a population receives some average irradiation, say 0.1 rem, in a short period of time, then the total dose in person-rems is simply 0.1 times the population. So long as no persons receive acute doses--in the lethal range--it does not make any statistical difference how the doses are distributed. In the no-threshold linear model, one person receiving ten rems or ten persons receiving one rem each have the same statistical result. Let us say a population receives a dose of one million person-rems in a period of a year or less. Then, the expectation is that $1.8 \times 10^{-4} \times 10^{6}$ or 180 fatal cancers will result. These cancers are not like the measles, however. They do not appear in two or three weeks. Rather, there will be a latent period of some years in which none of the cancers appears, followed by a rise in rate to some "plateau" that may last 30 years or even longer. During this 30-year period, some six extra cancer fatalities per year would result. Understand, these deaths would not be identifiable as related to the million-person-rem exposure. They are predicted by the model. If the population receives some radiation exposure at some constant rate over a long period of time, the predicted extra cancer fatality rate will come to a steady rate. If the dose rate is a steady 100,000 person-rems per year, the extra cancer fatality rate would rise to $1.8 \times 10^{-4} \times 10^{5} = 18$ per year.

Example 5.4

People often ask, "If low-level radiation is suspected of causing cancer, why couldn't the relationship be inferred by studying the cancer-rate variations in localities of different natural radiation background levels?" Let us assume we had two control populations of one million persons each. Population A receives 100 mrem/person · year more radiation than B. This is about the maximum difference one could expect to find in the natural background. What difference in the cancer rate would one expect to observe in these two populations?

Solution: First, we estimate the "background" cancer death rate by taking the U.S. average. In 1970, the value was 1620 deaths per million population. In one year, population A receives $10^{6} \times 0.1$ or 100,000 person-rems more dose than population B. Using the BEIR value of 180×10^{-6} cancer deaths per person-rem, we calculate 18 extra cancer deaths in population A. These expected deaths would occur over a period of many years and would be "lost" in the background of 1620 deaths each year. If our two populations were stable and had been subjected to the same background radiation level for many years, we would expect 18 extra cancer deaths in population A every year. Even this value is only 1% above the national average rate. When it is considered that such a study would be complicated by mobility within the populations and by differences in air pollution, water supply, and so on, it can be seen that a conclusive outcome is most improbable.

Genetic Effects

Genetic effects, in contrast to somatic, are those effects passed on to the offspring of a person receiving radiation. The genes are large molecules that pass on inheritable characteristics. They can be changed (mutated) by heat, ionizing radiation, and certain chemicals. Genetic defects are understandably of unique concern to people, because they represent a burden placed on future generations, sometimes for benefits enjoyed in the present. The BEIR report does point out, however, that if the no-threshold linear dose-response relationship is accepted as the basis for estimating cancer risks, the cancer risk so computed probably exceeds the risks involved in genetic damage.

Whole-body Radiation

It will be noted that Table 5.5 refers to whole-body radiation. The portion of the body under irradiation is one additional variable in evaluating the actual damage incurred. The term "whole body" means just what it says. Many radiation sources, particularly inhaled or ingested alpha- or beta-emitting isotopes, tend to be very specific to some part of the body. Thus, strontium-90 and radium-226 are bone-seekers. Iodine-135 goes to the thyroid. Each of these organs or body regions has its own dose-response curve.

The Dose Rate from Natural Radiation

One of the measures of the degree of seriousness of any radiation produced by human activities is the amount of radiation we all receive from natural sources. All life on earth has received radiation from natural sources since its inception. Typical values of whole-body radiation received from natural sources in the United States are given in Table 5.6.

Table 5.6 Typical Annual Dose Values from Natural Sources

SOURCE	WHOLE-BODY DOSE RATE (mrem/y)
Cosmic rays	45
Terrestrial gamma	60
Radioactivity in body	25
	130

There is a considerable variation, perhaps ± 50 mrem/y, in this natural radiation dose rate from place to place in the United States and in the world. The cosmic-ray dose, for example, increases approximately 1 mrem/y per 100 ft of elevation. The radioactivity in the body consists primarily of potassium-40 (half-life, 1.26 billion years) and heavy element alpha-emitters such as radon-222.

If one calculates the total annual exposure in the United States from natural sources, it is $220 \times 10^6 \times 0.13$ or 2.86×10^7 person-rems per year. Using the BEIR value of 1.8×10^{-4} cancer fatalities per person-rem, we calculate 5150 cancer fatalities per year. In fact, there are about 330,000. The simple explanation is that there are many other cancer-causing agents in our environment. In using the BEIR dose-response value, it is understood that the dose is that above background. The cancer fatalities so computed are called "extra" fatalities, to distinguish them from the "normal" or "spontaneous" cancer fatalities.

Radiation Protection Standards

The evaluation of the risks due to radiation has been carried on over the years by an interesting array of agencies. The oldest is apparently the International Commission on Radiological Protection (ICRP), founded in 1928. It has the role of evaluating radiation risks and recommending numerical standards. At the U. S. national level, there is the National Council on Radiation Protection and Measurements (NCRP), which has a similar function; therefore there is overlap. At the operational level, the Federal Radiation Council (FRC), was established in 1960, with the responsibility to give guidance for all federal agencies in the formulation of radiation standards. The agency most concerned with the implementation of standards has been the Atomic Energy Commission, whose responsibilities, in January 1975, were assumed by the Nuclear Regulatory Commission (NRC). In 1970, the FRC's responsibilities were transferred to the Environmental Protection Agency (EPA). EPA also assumed some of the responsibilities previously held by the AEC.

In nuclear-energy activities, the NRC is responsible for the translation of guidelines into quantitative operating standards. These standards are published in the Code of Federal Regulations (CFR), Title 10, Part 20. Such a designation is usually abbreviated to 10 CFR 20. Some of the standards are given in Table 5.7. For a proper perspective on the standards of the kind given in Table 5.7, it is necessary to understand the spirit in which they are set and implemented. This spirit was expressed in a 1960 FRC document: "There should not be any man-made radiation exposure without the expectation of benefit resulting from such exposure. . . . It is basic that exposure to radiation should result from a real determination of its need. . . . every effort should be made to encourage the maintenance of radiation doses as far below this guide as practicable."

Table 5.7 Examples of Standards for Limiting Doses

GROUP	KIND OF EXPOSURE	MAXIMUM PERMISSIBLE DOSE
Occupational	Whole-body	5 rem/y
Individual in the public	Whole-body	0.5 rem/y
Public at large	Whole-body	0.17 rem/y

This stated philosophy is known in the industry as the "as low as practicable" principle. (It is sometimes simply called "aslap.") The principle is made more concrete in specific cases. For example, in the regulations governing nuclear power reactors set forth in 10 CFR 50, Appendix I, the dose level at the plant boundary is not to exceed 5 mrem/y, a factor of 100 below the "maximum permissible." The same Appendix I goes on to state that the cost-benefit criterion shall be that an incremental expense to reduce exposure shall be made up to a value of \$1000/person-rem reduction. Using the BEIR dose-response value of 1.8×10^{-4} cancer fatalities per person-rem, this works out to a marginal value of $1000/1.8 \times 10^{-4}$ or \$5.6 million per life saved, assuming the validity of the no-threshold linear model.

The Maximum Permissible Concentration (MPC): Standards for Radioisotopes in the Environment

Even when we have a standard (Table 5.7) for maximum permissible dose, there is an evaluation problem. Much of the radiation exposure experienced by persons comes from radioisotopes. It is necessary, therefore, to translate the maximum permissible dose into a maximum permissible concentration (MPC) for each isotope in air and water. Each isotope is different, and its potential for adding to a person's radiation dose must be evaluated on the basis of the following characteristics:

Character (α, β, γ) and energy of radiation emitted

Half-life

Transport and history in the environment: air, water, plants, animals, and so on

History in the human body: biological half-life, selectivity for various organs, and so on

Evaluation of each radioisotope has been made by such organizations as the ICRP. The evaluation is based on models of a person breathing so much air at MPC, or drinking so much water at MPC, and so on. The basis for all the calculations is that if a given person drinks only water with a single isotope at the

Table 5.8 Maximum Permissible Concentrations of Some Radioisotopes in Plant Discharges as Effluent

ISOTOPE		MPC-AIR ($\mu Ci/cm^3$)	MPC-WATER ($\mu Ci/cm^3$)
Tritium (^3H):	Soluble	2×10^{-7}	3×10^{-3}
	Insoluble	2×10^{-7}	3×10^{-3}
Strontium-90:	Soluble	3×10^{-11}	3×10^{-7}
	Insoluble	2×10^{-10}	4×10^{-5}
Cesium-137:	Soluble	2×10^{-9}	2×10^{-5}
	Insoluble	5×10^{-10}	4×10^{-5}
Plutonium-239:	Soluble	6×10^{-14}	5×10^{-6}
	Insoluble	1×10^{-12}	3×10^{-5}

MPC, in the course of a year that person will have received the maximum permissible dose (Table 5.7). It follows, therefore, that there will be an MPC value for each isotope in air and water and also for persons occupationally exposed and for people in the public. The MPC also varies depending on whether the isotope is in a soluble or insoluble form, since that affects its behavior in the body. It follows from the manner of calculating MPC that each of several radioisotopes should not be at MPC. The sum of their fractional MPCs should not exceed a value of one. MPC values are tabulated for many isotopes in 10 CFR 20. A few values are given in Table 5.8.

Some Perspectives on Radiation Risks

The BEIR report gives data on the radiation from human sources to which the average person in the United States is subjected. These are given in Table 5.9.

Table 5.9 Average Dose Rate from Artificial Radiation in the United States

SOURCE	WHOLE-BODY DOSE RATE (mrem/y)
Medical and dental (diagnostic)	72
Weapons fallout	4
Occupational exposure	0.8
Nuclear power (1970)	0.003
Nuclear power (2000)	<1

The value projected for nuclear power assumes performance within regulatory standards and makes no allowance for accidental releases.

If we use the BEIR value of 1.8×10^{-4} fatal cancers per person-rem, the annual production of fatal cancers from medical and dental diagnostic X rays is $1.8 \times 10^{-4} \times 0.072 \times 220 \times 10^6$, or 2850. This number of late-effect fatalities exceeds the number estimated for a very serious reactor accident. If the no-threshold linear model is to serve as the basis for risk estimates in the case of routine and accidental releases in the nuclear industry, then it is hardly consistent not to apply it to other radiation-producing activities. The same principle also applies to genetic effects. To the extent that we are inconsistent in evaluating relative risks, we will misallocate resources devoted to the reduction of those risks.

SHIELDING

In the case of external radiation sources, the radiation dose can be reduced by the interposition of material between the source and the receiver. If the materials are placed there for that purpose, they are called shields. It often happens that the radiation source itself is embedded in some material that acts to reduce the dose received in the surroundings. This reduction is termed self-absorption or self-shielding.

Due to the number of variables involved—the configuration of the source, the configuration of the shielding, and materials of the source and shielding—the analysis of realistic shielding problems usually leads to very complicated equations involving special integrals. The results, however, are obtained by the synthesis of a few relatively simple analytical steps. We shall illustrate a few of these steps, selecting cases that give results in simple form. There are several books that give solutions to more complex shielding situations.[2]

The Shielding of a Point Source

The radiation field surrounding an unshielded point source is given by Eq. 5.4. If shielding intervenes between the source and a receiver

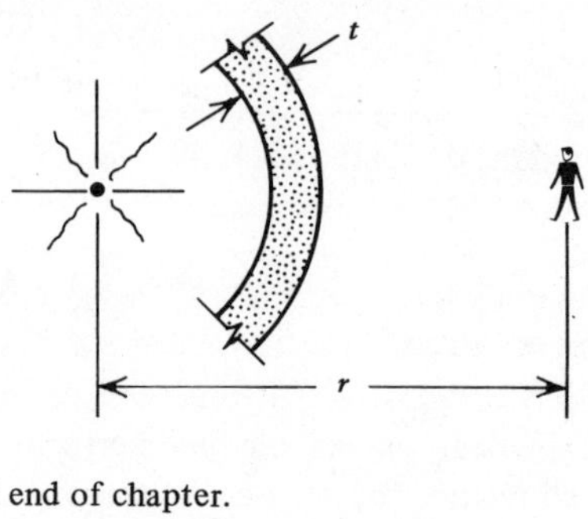

[2] See References 4, 5, 6 at end of chapter.

then

$$\phi_u(r) = \frac{\dot{S}}{4\pi r^2} e^{-\mu t} \tag{5.5}$$

where μ is the linear attenuation coefficient for gamma radiation in the shield material, as given by Eq. 4.55. Values can be obtained from Table 4.2, if the shield material density is known. If the source is a neutron emitter, then the total macroscopic cross section should be substituted for μ. If there is more than one material in the shield

$$\phi_u(r) = \frac{\dot{S}}{4\pi r^2} \exp - \sum_i \mu_i t_i \tag{5.6}$$

where the subscript i refers to different shield materials. The dose rate at r, which results from these uncollided photons or particles, can be obtained from Tables 5.3 or 5.4 or more extensive references.[3] There is other (secondary) radiation at the receiver, however, resulting from scattering of the primary radiation in the shield. We shall discuss the manner of dealing with this radiation shortly.

Example 5.5

If one curie of a radioisotope that emits a 1-MeV gamma photon on 75% of its decays is to be enclosed in a spherical iron container, what thickness should the container wall be in order to make the dose rate due to uncollided photons 1 mrem/h at 1 meter from the source?

Solution: In the terms of Eq. 5.6

$$\dot{S} = 0.75 \times 3.7 \times 10^{10} = 2.775 \times 10^{10} \text{ photons/s}$$

The mass absorption coefficient for 1-MeV photons in iron is 0.0599 cm^2/g (Table 4.2). If the density of iron is 7.8 g/cm^3

$$\mu = 0.0599 \times 7.8 = 0.467 \text{ cm}^{-1}$$

In order for the dose rate to be 1 mrem/h, the flux must be reduced to 561 photons/$cm^2 \cdot s$ (Table 5.3). Eq. 5.5 becomes

$$561 = \frac{2.775 \times 10^{10}}{4\pi \times 10^4} \exp(-0.467\, t)$$

$$t = 12.8 \text{ cm}$$

Note that if an isotope emits gamma photons of more than one energy, which is usually the case, the dose contribution from the photons of each energy can be computed separately. The individual contributions add to give the dose.

---

[3] See Reference 4 at end of chapter.

The Flux from a Plane Source

Although a point source is a fairly good approximation of any real source whose dimensions are less than a quarter of the distance to the receiver, the spatial distribution of the source must sometimes be treated explicitly. We shall consider a circular plane of radius R with the dose point located a distance a from

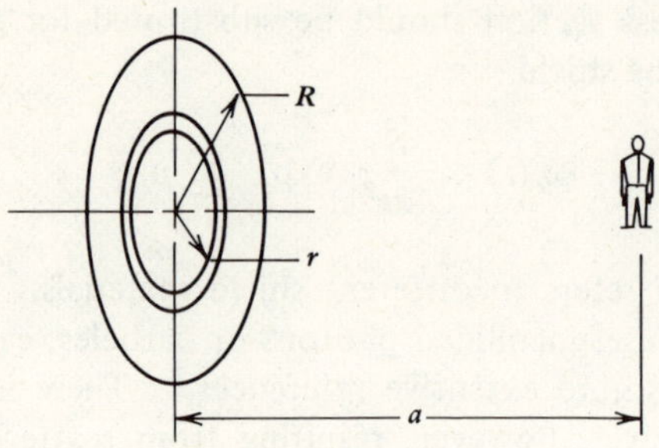

the center. The radiation source, $\dot{S}_A$ photons/cm^2 · s, is of uniform strength over the entire surface. The contribution to the flux from the differential plane area at r is

$$d\phi = \dot{S}_A \frac{2\pi r\, dr}{4\pi (a^2 + r^2)} = \frac{\dot{S}_A r\, dr}{2 (a^2 + r^2)} \tag{5.7}$$

The flux at the receiver is, therefore

$$\phi(a) = \frac{\dot{S}_A}{2} \int_0^R \frac{r\, dr}{a^2 + r^2} = \frac{\dot{S}_A}{4}\, ln \left(1 + \frac{R^2}{a^2}\right) \tag{5.8}$$

It will be noted that as R increases, the flux increases without limit. In a real situation, there would be some shielding ($\mu > 0$) in the medium between the source plane and the receiver. If a shield of constant thickness, t, and linear attenuation coefficient, μ, is placed between the plane source and the receiver

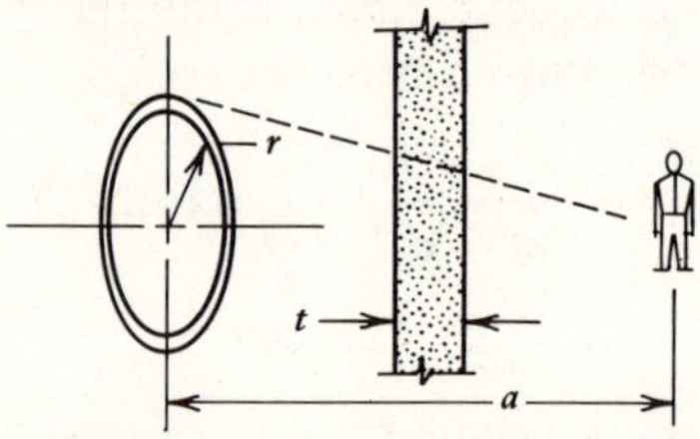

then the contribution to the flux at the receiver from the differential plane area at r is

$$d\phi_u = \frac{\dot{S}_A r\, dr}{2 (a^2 + r^2)} \exp - \left(\mu t \frac{\sqrt{a^2 + r^2}}{a}\right) \tag{5.9}$$

The results of the integration of Eq. 5.9, and the tables necessary to obtain numerical results, can be found in References 4, 5, and 6 at the end of this chapter.

The Flux from a Source with Self-Absorption

Another class of source–receiver relationship involves a volume-distributed source with self-absorption in the source. One example of such a circumstance would be a gamma-emitting radioisotope broadly distributed in the atmosphere. Consider a sphere of radius R and linear attenuation coefficient μ_s, with a uniform volume source $\dot{S}_v$ photons/cm$^3 \cdot$ s. The subscript s is used here to denote a property of the source and distinguish it from any external shielding. The

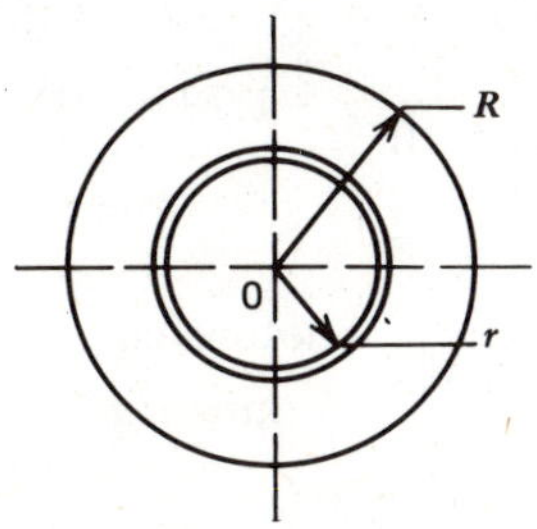

simplest dose point to analyze is that at the center of the sphere. In this case, the contribution to the flux from a differential spherical shell is

$$d\phi_u(0) = \dot{S}_v \frac{4\pi r^2\, dr}{4\pi r^2} e^{-\mu_s r} \tag{5.10}$$

which integrates to

$$\phi_u(0) = \dot{S}_v \int_0^R e^{-\mu_s r}\, dr = \frac{\dot{S}_v}{\mu_s}\left(1 - e^{-\mu_s R}\right) \tag{5.11}$$

Note that for $\mu_s R \gtrsim 3$, the sphere appears to be essentially infinite in extent. A solution can also be obtained[4] for the uncollided flux on the surface of the sphere.

$$\phi_u(R) = \frac{\dot{S}_v}{2\mu_s}\left[1 - \frac{1}{2\mu_s R}\left(1 - e^{-2\mu_s R}\right)\right] \tag{5.12}$$

Example 5.6
Consider that the 1-MeV gamma-emitting isotope of Example 5.5 is distributed uniformly in the atmosphere at a concentration of 10^{-6} Ci/m^3. What dose rate would result at ground level?

[4] See Reference 6 at end of chapter.

Solution: We can consider the receiver to be at the center of a hemisphere. Eq. 5.11 becomes

$$\phi_u(0) = \frac{\dot{S}_v}{2\mu_s}(1 - e^{-\mu_s R})$$

$$\dot{S}_v = 0.75 \times 10^{-6} \times 3.7 \times 10^{10} = 2.775 \times 10^4 \text{ photons/m}^3 \cdot \text{s}$$

$$= 2.775 \times 10^{-2} \text{ photons/cm}^3 \cdot \text{s}$$

From Table 4.2, the mass attenuation coefficient in air is 0.0636 cm^2/g. The density of air is 1.20×10^{-3} g/cm^3.

$$\mu = 0.0636 \times 1.20 \times 10^{-3} = 7.63 \times 10^{-5} \text{ cm}^{-1}$$

Assume that R is very large.

$$\phi_u(0) = \frac{2.775 \times 10^{-2}}{2 \times 7.63 \times 10^{-5}} = 1.79 \times 10^5 \text{ photons/cm}^2 \cdot \text{s}$$

From Table 5.3

$$\frac{\phi}{\dot{D}} = 561 \, \frac{\text{photons/cm}^2 \cdot \text{s}}{\text{mrem/h}}$$

$$\dot{D} = \frac{1.79 \times 10^5}{561} = 319 \text{ mrem/h}$$

We have assumed that $\mu_s R$ is large. How large would R have to be?

$$\mu_s R = 3$$

$$R = \frac{3}{7.63 \times 10^{-5}} = 3.93 \times 10^4 \text{ cm} = 393 \text{ m}$$

The Treatment of Secondary Radiation: Dose Buildup

We have noted that the flux, ϕ, which we have been calculating, is for those photons or neutrons that come directly from the source without interaction. We have been using the total cross section (μ or Σ), which includes scattering events. The radiation from these scatterings also contributes to the flux and the dose rate at the receiver. If the uncollided flux, ϕ_u, is combined with Tables 5.3 or 5.4 to calculate the dose rate, the answer will be low. It would be a lower bound. We could also use the same equation using only absorption cross sections and calculate an upper bound. Unfortunately, in many cases these two limits are too far apart to be useful, so a different approach is used.

The dose values have been computed[5] using sophisticated techniques which account for secondary radiation and the results expressed in terms of a dose buildup factor, B. The buildup factor is the ratio of the actual dose to the dose calculated on the basis of the uncollided flux.

$$\dot{D} = B \times \frac{\phi_u}{(\phi/\dot{D})} \tag{5.12}$$

The values of $\phi/\dot{D}$ come from Table 5.3, since the buildup factor approach is limited to photons. Values of B are given in Table 5.9 for the case of a point source in a few shielding materials.

Example 5.7
Compute the answer to Example 5.5, allowing for buildup.

Solution: The value of μt from Example 5.5 is

$$\mu t = 0.467 \times 12.8 = 6$$

From Table 5.9, the buildup factor, B, is 8.6 (by linear interpolation). The dose with the container 12.8 cm thick is, therefore, 8.6 mrem/h instead of 1 mrem/h. How much must the thickness be increased to obtain a dose rate of 1 mrem/h? By trial and error

$$t = 18 \text{ cm}$$

$$\mu t = 0.467 \times 18 = 8.4$$

$$B = 13$$

$$\phi_u = \frac{2.775 \times 10^{10}}{4\pi \times 10^4} e^{-8.4} = 49.7 \text{ photons/cm}^2 \cdot \text{s}$$

By Eq. 5.12

$$\dot{D} = 13 \times \frac{49.7}{561} = 1.15 \text{ mrem/h}$$

So the wall thickness should be increased from 12.8 to 18 cm.

As we have already seen, most shielding situations involve more than a point source surrounded by a shielding medium. The buildup factor approach can be extended considerably by expressing B as a function of μr in an empirical equation of the form

$$B = A e^{\alpha \mu r} + (1 - A) e^{-\beta \mu r} \tag{5.13}$$

[5] See Reference 7 at end of chapter.

Table 5.9 Dose Buildup Factors for an Isotropic Point Source[a]

	E_γ	μr						
MATERIAL	(MeV)	1	2	4	7	10	15	20
Water	0.255	3.09	7.14	23.0	72.9	166	456	982
	0.5	2.52	5.14	14.3	38.8	77.6	178	334
	1.0	2.13	3.71	7.68	16.2	27.1	50.4	82.2
	2.0	1.83	2.77	4.88	8.46	12.4	19.5	27.7
	3.0	1.69	2.42	3.91	6.23	8.63	12.8	17.0
	4.0	1.58	2.17	3.34	5.13	6.94	9.97	12.9
	6.0	1.46	1.91	2.76	3.99	5.18	7.09	8.85
Iron	0.5	1.98	3.09	5.98	11.7	19.2	35.4	55.6
	1.0	1.87	2.89	5.39	10.2	16.2	28.3	42.7
	2.0	1.76	2.43	4.13	7.25	10.9	17.6	25.1
	3.0	1.55	2.15	3.51	5.85	8.51	13.5	19.1
	4.0	1.45	1.94	3.03	4.91	7.11	11.2	16.0
	6.0	1.34	1.72	2.58	4.14	6.02	9.89	14.7
Lead	0.5	1.24	1.42	1.69	2.00	2.27	2.65	2.73
	1.0	1.37	1.69	2.26	3.02	3.74	4.81	5.86
	2.0	1.39	1.76	2.51	3.66	4.84	6.87	9.00
	3.0	1.34	1.68	2.43	3.75	5.30	8.44	12.3
	4.0	1.27	1.56	2.25	3.61	5.44	9.80	16.3
	6.0	1.18	1.40	1.97	3.34	5.69	13.8	32.7

[a]Valves taken from the *Radiological Health Handbook*, U.S. Department of Health, Education, and Welfare, January 1970.

This expression can then be inserted into differential equations such as Eqs. 5.7 or 5.10, and integrated to obtain an augmented value of the flux, which can then be converted to dose. Again, the interested reader is referred to more extensive treatments of this subject in References 4, 5, and 6.

REFERENCES

1. R. H. Thomas, "The Conversion of Particle Fluence to Dose Equivalence," *Health Physics and Occupational Health*, Stanford University, 1969.
2. "The Effects on Populations of Exposure to Low Levels of Ionizing Radiation," NAS-NRC, November 1972.
3. "Reactor Safety Study," U.S. Nuclear Regulatory Commission, October 1975.
4. K. Z. Morgan, J. E. Turner, *Principles of Radiation Protection*, Wiley, New York, 1967.
5. E. P. Blizard (editor), *Reactor Handbook*, Volume III, Part B, "Shielding," Interscience Publishers, New York, 1972.

6. R. G. Jaeger (editor), *Engineering Compendium on Radiation Shielding*, Springer-Verlag, New York, 1968.
7. H. Goldstein, *Fundamental Aspects of Reactor Shielding*, Addison-Wesley, Reading, Mass., 1959.
8. N. M. Schaeffer, *Reactor Shielding for Nuclear Engineers*, U.S. AEC Document TID 25951, 1973.

PROBLEMS

5.1 In the explosion of a certain fission bomb in the atmosphere, 10^{25} fissions occur over a period of less than one second. If six 1-MeV prompt gamma photons are emitted per fission, compute the gamma dose (rem) received by a person standing 1000 meters away and directly exposed to the bomb. (a) Neglect shielding effect of air, and (b) consider shielding effect of air.

5.2 A gamma photon enters normal to a slab of material of thickness t. The linear attenuation coefficient, μ, is 0.4 cm^{-1} and the coefficient of Compton scattering, μ_{cs}, is 0.3 cm^{-1}.
(a) What is the probability that the photon will go through the material without a reaction?
(b) What is the probability that the photon will undergo a Compton scattering?

5.3 Ordinary concrete has a density of 2.35 g/cm^3. Its composition may be approximated as follows:

i	wt %	$(\mu/\rho)_i$, cm^2/g
Hydrogen	1	0.295
Oxygen	50	0.151
Silicon	32	0.172
Calcium	17	0.235

The mass attenuation coefficients for 0.1 MeV gamma photons for each constituent is shown. Calculate the linear attenuation coefficient, μ, for 0.1-MeV photons in this concrete.

5.4 If one million curies of ^{137}Cs are to be stored in a cubical room, 10 feet on an inside side, what thickness concrete would you recommend for the room wall if people are to work regularly outside that wall? (Assume the ^{137}Cs will be in the center.) If the source is moved to a side wall, by what factor would you estimate the dose would increase at a point on the outside wall just opposite the source?

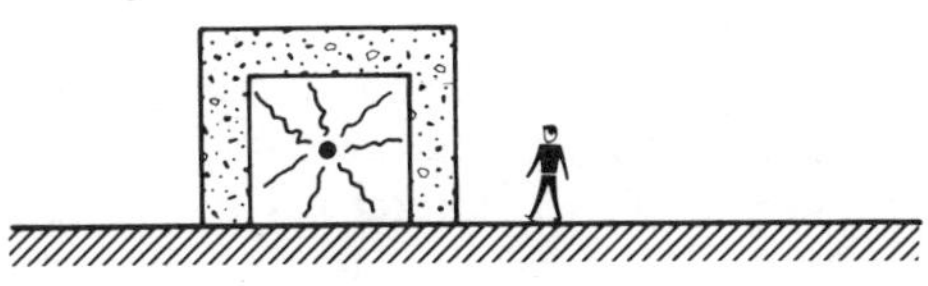

A handbook gives the following mass attenuation coefficients for a standard concrete (density = 2.7 g/cm^3).

E_γ (MeV)	0.1	0.2	0.5	1.0	2.0
μ/ρ (cm^2/g)	0.169	0.124	0.087	0.0635	0.0445

5.5 A certain pump in a nuclear power plant is located in an average thermal neutron flux of 10^9 neutrons/cm$^2 \cdot$ s. The pump contains some 200 pounds of steel containing 0.1 w/o cobalt. After 15 years of such neutron exposure, how many curies of cobalt activity would be in the pump? Could a person work on this pump for maintenance purposes? Support your answer with numbers.

5.6 A concrete made from Portland cement has a density of 2.2 g/cm^3 and a composition as follows:

	w/o	μ (E_γ = 1 MeV)/ρ (cm^2/g)
H	1.0	0.126
O	52.9	0.0636
Si	33.7	0.0635
Al	3.4	0.0614
Fe	1.4	0.0598
Ca	4.4	0.0637
Na	1.6	0.0608
K	1.3	0.0620

Estimate the linear attenuation coefficient, μ, for this concrete. What thickness of this concrete would you recommend for shielding a 100 Ci emitter of 1-MeV gammas in order that persons could work regularly in the immediate vicinity?

5.7 Iron has a density of 7.86 g/cm^3 and an atomic weight of 55.85. For a neutron traveling at 2200 m/s, iron has microscopic cross sections for scattering and absorption of 11 and 2.62 barns, respectively. For neutrons traveling at this speed, how long (how many seconds) after the neutrons enter a piece of iron, on the average, would you expect an interaction with a nucleus? For neutrons of this energy entering normal to a sheet of iron one centimeter thick, what percentage would you expect to go through without interacting?

5.8 The mass attenuation coefficient, μ/ρ, for 1-MeV photons in iron is 0.06 cm^2/g, and the density of iron is 7.8 g/cm^3. What thickness of iron must be placed between a source and receiver 100 cm apart to reduce the flux at the receiver to 1000 photons (1 MeV)/cm$^2 \cdot$ s? The source emits 10^{12} photons (1 MeV)/s.

5.9 In a recent news release, the AEC announced that about 0.1 g of tritium (^{3}H) is contained in the project Rulison cavity. This refers to a Plowshare project in Colorado in which natural gas is released from underground entrapment by exploding a nuclear device. How many curies is represented by 0.1 g? The allowable concentration for release of tritium to the public environment is 2×10^{-7} μCi/ml (10 CFR20). How many cubic feet of gas would be required to reduce the 0.1 g of tritium to this concentration?

5.10 What is the flux (or current) of gamma photons at a distance of one meter from a 1-g point source of ^{144}Ce (with its ^{144}Pr daughter at equilibrium concentration)? What is the dose rate in rem/h at this point?

5.11 Consider a point source of 1-MeV gamma photons located in a semi-infinite slab of H_2O, one foot from the surface. What fraction of the emitted gammas get out of this surface without undergoing any interaction with the water?

5.12 The maximum permissible concentration (10 CFR 20) for ^{137}Cs in water passing into the public domain is 2×10^{-5} μCi/ml. Suppose one has a detector through which water flows and in which 100 ml is held up at any time. If the detector can count 10% of the disintegrations, what would be the count rate (CPM) if the water contained only ^{137}Cs at the MPC?

5.13 Considering only the neutrons emitted by the spontaneous fission of ^{240}Pu, compute the dose rate (rem/h) 1 m away from 1 g of ^{240}Pu with no shielding.

5.14 Air has a density of 0.0013 g/cm^3 and a mass attenuation coefficient for 1-MeV gamma photons of 0.035 cm^2/g. If you are standing 1000 meters away from a source that emits 10^8 1-MeV photons/s, what current of these photons (1-MeV photons/cm^2 · s) is impinging on you? In order to calculate the dose rate you are receiving, you would have to introduce the buildup factor. Explain why and how the buildup factor is used.

5.15 One gram of ^{90}Sr, which may be treated as a point source of gamma radiation, is to be put in a hollow spherical lead (density 11.35 g/cm^3) container with an inside diameter of 10 cm. How thick must the wall be in order that the absorbed dose rate for a person 1 m from the container's outer surface is 10 mrad/h?

5.16 Consider that one curie of a hypothetical isotope that emits a 2-MeV gamma photon each decay is to be shielded with a concrete wall of density 2.7 g/cm^3. What thickness must the concrete be to ensure that 99.9% of the photons that impinge normal to the wall undergo an interaction?

5.17 Consider that a room $20' \times 20' \times 10'$ has air containing the MPC_a for insoluble ^{239}Pu (that is, 1×10^{-12} μCi/cm^3). How many grams of ^{239}Pu are in the air of the room? What is the disintegration rate of ^{239}Pu in the room?

5.18

Radioactive ^{100}X has the above decay scheme. One kg of it is to be stored in a spherical shield of lead (density 11.3 g/cm^3). What wall thickness of Pb is required to reduce the dose at a receiver 10 meters away to 1 mrem/h? Neglect buildup in the lead and attenuation in the air.

5.19 Suppose you worked for 40 hours a week, 50 weeks a year, at an average distance of 1 meter (no shielding) from a 1-millicurie radioactive source emitting one 1-MeV gamma photon and one 2-MeV gamma photon on each disintegration. Would your exposure (a) exceed dose level standards for radiation workers (that is, occupational standards), (b) exceed the maximum exposure for someone in the public, and (c) be high enough so that you might be expected to exhibit clinical symptoms at the end of one year?

5.20

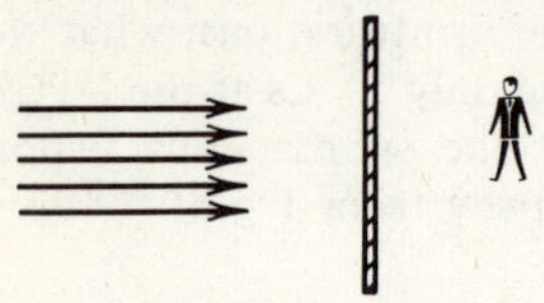

A collimated neutron beam of large cross section impinges on a receiver. The beam current is 10^5 neutrons/cm$^2 \cdot$s. A 2mm-thick sheet of cadmium metal is placed in front of the receiver. Estimate the dose rate (mrem/h) if the neutron energy is (a) 0.025 eV and (b) 2 MeV. Consider only the dose contribution from unscattered neutrons.

Chapter 6

Radioactivity and decay thermal power

In Chapter 3 we dealt briefly with radioisotopes as large-scale energy sources. Although in this respect we found radioisotopes wanting, they nevertheless play a very important role in all nuclear energy systems. In this chapter, therefore, we treat the subject of radioactivity in greater detail.

CHAIN DECAY

It often happens that the daughter nuclide formed in a radioactive decay is itself radioactive. A succession of decays then results; this is called chain decay. In such a case, the daughter activity is often of interest. We represent a chain

$$X_1 \xrightarrow{\lambda_1} X_2 \xrightarrow{\lambda_2} X_3 \dashrightarrow \qquad (6.1)$$

where X_1 is the chain starter. It is not the decay product of any isotope present and is not otherwise being produced. The population and activity of X_1 nuclei, therefore, follows Eq. 3.3 and 3.4, respectively. The population of X_2 nuclei is described by the differential equation

$$\frac{dN_2}{dt} = \lambda_1 N_1 - \lambda_2 N_2 \qquad (6.2)$$

$$N_1(t) = N_1(0)\, e^{-\lambda_1 t} \qquad (3.3)$$

Substituting:

$$\frac{dN_2}{dt} = \lambda_1 N_1(0)\, e^{-\lambda_1 t} - \lambda_2 N_2 \qquad (6.3)$$

This is a linear, first-order, ordinary differential equation and has the solution

$$N_2(t) = \frac{\lambda_1 N_1(0)}{\lambda_2 - \lambda_1}\, [e^{-\lambda_1 t} - e^{-\lambda_2 t}] + N_2(0)\, e^{-\lambda_2 t} \qquad (6.4)$$

Example 6.1
Yttrium-90 is a radioactive daughter of strontium-90. Consider that strontium is chemically separated from yttrium at a given time. For each curie of ^{90}Sr activity, how much activity of ^{90}Y will there be in the sample after one day?

Solution: The decay chain is

$$^{90}\text{Sr} \xrightarrow{28.1\,\text{y}} {}^{90}\text{Y} \xrightarrow{64\,\text{h}} {}^{90}\text{Zr}$$

$$X_1 \xrightarrow{\lambda_1} X_2 \xrightarrow{\lambda_2} X_3$$

The chemical separation means that the initial value, $N_2(0)$, is zero. It will suffice to express the values in hour units.

$$\lambda_1 = \frac{\ln 2}{28.1 \times 365 \times 24} = 2.816 \times 10^{-6}\ \text{h}^{-1}$$

$$\lambda_2 = \frac{\ln 2}{64} = 1.083 \times 10^{-2}\ \text{h}^{-1}$$

$$\lambda_1 N_1(0) = 3.7 \times 10^{10}\ \text{dps (one Ci)}$$

$$\lambda_2 N_2(24\text{h}) = 3.7 \times 10^{10} \times \frac{1.083 \times 10^{-2}}{1.083 \times 10^{-2} - 2.816 \times 10^{-6}} [1.00 - 0.7711]$$

$$\lambda_2 N_2(24\text{h}) = 8.47 \times 10^{9}\ \text{dps}$$

$$= 0.23\ \text{Ci}$$

Example 6.1 illustrates the case of the daughter with a much shorter half-life than the parent. Note that, in this case, Eq. 6.4 becomes very similar to Eq. 3.7, except that $\lambda_1 N_1(0)$ has replaced $\dot{P}$, the production rate. But, of course, $\lambda_1 N_1$ is the production rate of X_2. If the half-life of X_1 is relatively long, then in the time frame of a few half-lives of X_2, $\lambda_1 N_1$ appears constant. The activity of X_2 approaches the activity of the parent, going 50% of the way in one X_2 half-life, as depicted in Figure 3.5. The daughter's activity approaches that of the parent, a condition called secular equilibrium. Although Eq. 6.4 gives exact results, it is often useful to know that a short-lived daughter will reach the same activity as the parent.

It is always necessary to keep in mind the radioactivity of any daughters. Strontium-90 is a good case in point. People speak of a strontium-90 radioisotope power source. Yet it is the yttrium-90 decay that gives most of the energy. Also, it gives off most of the gamma photons and, therefore, sets the design criteria for shielding. Of course, strontium-90 is the starting material and is always the most abundant isotope, since N_1/N_2 is equal to λ_2/λ_1.

Gamma Emission as Chain Decay

It may be a small point, but it should be noted that all gamma emission in the radioactive process is actually an example of chain decay. The excited levels of the daughter decay by gamma emission (isomeric transition) as a subsequent step to the alpha or beta emission of the parent. The half-life of the excited daughter states is generally so short that they are instantly in secular equilibrium with the parent, and we treat the two decays as a single event. In Figure 3.4, for example, we see that the 1.333-MeV level of ^{60}Ni has a half-life of 0.7 picoseconds. There are cases, such as ^{137}Cs decay, in which the excited level of the daughter (^{137}Ba) has a long half-life. The excited level activity can be calculated by Eq. 6.4.

Example 6.2
One kilogram of chemically purified plutonium contains 10 w/o ^{241}Pu. What weight and activity of ^{241}Am will it contain after two years?

Solution: The decay chain of interest is

$$^{241}Pu \xrightarrow{13.2\,y} {}^{241}Am \xrightarrow{458\,y} {}^{237}Np$$

$$N_1 \xrightarrow{\lambda_1} N_2 \xrightarrow{\lambda_2} N_3$$

The initial activity of ^{241}Pu is calculated:

$$\lambda_1 = \frac{\ln 2}{13.2 \times 3.156 \times 10^7} = 1.664 \times 10^{-9}\ s^{-1}$$

$$N_1(0) = \frac{1000 \times 0.1}{241} \times 6.023 \times 10^{23} = 2.499 \times 10^{23}\ \text{nuclei}$$

$$\lambda_1 N_1(0) = 4.158 \times 10^{14}\ \text{dps} = 1.124 \times 10^4\ \text{Ci}$$

Solve Eq. 6.4 for ^{241}Am activity.

$$\lambda_1 = \frac{\ln 2}{13.2} = 5.25 \times 10^{-2}\ y^{-1}$$

$$\lambda_2 = \frac{\ln 2}{458} = 1.513 \times 10^{-3}\ y^{-1}$$

$$\lambda_2 N_2(2y) = \frac{1.513 \times 10^{-3}}{1.513 \times 10^{-3} - 5.25 \times 10^{-2}} \times 4.158 \times 10^{14}$$

$$\times\ [0.9003 - 0.9970]$$

$$= 1.193 \times 10^{12}\ \text{dps}\ (32.2\ \text{Ci})$$

The weight of ^{241}Am is computed:

$$\lambda_2 = \frac{\ln 2}{458 \times 3.156 \times 10^7} = 4.795 \times 10^{-11}\ \text{s}^{-1}$$

$$N_2 = \frac{1.193 \times 10^{12}}{4.795 \times 10^{-11}} = 2.488 \times 10^{22}\ \text{nuclei}$$

$$m_2 = \frac{2.488 \times 10^{22}}{6.023 \times 10^{23}} \times 241 = 9.95\ \text{g}$$

After the two-year period, the "plutonium" is about 1 w/o ^{241}Am. Although the activity (curies) of the ^{241}Am is far less than that of the original ^{241}Pu, the gamma emission (see LHP) of the ^{241}Am is much greater (^{241}Pu has no gamma emission). This fact can cause problems in the fuel cycle, so it is desirable to perform operations with plutonium as soon after purification as possible.

The Analysis of Longer Decay Chains

The previous analysis of the activity of the second member of a decay chain can be extended to the third member, and so on. For example

$$\frac{dN_3}{dt} = \lambda_2 N_2 - \lambda_3 N_3 \tag{6.5}$$

We can substitute for N_2 by Eq. 6.4. Rather than perform that operation, however, we shall skip directly to the general result for the ith member of a chain:

$$N_i(t) = \lambda_1 \lambda_2 \cdots \lambda_{i-1} N_1(0) \sum_{j=1}^{i} \frac{\exp(-\lambda_j t)}{\prod_{\substack{k=1 \\ k \neq j}}^{i} (\lambda_k - \lambda_j)} \tag{6.6}$$

The interpretation of this formulation is illustrated in Example 6.3. The formulation of Eq. 6.6 does assume that at time zero, only X_1 is present. If other isotopes in the chain are initially present, then the activities that follow could be calculated by making each such isotope X_1 in a series of separate applications of Eq. 6.6, and adding the results for time t. (This is a case of superposition.)

Example 6.3
Some natural uranium is completely separated from all other elements at some given time. Calculate the concentration of ^{234}U present, assuming it to be in secular equilibrium with ^{238}U. Compare this value with the natural abundance

of 0.0057% reported in LHP. If the purified uranium is stored indefinitely, estimate the time it would take for ^{226}Ra to reach half its equilibrium value.

Solution: The relevant decay chain is

$$^{238}\text{U} \xrightarrow[\alpha]{4.51 \times 10^9\text{ y}} {}^{234}\text{Th} \xrightarrow[\beta^-]{24.1\text{ d}} {}^{234}\text{Pa} \xrightarrow[\beta^-]{6.75\text{ h}}$$

$$^{234}\text{U} \xrightarrow[\alpha]{2.47 \times 10^5\text{ y}} {}^{230}\text{Th} \xrightarrow[\alpha]{8.0 \times 10^4\text{ y}} {}^{226}\text{Ra} \xrightarrow[\alpha]{1602\text{ y}}$$

If ^{234}U and ^{238}U are in secular equilibrium, their λN values are equal. Therefore

$$\frac{N_{234}}{N_{238}} = \frac{2.47 \times 10^5}{4.51 \times 10^9} = 5.48 \times 10^{-5}$$

This value is in good agreement with the 0.0057% abundance value. We ignore the slight effect of the presence of 0.71% ^{235}U in the natural uranium.

If the uranium is completely purified, then the various daughters–thorium, protactinium, radium, and so on–are all removed. They will again begin to build up, of course. We can say that within a year, ^{234}Th and ^{234}Pa will again be essentially at their equilibrium concentration. Since ^{234}U is already in secular equilibrium with ^{238}U, it will stay there; the half-lives of the two intermediate isotopes are so short that they cause no interruption in ^{234}U production. In effect, therefore, the decay rate of ^{234}U follows ^{238}U continuously. That activity, based on one kg of ^{238}U, is

$$\lambda = \frac{\ln 2}{4.51 \times 10^9 \times 3.156 \times 10^7} = 4.87 \times 10^{-18}\text{ s}^{-1}$$

$$\lambda N = 4.87 \times 10^{-18} \times \frac{1000}{238} \times 6.023 \times 10^{23} = 1.232 \times 10^7\text{ dps}$$

$$(0.333\text{ mCi/kg-}^{238}\text{U})$$

This activity decays with a 4.5 billion year half-life. By inspection of the decay chain, we expect ^{230}Th activity to reach half this activity in 80,000 years. The activity of ^{238}U and ^{234}U will not have changed significantly in this time. The ^{226}Ra activity will be following the ^{230}Th activity by a relatively short period of time. Let us guess that ^{226}Ra will reach half secular equilibrium in 82,000 years and calculate its activity at that time. The *effective* chain is

$$^{238}\text{U} \xrightarrow{4.51 \times 10^9\text{ y}} {}^{230}\text{Th} \xrightarrow{8 \times 10^4\text{ y}} {}^{226}\text{Ra} \xrightarrow{1602\text{ y}}$$

$$X_1 \xrightarrow{\lambda_1} X_2 \xrightarrow{\lambda_2} X_3 \xrightarrow{\lambda_3}$$

Applying Eq. 6.6 to this case

$$\lambda_3 N_3(t) = \lambda_2 \lambda_3 \, \lambda_1 N_1(0) \left[\frac{e^{-\lambda_1 t}}{(\lambda_2 - \lambda_1)(\lambda_3 - \lambda_1)} + \frac{e^{-\lambda_2 t}}{(\lambda_1 - \lambda_2)(\lambda_3 - \lambda_2)} + \frac{e^{-\lambda_3 t}}{(\lambda_1 - \lambda_3)(\lambda_2 - \lambda_3)} \right]$$

$$\lambda_1 = \frac{\ln 2}{4.51 \times 10^9} = 1.537 \times 10^{-10}\ \text{y}^{-1}$$

$$\lambda_2 = \frac{\ln 2}{8 \times 10^4} = 8.664 \times 10^{-6}\ \text{y}^{-1}$$

$$\lambda_3 = \frac{\ln 2}{1602} = 4.327 \times 10^{-4}\ \text{y}^{-1}$$

$$\lambda_3 N_3 (82{,}000\ \text{y}) = 8.664 \times 10^{-6} \times 4.327 \times 10^{-4} \times 1.232 \times 10^7$$
$$\times\ [2.668 \times 10^8 - 1.338 \times 10^8 + 0]$$
$$= 6.14 \times 10^6\ \text{dps}\ (0.166\ \text{mCi})$$

This is very close to half the ^{234}U and ^{238}U activity. Although Eq. 6.6 can be solved exactly, substituting in each and every isotope in the chain, it does simplify matters to make approximations of the type illustrated here.

ISOTOPE CHANGES IN NUCLEAR REACTORS

We have mentioned in Chapter 3 that some important radioisotopes are formed in reactors (that is, in a neutron flux) through a succession of (n, γ) reactions and radioactive decays. The conversion of ^{239}Pu from ^{238}U is a case in point. Consider the general case of a radionuclide. It can be formed either as the product of a (n, γ) reaction or, indeed, other neutron reactions; it may also be the daughter of a radioactive parent. It can disappear through neutron absorption or its own radioactive decay. One representation of a group of such processes is

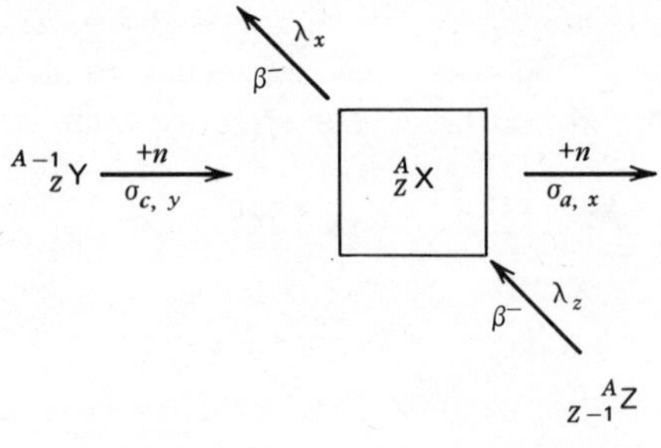

The change in population of ${}^{A}_{Z}X$ nuclei with time can be represented

$$\frac{dN_x}{dt} = \lambda_z N_z + N_y \sigma_{c,y} \phi - \lambda_x N_x - N_x \sigma_{a,x} \phi \tag{6.7}$$

If N_x is the only term that varies significantly over time, then the solution of Eq. 6.7 is analogous to that of Eq. 3.7.

$$N_x(t) = \frac{(\lambda_z N_z + N_y \sigma_{c,y} \phi)}{\lambda_x + \sigma_{a,x} \phi} [1 - \exp - (\lambda_x + \sigma_{a,x} \phi) t] \tag{6.8}$$

More general cases, where several N values are changing with time, require the simultaneous solution of several equations of the form of Eq. 6.7. We will not go into further detail, however. It is interesting to note that when a given nuclide is disappearing both by radioactive decay and by neutron absorption, the measure of the relative importance of the two modes is simply the relative value of λ and $\sigma_a \phi$. The value of λ is unchanging with time, while the value of $\sigma_a \phi$ depends on the maintenance of the neutron flux value by means of a reactor or accelerator. Consequently, $\sigma_a \phi$ is time-varying. In the application of Eq. 6.7 and 6.8, the user must consider whether a time-averaged $\sigma_a \phi$ value is suitable or whether the variations of $\sigma_a \phi(t)$ must be handled more explicitly.

As we did for the radioactive decay chain in obtaining a general equation (Eq. 6.6) for any isotope in the chain, we can do also in the present case of a chain consisting of neutron reactions and radioactive decays. We write a general equation

$$\frac{dN_i}{dt} = N_{i-1} \sigma_{(i-1)i} \phi - N_i \sigma_i \phi \tag{6.9}$$

where $\sigma_{(i-1)i}$ is the neutron reaction cross section for the transformation of isotope $i - 1$ to i. *Alternatively*, $\sigma_{(i-1)i} \phi$ could be replaced by λ_{i-1}, the radioactive decay constant of isotope $i - 1$, if that is the route from $i - 1$ to i. The quantity σ_i is a measure of the removal of isotope i by any mechanism—neutron absorption or radioactive decay, or both.

$$\sigma_i = \sigma_{a,i} + \frac{\lambda_i}{\phi} \tag{6.10}$$

We also define

$$d\tau = \phi(t)\, dt \tag{6.11}$$

We can now write Eq. 6.9

$$\frac{dN_i}{d\tau} = N_{i-1} \sigma_{(i-1)i} - N_i \sigma_i \tag{6.12}$$

Isotope 1, by definition, is the chain starter. If isotope 1 has a formation reaction, it is not included.

$$\frac{dN_1}{d\tau} = -N_1\,\sigma_1$$

$$\frac{dN_2}{d\tau} = N_1\,\sigma_{12} - N_2\,\sigma_2$$

The solution of Eq. 6.12 for the ith isotope is

$$N_i(\tau) = \sigma_{12}\,\sigma_{23}\cdots\sigma_{(i-1)i}\,N_1(0)\sum_{j=1}^{i}\frac{\exp(-\sigma_j\tau)}{\prod_{\substack{k=1\\k\neq j}}^{i}(\sigma_k-\sigma_j)} \tag{6.13}$$

This equation may look very complex, but it can be handled with some practice. Its use is illustrated in examples to follow. In the examples, we shall use only cross section values at 0.0253 eV. In practical cases the neutron spectrum is more complex (Chapters 4, 7), and some spectrum-averaged cross section must be used.

Example 6.4

In a thermal-neutron reactor that has only ^{235}U as a fissile material, some ^{10}B is mixed with the ^{235}U. When the ^{235}U has been 50% consumed (often expressed as 50% burnup), what percentage of the ^{10}B has been consumed?

Solution: Since it is a thermal-neutron reactor, we shall use the 0.0253 eV cross sections of Table A.3. The cross section for disappearance is the absorption cross section. In the case of ^{235}U

$$\sigma_a = \sigma_c + \sigma_f = 98.6 + 582.2 = 680.8 \text{ barns}$$

In the case of ^{10}B

$$\sigma_a \text{ (of the } (n,\alpha) \text{ reaction)} = 3837 \text{ barns}$$

In the case of each of these isotopes, there is no formation by neutron reactions or radioactive decay. Each is the "first" isotope in a chain. In terms of Eq. 6.12

$$\frac{dN}{d\tau} = -N\sigma_a$$

Integrating

$$\frac{N(\tau)}{N(0)} = \exp(-\sigma_a\tau)$$

In the case of ^{235}U

$$\frac{N(\tau)}{N(0)} = 0.5 = \exp(-680.8\tau)$$

$$\tau = 1.018 \times 10^{-3}$$

The same value of τ applies to ^{10}B. So for ^{10}B

$$\frac{N(\tau)}{N(0)} = \exp\left(-3837 \times 1.018 \times 10^{-3}\right)$$

$$= 0.0201$$

which means that 98% of ^{10}B has been consumed.

Example 6.5
In the conversion of ^{232}Th (natural thorium) to fissile ^{233}U, the reactions are

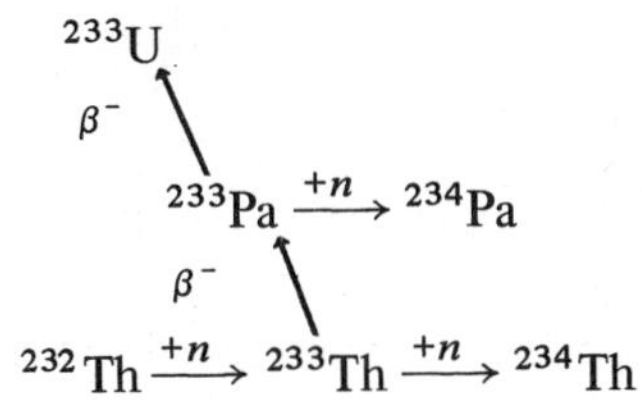

The neutron absorption reactions in ^{233}Th and ^{233}Pa reduce the ^{233}U actually formed. Compute the fraction of ^{233}Th actually converted to ^{233}U as a function of the neutron flux, ϕ.

Solution: The relevant thermal neutron cross sections and radioactive decay constants are:

	HALF-LIFE	λ (s^{-1})	σ_a (BARNS)
^{233}Th	22.2 min	5.20×10^{-4}	1515
^{233}Pa	27 day	2.97×10^{-7}	41

The quantity $\lambda/\sigma_a\phi$ gives the ratio of radioactive decays to neutron absorptions. Values of $\lambda/\sigma_a\phi$ as a function of ϕ, are:

	$\phi(n/\text{cm}^2 \cdot \text{s})$		
	10^{13}	10^{14}	10^{15}
^{233}Th	3.43×10^4	3.43×10^3	343
^{233}Pa	724	72.4	7.24

From these values we see that the reaction that competes most strongly with ^{233}U formation is neutron absorption in ^{233}Pa. At a flux of 10^{15} $n/\text{cm}^2 \cdot \text{s}$ there are 7.24 beta decays for each neutron capture in ^{233}Pa. The fraction of ^{233}Th going to ^{233}U is 7.24/8.24, or 0.88. In a reactor based on the thorium fuel cycle, therefore, it is important that the neutron flux not be too high, because the effect is to reduce the conversion of thorium to ^{233}U. In the conversion of ^{238}U to ^{239}Pu, the half-lives of the intermediate isotopes are so short that loss of the intermediates by neutron absorption is not significant.

Example 6.6

Pure ^{239}Pu is placed in a reactor. When the ^{239}Pu is 50% consumed, how much ^{240}Pu and ^{241}Pu will there be? The thermal neutron flux is 10^{12} $n/\text{cm}^2 \cdot \text{s}$.

Solution: The reactions of interest are

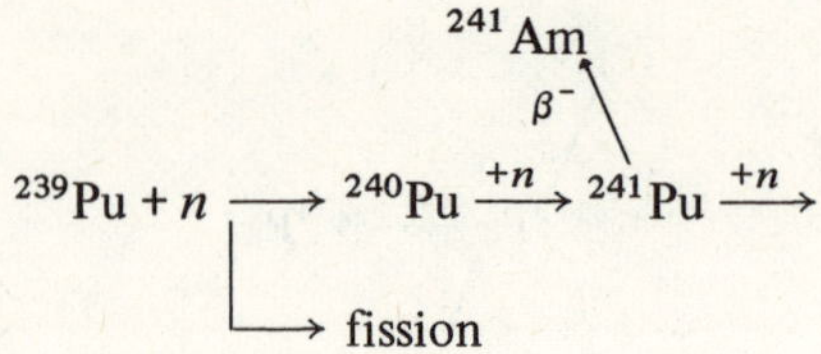

The decay and neutron reaction data (Appendix A.3) are:

	HALF-LIFE	λ (s^{-1})	σ_c (BARNS)	σ_a (BARNS)
^{239}Pu	24,400 y		268.8	1011.3
^{240}Pu	6,580 y		289.5	289.5
^{241}Pu	13.2 y	1.66×10^{-9}	368	1377

For ^{241}Pu, the ratio of beta decay to neutron absorption, $\lambda/\sigma_a\phi$, is 1.208. Therefore, both mechanisms for ^{241}Pu must be considered. The σ value for ^{241}Pu is (Eq. 6.10)

$$\sigma = \sigma_a + \frac{\lambda}{\phi} = 1377 + 1664 = 3041$$

We shall apply Eq. 6.13 to the three isotopes in order.

$$N_{239} \equiv N_1(\tau) = N_1(0) \exp(-1011.3\,\tau)$$

$$N_1(\tau)/N_1(0) = 0.5$$

$$\tau = 6.854 \times 10^{-4} \text{ neutrons/barn}$$

$$N_{240} \equiv N_2(\tau) = 268.8\,N_1(0)\left[\frac{\exp(-1011.3 \times 6.854 \times 10^{-4})}{1011.3 - 289.5} + \frac{\exp(-289.5 \times 6.854 \times 10^{-4})}{1011.3 - 289.5}\right]$$

$$\frac{N_2(\tau)}{N_1(0)} = 268.8\,(-6.927 \times 10^{-4} + 1.136 \times 10^{-3})$$

$$= 0.119$$

$$N_{241} \equiv N_3(\tau) = 268.8 \times 289.5\,N_1(0) \left[\frac{\exp(-1011.3 \times 6.854 \times 10^{-4})}{(289.5 - 1011.3)(3041 - 1011.3)} \right.$$

$$+ \frac{\exp(-289.5 \times 6.854 \times 10^{-4})}{(1011.3 - 289.5)(3041 - 289.5)}$$

$$\left. + \frac{\exp(-3041 \times 6.854 \times 10^{-4})}{(1011.3 - 3041)(289.5 - 3041)} \right]$$

$$\frac{N_3(\tau)}{N_1(0)} = 0.0073$$

For every kg of ^{239}Pu originally in the reactor there will be

	$\frac{N(\tau)}{N_1(0)}$	WEIGHT (kg)	Pu COMPOSITION (w/o)
^{239}Pu	0.5	0.500	79.8
^{240}Pu	0.119	0.119	19.0
^{241}Pu	0.0073	0.0073	1.2

In order to know the calendar time, or real time, over which these isotopic changes occur, one must know the average flux or some other information about the system.

Note that in this calculation it is assumed that no ^{240}Pu or ^{241}Pu was present at time zero. The presence of either of these isotopes at the start could be handled by making it isotope 1 in Eq. 6.13, calculating its quantity and the quantity of higher isotopes in the chain at τ, and adding these quantities to those calculated with ^{239}Pu as the chain starter.

Xenon Poisoning of Fission Reactors

In Chapter 3 we discussed the fact that many kinds of isotopes are formed in the fission process. These isotopes are known collectively as fission products. Most of them are radioactive. There are well over one hundred such isotopes, but only a few stand out from the crowd by virtue of some special characteristic or combination of characteristics. One of these is the isotope xenon-135. Its special characteristic is its neutron capture cross section, which at 0.0253 eV is

2.65 million barns. This capture cross section is about 4000 times the absorption cross section of ^{235}U. The result is that, in a reactor, ^{135}Xe competes strongly with fissile material for neutrons; a relatively low concentration of ^{135}Xe can actually break up the fission chain reaction (Figure 3.8) and prevent a reactor from operating. Strong (high cross section) neutron absorbers of this nature are called "poisons"–hence the term "xenon poisoning."

An important characteristic of xenon poisoning is its time-dependence, which is determined by the half-lives of members of the decay chain of which ^{135}Xe is a member. The decay chain is

$$^{135}Sb \xrightarrow[1.7\,s]{\beta^-} {}^{135}Te \xrightarrow[19.2\,s]{\beta^-} {}^{135}I \xrightarrow[6.58\,h]{\beta^-} {}^{135}Xe \xrightarrow[9.17\,h]{\beta^-} {}^{135}Cs$$

The quantity of each isotope in the chain can be described by an equation very similar to Eq. 6.9.

$$\frac{dN_i}{dt} = \gamma_i \Sigma_f \phi + \lambda_{i-1} N_{i-1} - N_i \sigma_{a,i} + \lambda_i N_i \tag{6.14}$$

The source term, $\gamma_i \Sigma_f \phi$, needs some explanation. We have already shown (as in Example 4.12) that $\Sigma_f \phi$ is the fission rate in fissions/cm$^3 \cdot$s. The term γ_i is the fission yield (Table 3.1), or the fraction of fissions in which isotope i is formed. Therefore $\gamma_i \Sigma_f \phi$ is the rate of formation of isotope i in the fission process.

The complete solution to Eq. 6.14 could be written in an equation similar to Eq. 6.13, with ^{135}Sb being the first in the chain. As a practical matter, it is usually wiser to inspect the values of various terms in the set of Eq. 6.14 to see which can be omitted. By virtue of the very short half-lives of ^{135}Sb and ^{135}Te, we can see that these isotopes will "saturate" (see Figure 3.5) at a decay rate equal the production rate, $\gamma_i \Sigma_f \phi$, in a matter of seconds. Furthermore, the ratio $\lambda_i / \sigma_{a,i} \phi$ for these isotopes, and ^{135}I as well, will be so high that neutron absorption can be neglected. These simplifications allow us to omit any explicit treatment of ^{135}Sb and ^{135}Te and to allow for their existence by adding their yield values, γ_i, to the γ_i for iodine. We now have reduced the chain to two members:

$$^{135}I: \qquad \frac{dN_1}{dt} = \gamma_1 \Sigma_f \phi - \lambda_1 N_1 \tag{6.15}$$

$$^{135}Xe: \qquad \frac{dN_2}{dt} = \gamma_2 \Sigma_f \phi + \lambda_1 N_1 - N_2 \sigma_{a,2} \phi - \lambda_2 N_2 \tag{6.16}$$

These equations have the solutions for the case of constant flux:

$$^{135}I: \qquad N_1(t) = \frac{\gamma_1 \Sigma_f \phi}{\lambda_1} [1 - \exp(-\lambda_1 t)] + N_1(0) \exp(-\lambda_1 t) \tag{6.17}$$

$$^{135}\text{Xe:}\quad N_2(t) = \frac{(\gamma_1 + \gamma_2)\Sigma_f\phi}{\lambda_2 + \sigma_{a,2}\phi}[1 - \exp - (\lambda_2 + \sigma_{a,2}\phi)t]$$
$$+ \frac{\gamma_1\Sigma_f\phi - \lambda_1 N_1(0)}{\lambda_2 + \sigma_{a,2}\phi - \lambda_1}[\exp - (\lambda_2 + \sigma_{a,2}\phi)t - \exp(-\lambda_1 t)]$$
$$+ N_2(0)\exp - (\lambda_2 + \sigma_{a,2}\phi)t \qquad (6.18)$$

Both iodine-135 and xenon-135 have half-lives of several hours so that, at constant flux, we expect their concentrations, N, to saturate in a matter of days. The saturation values are obtained from Eq. 6.17 and 6.18 by allowing time to increase indefinitely.

$$^{135}\text{I:}\quad N_1(\infty) = \frac{\gamma_1\Sigma_f\phi}{\lambda_1} \qquad (6.19)$$

$$^{135}\text{Xe:}\quad N_2(\infty) = \frac{(\gamma_1 + \gamma_2)\Sigma_f\phi}{\lambda_2 + \sigma_{a,2}\phi} \qquad (6.20)$$

These equations are, in fact, simply variations on the general isotope-change equations such as Eq. 6.13. The xenon-135 case if of special interest, however, because (1) the thermal-neutron cross section of xenon-135 is so high that its presence in a reactor has a pronounced influence, and (2) the combination of γ and λ values that apply in Eq. 6.17 and 6.18 produces some unusual and counterintuitive transients of xenon concentration. For example, when a reactor is shut down, the xenon concentration may rise. This effect is flux-level-dependent, however. The exact history of xenon concentration for any set of conditions can be obtained from a suitable solution to Eq. 6.15 and 6.16.

In order to illustrate xenon concentration transients, we shall solve a specific case. Duderstadt[1] gives the necessary properties for the fission of ^{235}U:

$$\gamma_1 = 0.06386$$
$$\gamma_2 = 0.00228$$
$$\lambda_1 = 0.1035\ \text{h}^{-1} = 2.875 \times 10^{-5}\ \text{s}^{-1}$$
$$\lambda_2 = 0.0753\ \text{h}^{-1} = 2.092 \times 10^{-5}\ \text{s}^{-1}$$
$$\sigma_{a,2} = 2.65 \times 10^6\ \text{barns} = 2.65 \times 10^{-18}\ \text{cm}^2$$

The macroscopic fission cross section, Σ_f, is taken as 0.1 cm^{-1}. We start with a clean reactor (that is, no ^{135}I or ^{135}Xe present) and a flux of 2×10^{14}. The reactor is operated at this level until iodine and xenon saturate. The flux level is

[1] See Reference 1 at end of chapter.

then lowered abruptly to 1×10^{14} n/cm^2 · s. Again, the iodine and xenon are allowed to saturate. Then the reactor is shut down. The history of the ^{135}Xe concentration is given in Table 6.1. The results are also plotted in Figure 6.1. One should remember that these results are confined to thermal neutron reactors because, at fast-neutron (1 MeV) energies, ^{135}Xe does not exhibit an unusually high cross section.

From Table 6.1 and Figure 6.1, we see that ^{135}Xe builds up toward its saturation value (Eq. 6.20) in a matter of hours. At 20 hours, for example, it is 87% of the way. When the flux is cut in half, we reset the time to zero but keep the previous saturation values of ^{135}I and ^{135}Xe for the new initial conditions. The interesting occurrence is that the xenon concentration initially *rises*, peaking at about 2 hours after the power decrease and then declining toward its saturation

Table 6.1 A History of Xenon-135 Concentration in a Reactor

TIME (h)	NEUTRON FLUX (n/cm^2 · s)	FISSION POWER (W/cm^3)	^{135}I CONCENTRATION (nuclei/cm^3)	^{135}Xe CONCENTRATION (nuclei/cm^3)
0	2×10^{14}	640	0	0
1				2.017×10^{14}
10				1.532×10^{15}
20				2.092×10^{15}
30				2.291×10^{15}
—				—
∞			4.44×10^{16}	2.401×10^{15}
0	1×10^{14}	320	4.44×10^{16}	2.401×10^{15}
0.5				3.239×10^{15}
1				3.696×10^{15}
2				4.026×10^{15}
5				3.779×10^{15}
10				3.195×10^{15}
20				2.626×10^{15}
—				—
∞			2.22×10^{16}	2.313×10^{15}
0	0	0	2.22×10^{16}	2.313×10^{15}
1				4.248×10^{15}
5				8.948×10^{15}
10				1.053×10^{16}
15				9.839×10^{15}
20				8.311×10^{15}
40				2.827×10^{15}
50				1.482×10^{15}
∞			0	0

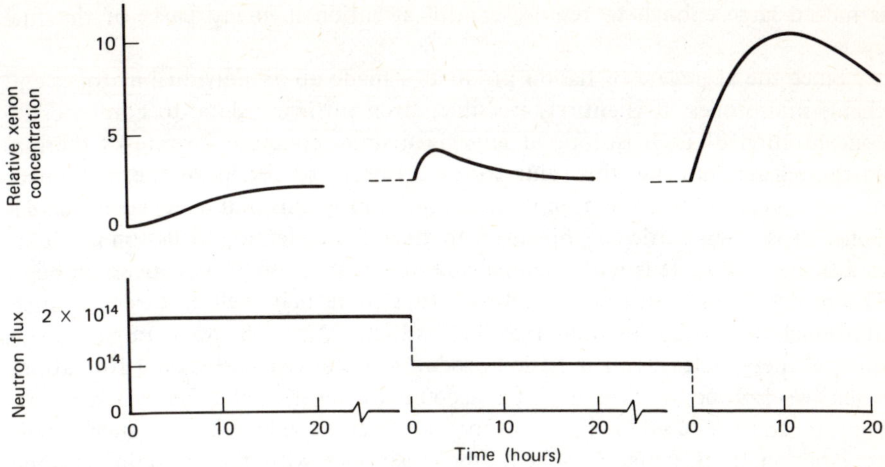

Figure 6.1 Typical xenon concentration behavior in a reactor as a function of thermal neutron flux.

value. Following shutdown, the rise in ^{135}Xe is even more dramatic. It goes to over four times it shutdown value at about 10 hours after shutdown before beginning to decline. Even 40 hours after shutdown, ^{135}Xe is above its shutdown concentration.

The transient behavior of ^{135}Xe concentration has very important implications for reactor design and operation. We will not pursue the matter further here. It is important for the reader to understand that these mathematical treatments of the transient behavior of radioisotope concentrations have many applications. The xenon poisoning case is only one. There are applications to the nuclear fission fuel cycle. These same kinds of equations apply to isotopes contained in waste from nuclear reactors.[2] In the latter case, the characteristic times are often in centuries instead of hours. The approximations that can be made vary from case to case, but each approximation comes down to finding satisfactory solutions to simultaneous equations of the form of Eq. 6.7.

FISSION PRODUCT DECAY THERMAL POWER

In Chapter 3 we discussed the thermal power released by the decay of individual radioisotopes and its variation with time. The collection of radioisotopes that constitute fission products obviously must release thermal energy as the individual radionuclides decay. It so happens that this energy, although not large enough to make fission products likely candidates as large-scale energy sources,

[2] See Reference 2 at end of chapter.

is indeed large enough to require careful attention in many parts of the fuel cycle.

Since the aggregate of fission products is made up of individual isotopes and chains of isotopes, it is entirely possible, given sufficient data, to compute the concentration of each isotope at any given time, compute its rate of thermal energy release, and sum this value for each isotope to obtain the thermal power of the aggregate. There are, in fact, computer programs that make such calculations. It is often sufficient, however, to treat the collection of fission products as a single entity. It is with such calculations that we shall be concerned here. The reader should remember, however, that there may well be cases in which individual isotope calculations may be justified. A case in point might be the thermal energy release rate in fission products several years after their formation, when two isotopes, ^{90}Sr and ^{137}Cs, account for most of the energy. When attention can be focused on a few isotopes, allowance can be made for such details as variation in the fission yield of those isotopes with the fissioning isotope and with the reactor neutron spectrum. The calculations to be discussed here do not have such detail.

Before discussing the calculations, it will help to review the state of fission products at various stages in the nuclear fuel cycle. Fission products are, of course, created in the reactor fuel. In the typical solid-fuel reactor, these isotopes for the most part remain in place, tightly associated with the fuel. Fission products in the gaseous state, such as krypton and xenon isotopes, do tend to migrate, but we shall ignore this effect. There are primary fission products: those nuclides formed directly in the fission reaction, and secondary products that are either part of the decay chain of a primary or the result of (n, γ) reactions in fission products. We regard all of these isotopes as a single entity, called "fission products." Fission products release their thermal energy within the fuel just as the fissioning isotopes do. While the reactor is in operation, the decay thermal power reaches 7 to 8% of the fission power. When the reactor is shut down and the fission reaction ceases, however, the fission-product-decay energy release goes on, declining over time according to fixed decay laws quite outside the control of the reactor operator. It is the inability to affect the decay thermal power that makes it a phenomenon to be reckoned with. Satisfactory means of removing this thermal energy must be available at all times, or else the fuel will overheat and eventually melt.

When the fuel is removed from the reactor, it is quickly transferred to a large pool of water that provides the needed cooling. Later (see Figure 3.11) it must be shipped to reprocessing, and cooling must be provided in transit. At the reprocessing plant, the fission products are to be separated by chemical means from the other fuel constituents, principally uranium and plutonium. The separated fission products, mixed with some uranium and plutonium because the separation is not perfect, are now referred to as high-level waste. By whatever name they are called, however, their decay heat release continues, and the necessity to make adequate design provision for its removal continues. As we shall see, the rate of decay thermal energy release declines with time and, accordingly, presents a less difficult problem as time passes.

Calculation of Decay Heat Rate

Figure 6.2 gives the model for the calculation of the rate of fission product energy release (the decay heat rate). A reactor. or any unit of fuel initially without fission products, is operated at constant thermal fission power, P_0, for time t_0, at which time the fission power drops to zero. The question is asked: what is the decay heat rate at time t_s after shutdown? It has been found that the average decay heat rate at time t following fission is of the form

$$\delta P = Ct^{-(a+1)} \text{ MeV/fission} \cdot \text{s} \tag{6.21}$$

That is to say, at time t, following an *average* fission, the decay heat rate is δP MeV/s. As can be seen from Figure 6.2, in the case of a finite operating time, t_0, the value of t is a variable, so we must integrate. The fission rate during t_0 is $P_0/200$ fissions/s, assuming P_0 in MeV units. The decay heat rate, P, is

$$\frac{P}{P_0} = \frac{C}{200} \int_{t_s}^{t_0+t_s} t^{-(a+1)}\, dt \tag{6.22}$$

We designate $P(t_0, t_s)$ as the decay heat rate following reactor operation of length t_0 and a shutdown of length t_s.

$$\frac{P(t_0, t_s)}{P_0} = \frac{C}{200\,a} [t_s^{-a} - (t_0 + t_s)^{-a}] \tag{6.23}$$

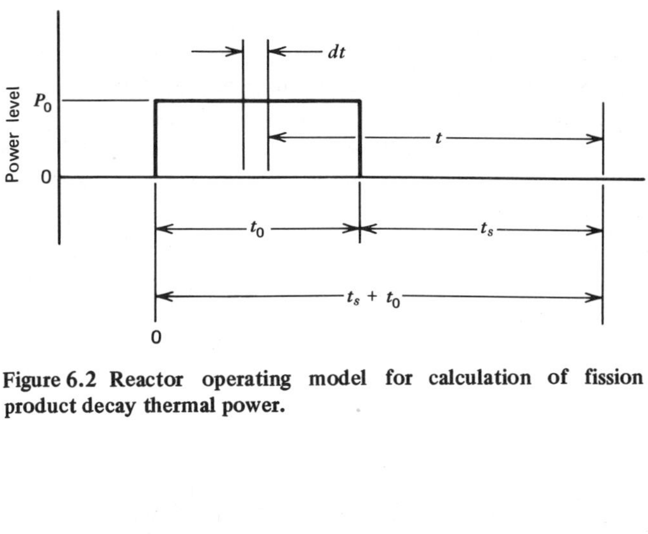

Figure 6.2 Reactor operating model for calculation of fission product decay thermal power.

Table 6.2 Thermal Power from Decay of Aggregate Fission Products

TIME FROM SHUTDOWN			$\frac{P(10d, t_s)}{P_0}$	$\frac{P(\infty, t_s)}{P_0}$	
(s)	(d)	(y)	ANS[3]	ANS[3]	$0.279\, t_s^{-0.333}$
0.1			0.0701	0.0728	
1			0.0605	0.0630	
10			0.0508	0.0535	
20			0.0448	0.0475	
50			0.0380	0.0407	
100			0.0335	0.0362	
200			0.0291	0.0318	
500			0.0223	0.0250	
1000			0.0180	0.0208	
2000			0.0144	0.0171	
5000			0.0105	0.0132	
10^4	0.1		0.00828	0.0110	0.0130
10^5	1.2		0.00339	0.00603	0.00603
10^6	11.6		0.00046	0.00264	0.00280
10^7	116	0.3			0.00130
10^8		3.2			0.000604
10^9		32			0.000281

Note that if the reactor operated for an indefinitely long period

$$\frac{P(\infty, t_s)}{P_0} = \frac{C}{200\,a} t_s^{-a} = A t_s^{-a} \tag{6.24}$$

where we grouped the constants as A. Equation 6.23 can be written

$$\frac{P(t_0, t_s)}{P_0} = P(\infty, t_s) - P(\infty, t_0 + t_s) \tag{6.25}$$

This relationship means that if we have the function $P(\infty, t_s)/P_0$, we can calculate any $P(t_0, t_s)/P_0$.

Over the years, several evaluations have been made of the fission product decay heat rate. Some results are given in Table 6.2. The ANS values[3] were fitted to the expression At^{-a} for times greater than 10^5 seconds. The results are also given in Table 6.2. The ANS values for fuel irradiated for only 10 days and for an indefinitely long period are shown. It can be seen that for short times after shutdown there is not much difference. The reason is that the thermal power is coming primarily from short-lived fission products, and these reach equilibrium or saturation values in a very short time. As the shutdown

[3] See Reference 3 at end of chapter.

time increases, the difference in thermal power between the short-exposure fuel and long-exposure fuel becomes great. The computation of decay heat rates is illustrated in Examples 6.7 to 6.10.

Example 6.7

Spent uranium fuel from a power reactor has a burnup or exposure of 25 MWD(t)/kg-U (megawatt days thermal per kilogram of uranium) after a three-year period in a reactor. One hundred days after removal from the reactor the fuel is to be shipped to a reprocessing plant (Figure 3.11). What is the thermal power from fission product decay per kg-U?

Solution: Although we know that, during the three years this fuel was in the reactor, it probably operated at many different power levels, it will suffice to take an average power, or

$$P_0 = \frac{25}{3 \times 365} = 2.28 \times 10^{-2}\,\frac{\text{MW}(t)}{\text{kg-U}} = 22.8\,\frac{\text{kW}(t)}{\text{kg-U}}$$

We will use the relation

$$\frac{P(t_0, t_s)}{P_0} = \frac{P(\infty, t_s)}{P_0} - \frac{P(\infty, t_0 + t_s)}{P_0}$$

where

$$\frac{P(\infty, t_s)}{P_0} = 0.279\, t_s^{-0.333}$$

$$t_s = 8.64 \times 10^6 \text{ s}$$

$$t_0 = 9.46 \times 10^7 \text{ s}$$

$$t_0 + t_s = 1.032 \times 10^8 \text{ s}$$

Substituting in the above equation

$$\frac{P(\infty, t_s)}{P_0} = 1.37 \times 10^{-3}$$

$$\frac{P(\infty, t_0 + t_s)}{P_0} = 5.98 \times 10^{-4}$$

$$\frac{P(t_0, t_s)}{P_0} = 7.68 \times 10^{-4}$$

$$P(t_0, t_s) = 7.68 \times 10^{-4} \times 22.8 = .0175 \text{ kW/kg-U}$$

$$= 17.5 \text{ W/kg-U}$$

Example 6.8
Estimate the thermal power generated by fission products one month after reactor startup, with the reactor still running.

Solution: The fission power is designated P_0. In terms of the equation

$$\frac{P(t_0, t_s)}{P_0} = \frac{P(\infty, t_s)}{P_0} - \frac{P(\infty, t_0 + t_s)}{P_0}$$

we are interested in

$$\frac{P(t_0, 0)}{P_0} = \frac{P(\infty, 0)}{P_0} - \frac{P(\infty, t_0)}{P_0}$$

$$t_0 = 2.59 \times 10^6 \text{ s}$$

We shall take $P(\infty, 0)/P_0$ equal to 0.0728 from Table 6.2, for $t_s = 0.1$ s. Using the relation

$$\frac{P(\infty, t_s)}{P_0} = 0.279\ t_s^{-0.333}$$

we obtain

$$\frac{P(\infty, t_0)}{P_0} = 2.04 \times 10^{-3}$$

and

$$\frac{P(t_0, 0)}{P_0} = 0.0728 - 0.0020 = 0.0708$$

The power from fission product decay is 7% of the fission power. It is clear that the decay power concurrent with reactor operation rises quickly toward a saturation value in the range of 7 to 8%.

Example 6.9
Consider the case of a reactor fuel which operated for 30 months at 30 kW(t)/kg-U and then was reduced in power density to 15 kW(t)/kg-U for six months. The fuel was then removed from the reactor. Estimate the decay heat rate in this fuel three months after removal.

Solution: The decay power from the two different periods of operation can be calculated separately and the results added. For the 30-month operation, we are interested in $P(t_0, t_s)/P_0$, where

$$t_0 = 30 \text{ months} = 2.15 \times 10^9 \text{ s}$$

$$t_s = 9 \text{ months} = 5.37 \times 10^7 \text{ s}$$

Using the relationships

$$\frac{P(\infty, t_s)}{P_0} = 0.279\, t_s^{-0.333}$$

$$\frac{P(t_0, t_s)}{P_0} = \frac{P(\infty, t_s)}{P_0} - \frac{P(\infty, t_0 + t_s)}{P_0}$$

$$\frac{P}{P_0} = 7.44 \times 10^{-4} - 2.16 \times 10^{-4} = 5.28 \times 10^{-4}$$

the decay power from the 30-month operation is

$$P = 5.28 \times 10^{-4} \times 30 = 1.58 \times 10^{-2}\ \text{kW}(t)/\text{kg-U}$$

For the 6-month operation

$$t_0 = 6\ \text{months} = 3.58 \times 10^7\ \text{s}$$

$$t_s = 3\ \text{months} = 1.79 \times 10^7\ \text{s}$$

$$\frac{P}{P_0} = 1.073 \times 10^{-3} - 7.44 \times 10^{-4} = 3.29 \times 10^{-4}$$

The decay power from the 6-month operation is

$$P = 3.29 \times 10^{-4} \times 15 = 4.93 \times 10^{-3}\ \text{kW}(t)/\text{kg-U}$$

The decay power from the two operations simply add.

$$P = 1.58 \times 10^{-2} + 4.9 \times 10^{-3} = 0.0207\ \text{kW}(t)/\text{kg-U}$$

Example 6.10

A uranium dioxide (UO_2) fuel has operated at a thermal power density of 30 kW(t)/kg-U for a long time. One week after reactor shutdown the fuel is to be removed from the reactor. If the fuel is totally insulated for one hour during this operation, estimate the temperature rise that would occur. Use a value of 1000 J/kg · K as the heat capacity of UO_2.

Solution: We wish to know the total thermal energy release in the hour in question. Using the relationship

$$\frac{P(\infty, t_s)}{P_0} = 0.279\, t_s^{-0.333}$$

$$t_s = 6.048 \times 10^5\ \text{s}$$

$$\frac{P(\infty, t_s)}{P_0} = 0.00331$$

$$P = 0.00331 \times 30 = 0.0994\ \text{kW/kg-U} = 99.4\ \text{W/kg-U}$$

The power does not decline significantly over the hour period. The total thermal energy release in the hour is

$$3600 \times 99.4 = 3.58 \times 10^5 \text{ J/kg-U}$$

The temperature rise would be the total thermal release divided by the heat capacity.

$$\Delta T = \frac{3.58 \times 10^5}{1000} = 358°\text{C}$$

RADIOISOTOPE POWER SOURCES

Although our analysis in Chapter 3 indicated that radioisotopes could never play the role of large-scale energy sources, they nevertheless have found some important uses in special applications. The space program has provided the principal applications. The Apollo missions, for example, used radioisotope sources to provide power to the automatic measuring stations left behind on the surface of the moon. These sources have a power of about 60 W(*e*). They employ a thermoelectric converter, which has the advantage of simplicity (no moving parts) but achieves only about 4% conversion efficiency. The thermal power, therefore, is about 1500 W(*t*). At a much lower power level (~1 mW(*e*)), isotope power sources have found some application in cardiac pacemakers. Economic and other considerations indicate that the feasible range of electric power from isotope sources is 1 mW(*e*) to 1 kW(*e*).[4]

In spite of the fact that there are many radioisotopes, very few are candidates to drive power sources. Table 6.3 lists the principal candidates; even among this select group, few are likely to be chosen. The reason is that the requirements for an acceptable power source are usually very stringent.

Requirements for a Power Source Radioisotope

In Chapter 3, we demonstrated the way in which the thermal power of a radioisotope is calculated, given its quantity, its half-life, and its decay scheme (Figure 3.4). We shall not repeat these calculations here. However, a number of topics have been covered since Chapter 3 that also bear on the analyses of radioisotope sources. We shall summarize these points briefly.

Some of the isotopes of Table 6.3 decay to stable daughters, while others decay to radioactive daughters. In the latter case, the energy and the radiation from all members of the chain must be calculated. In the case of ^{90}Sr and ^{144}Ce,

[4]See Reference 4 at end of chapter.

TABLE 6.3 Radioisotopes of Interest as Power Sources

ISOTOPE	HALF-LIFE	SPECIFIC ACTIVITY (Ci/g)	SPECIFIC POWER (W(*t*)/g)	% ENERGY FROM ALPHA	BETA	GAMMA	SOURCE
Cobalt-60	5.26 y	1133	17.4	—	3.6	96.4	Cobalt (n, γ)
Strontium-90	28.1 y	141	0.95	—	100	neg	Fission product
Cesium-137	30 y	87	0.42	—	25.4	74.6	Fission product
Cerium-144	284 d	3191	0.33	—	95.1	4.9	Fission product
Promethium-147	2.62 y	928	0.33	—	100	0	Fission product
Thulium-170	130 d	5900	12.1	—	99.0	1.0	Thulium (n, γ)
Polonium-210	138 d	4500	141	100	—	neg	Bismuth (n, γ)
Plutonium-238	86 y	17.5	0.56	100	—	neg	^{237}Np (n, γ)
Curium-242	163 d	3310	120	100	—	neg	Reactor spent fuel
Curium-244	17.6 y	83.3	2.8	100	—	neg	Reactor spent fuel

for example, most of the energy release comes from the decay of the relatively short-lived daughters.

The need for shielding[5] to reduce radiation in the surroundings will, of course, vary with the application. For a remotely located power source, shielding might be required only during installation. For an implanted cardiac pacemaker, it is obvious that little radiation can be tolerated in the surroundings. In general, shielding is necessary, so a very important characteristic of a candidate isotope is that it emit a minimum of penetrating radiation. Most of the isotopes in Table 6.3, therefore, are alpha or beta emitters. Those that emit gamma photons are at a serious disadvantage in proportion to the number and energy of the photons, that is, in proportion to the shielding problem presented. While it is true, of course, that the energy of a captured photon is converted to thermal energy, this energy may be so widely distributed spatially that it is not worthwhile to attempt to recover or convert it. It should also be noted that several of the alpha emitters listed in Table 6.3 do have some gamma emission, even though the gamma energy relative to the alpha energy is negligible.

Gamma radiation is not the only concern in shielding a radioisotope power source. The beta electrons give rise to the electromagnetic radiation known as bremsstrahlung. The fraction of an electron's energy that appears as bremsstrahlung may be estimated as $ZE/800$, where E is the electron kinetic energy in MeV, and Z is the atomic number of the medium in which the electron is being decelerated. Bremsstrahlung has a spectrum with an upper limit equal to the electron energy, but the average photon energy is a small fraction of this. In most cases of shielding beta emitters, the gamma radiation is of much more concern than bremsstrahlung. However, because in isotope power sources one is seeking pure beta emitters, the bremsstrahlung can be relatively important.

Another form of radiation that must be considered in these heavy isotopes, ^{238}Pu, ^{242}Cm, and ^{244}Cm, is the neutrons from spontaneous fissions. Each of these isotopes undergoes spontaneous fission. The rate is expressed as a half-life in such references as LHP. The calculation is illustrated in Example 6.11.

Another important consideration of isotopes for power sources is their dilution or contamination with other isotopes, which may be an inevitable result of the way in which the isotopes are produced. In some cases, the contaminating isotopes may prove to be a serious source of gamma radiation; in others, the isotope simply lowers the specific power. ^{238}Pu, for example, can have trace quantities of 2.85-year ^{236}Pu. This isotope starts a decay chain that can lead to the buildup of some serious gamma emission over a period of a few years. In the case of an isotope such as ^{60}Co, produced in the ^{59}Co(n, γ) reaction, the ^{60}Co is greatly diluted by its stable isotope.

[5] Reference 5 provides an excellent detailed treatment of the energetics and the shielding of several radioisotopes of interest as power sources.

Given all of the requirements and potential problems, it is not surprising that few isotopes have achieved success as special-purpose energy sources. The two principal ones are ^{90}Sr and ^{238}Pu. Schalch et al.[6] discuss several applications of these. Between these two, ^{238}Pu has the advantage of requiring far less shielding,[7] but ^{90}Sr costs far less–about \$42/g or \$45/W(t)–compared to \$1250/g or \$2200/W(t) for ^{238}Pu[8].

Example 6.11
Calculate the rate of spontaneous fission neutrons emitted by a ^{238}Pu source with a thermal power of 0.1 W(t).

Solution: From Table 6.3 we obtain the specific power of ^{238}Pu as 0.56 W(t)/g. In the present case, therefore, we need

$$\frac{0.1}{0.56} = 0.18 \text{ g-}^{238}\text{Pu}$$

LHP gives the half-life for spontaneous fission (SF) as 4.9×10^{10} y. The rate of spontaneous fission is

$$\lambda_{SF}N = \frac{\ln 2}{4.9 \times 10^{10} \times 3.156 \times 10^7} \times \frac{0.18}{238} \times 6.023 \times 10^{23}$$

$$= 203 \text{ fissions/s}$$

If we assume 2.5 neutrons/fission, the neutron emission rate is

$$2.5 \times 203 = 508 \text{ neutrons/s}$$

REFERENCES

1. J. J. Duderstadt and L. J. Hamilton, *Nuclear Reactor Analysis*, Wiley, New York, 1976.
2. T. H. Pigford, "Environmental Aspècts of Nuclear Energy Production," *Annual Review of Nuclear Science*, Vol. 24, 1974.
3. G. J. Scatena, G. L. Upham, "Power Generation in a BWR following Normal Shutdown or Loss-of-Coolant Accident Conditions," NEDO-10625, March 1973.
4. D. Schalch, A. Scharmann, K. J. Euler, H. Mayer, "Radionuclide Batteries," *Kerntechnik* 17, 57–73 (1975).
5. E. D. Arnold, "Handbook of Shielding Requirements and Radiation Characteristics of Isotope Power Sources for Terrestrial, Marine, and Space Applications," ORNL-3576, 1964.

[6] See Reference 4 at and of chapter.
[7] See Reference 5 at end of chapter.
[8] See Reference 4.

PROBLEMS

6.1 Given 1 microgram (10^{-6} grams) of pure strontium-90, what is the activity in curies? After 32 hours, what is the activity of yttrium-90 in the sample? How many grams of yttrium-90 are present?

6.2 After yttrium-90 has reached saturation in the above sample, what will be the heat generated from beta particles (in watts)?

6.3 A chain decay

$$A \xrightarrow{t_{1/2}=1\,s} B \xrightarrow{t_{1/2}=10\,s} C \text{ (stable)}$$

takes place. At a given time there are 10^{10} atoms/cm^3 of A present and 10^{12} atoms/cm^3 of B present. What is the rate of change in concentration of B at this instant?

6.4 One gram of uranium-235 and 0.1 gram of boron-10 are intimately mixed. The material is placed in a thermal neutron (2200 m/s) flux of 10^{13} neutrons/cm^2 · s for one year. What mass of each is left at the end of this time?

6.5 Plutonium-239 is formed in a reactor by a series of reactions beginning with uranium-238. The reactions are as follows:

$$^{238}U + n \rightarrow {}^{239}U$$
$$^{239}U \rightarrow \beta^- + {}^{239}Np$$
$$^{239}Np \rightarrow \beta^- + {}^{239}Pu$$
$$^{239}Pu + n \rightarrow {}^{240}Pu$$
$$^{239}Pu + n \rightarrow \text{fission}$$

Assume that other possible reactions are not significant. Assume that the neutron flux, ϕ, is constant. Write the set of differential equations that must be solved to obtain $N_{49}(t)$, the number of plutonium-239 atoms at any time t. From information in the text and in the chart of the nuclides, give numerical values to as many parameters in these equations as you can. For the sake of uniformity, use the following number designations:

Nuclide:	^{238}U	^{239}U	^{239}Np	^{239}Pu	^{240}Pu
Designation:	N_{28}	N_{29}	N_{39}	N_{49}	N_{40}

6.6 A decay chain goes

$$A \xrightarrow{t_{1/2}=100\,s} B \xrightarrow{t_{1/2}=10\,s} C \xrightarrow{t_{1/2}=0.1\,s}$$

At a given instant, concentrations are

$$N_A = 10^{10} \text{ nuclei/cm}^3$$
$$N_B = 10^{10} \text{ nuclei/cm}^3$$
$$N_C = 0$$

(a) What are the radioactive decay rates of A and B?
(b) What is the net rate of change of the population of B?
(c) Approximately what will be the population of C nuclei one second after the above situation?

6.7 Plans for radioisotope power sources for space auxiliary power call for a 25 kW (thermal) ^{238}Pu system by the mid-1970s.

(a) How many grams of ^{238}Pu will be required? Assume the 25 kW is an initial operating power.

(b) ^{238}Pu decays to ^{234}U, which is itself radioactive. Will the heat power contribution from ^{234}U decay become significant? After one year of operation, at what rate is heat being generated by ^{234}U decay?

6.8 Radium-226 has a half-life of 1620 years. How do you explain the fact that it is found in nature, in view of such a short life?

6.9 In a radioactive decay chain

$$\mathrm{A} \xrightarrow{t_{1/2,\mathrm{A}}} \mathrm{B} \xrightarrow{t_{1/2,\mathrm{B}}} \mathrm{C}\ (\text{stable})$$

a sample contains one curie of pure, isolated isotope A at time zero. Approximately how many curies of B will there be 10 minutes later for the cases

(a) $t_{1/2,\mathrm{A}} = 10$ minutes; $t_{1/2,\mathrm{B}} = 1$ year

(b) $t_{1/2,\mathrm{A}} = 1$ year; $t_{1/2,\mathrm{B}} = 10$ minutes

6.10 A sample of material is placed in a fission reactor. A certain radioisotope having a half-life of 2 days is being produced in this sample by an (n, γ) reaction at the rate of 10^{12} reactions/s. What will be the activity (disintegrations/s) of this isotope at the end of 3 days in the reactor?

6.11 Under bombardment by 14-MeV neutrons, molybdenum isotopes undergo $(n, 2n)$ reactions. In the case of natural molybdenum, describe the radioactivity produced in these reactions. What stable nuclides ultimately are produced?

6.12 The isotope ^{144}Ce is a fission product and a potential isotope heat source. Assume that cerium is chemically purified and stored for some time.

(a) Calculate the heat generated in W/g-^{144}Ce due to beta particles and due to gamma photons.

(b) Tabulate the gamma emission spectrum as follows:

E_γ (MeV)	PHOTONS PER DISINTEGRATION
———	———
———	———

(c) It is projected that by 1980, some 100,000 MW(e) of light-water power reactors will be installed in the United States. What production of ^{144}Ce would there be?

6.13

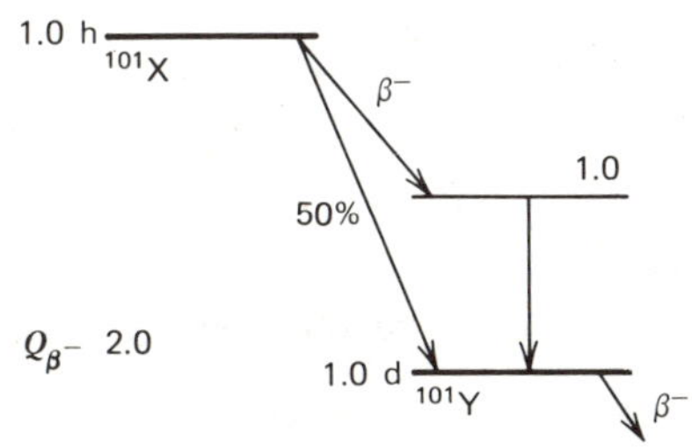

(a) If ^{101}X is a fission product formed in 1% of fissions, how many curies would be present in a reactor of 3000 megawatts thermal after one year of operation?
(b) Assuming that ^{101}Y is not a direct fission product but formed only by the decay of ^{101}X, how many curies of ^{101}Y are present under the conditions of the preceding question?

6.14 Some plutonium discharged from a reactor is chemically segregated and purified so that only plutonium isotopes are present. This plutonium is then sent to storage. Initially, the batch contains 1 kg of the isotope ^{241}Pu. After 2 years' storage, how many curies of ^{241}Am will be present in the batch as a result of ^{241}Pu decay?

6.15 Analysis of some Apollo-11 rock samples showed a ratio of $^{207}\text{Pb}/^{235}\text{U} = 95$. One line of reasoning is that all the ^{207}Pb now on the moon comes from the decay of ^{235}U. The naturally occurring lead isotopes are in different percentages on the moon than on the earth. Write the decay chain with appropriate half-lives starting

$$^{235}\text{U} \xrightarrow{7.1 \times 10^8\ \text{y}} {}^{231}\text{Th} \xrightarrow{25.5\ \text{h}} \text{etc.} \cdots \longrightarrow {}^{207}\text{Pb (stable)}$$

Assuming this theory is correct, how old is the moon?

6.16 If the (n, γ) cross section for ^{137}Cs is 0.11 barn and some ^{137}Cs is placed in a neutron flux of 10^{13} neutrons/cm$^2 \cdot$ s, what would be the initial rate of disappearance of ^{137}Cs (% per year)? How does this compare with the radioactive decay rate?

6.17 A commercial operation detects metals in ore samples by means of neutron-activation analysis. For platinum, the method is to irradiate a sample in a thermal neutron flux of 10^{13} neutrons/cm$^2 \cdot$ s for 30 minutes. The activated isotope that is measured is ^{199}Au, which results from the (n, γ) reaction in ^{198}Pt followed by the beta decay of ^{199}Pt. The capture cross section of ^{198}Pt for thermal neutrons is 4.0 barns. The gamma photon that is counted is the 0.158-MeV gamma from ^{199}Au. Assume that they wait 10 hours after the irradiation to let the ^{199}Pt die out and that the efficiency of counting the 0.158-MeV gamma is 70%. They claim a sensitivity of—that is, an ability to detect as little as—0.1 μg *natural* platinum. How many counts per hour should they get from the activation of this quantity? Note: in the interest of simplicity, you may assume that ^{199}Pt decays immediately to ^{199}Au; in other words, the (n, γ) reaction in ^{198}Pt goes directly to ^{199}Au. The error so introduced is small; the simplification is great.

6.18 A single fuel assembly removed from a pressurized water reactor (PWR) has reached a burnup (specific energy yield would be a more descriptive term) of 30,000 MWd(t)/MTU over a period of three calendar years. The fuel assembly contained 500 kg of uranium when charged to the reactor. The fuel assembly is stored at the utility for 150 days following discharge and then is shipped by rail to reprocessing. How much thermal power

(W) is being generated by this fuel assembly at this time? Can you make some quantitative judgment regarding how high a temperature such an assembly might reach if you just stood it up against your garage on a nice day?

Chapter 7

The fission power reactor system

In Chapter 3 we discussed the fission reactor briefly. In Chapter 4 we described some of the neutron reactions that go on in a reactor. In the present chapter we shall apply this information to develop an understanding of fission power systems. More specifically, we shall be discussing (1) the reactor core, that region in which the fission reaction takes place, (2) the accompanying energy conversion system necessary to convert the released fission energy to electrical energy, and (3) the fuel cycle or the support system necessary to keep power reactors fueled.

There are perhaps three reasons for building a fission reactor. The most important reason by far is to convert fission energy to thermal energy, usually with the intention of converting the thermal energy to electric energy. Reactors designed for this purpose are called power reactors. Reactors are also built for research and testing purposes. There are a variety of research needs that a reactor can fill; the principal one is for neutron or gamma irradiation facilities, or both. It can also be used to demonstrate certain characteristics common to fission chain reactors. A test reactor is usually designed to provide a high neutron flux or environment in which to test materials and subsystems eventually intended for use in power reactors. Research reactors usually operate at relatively low power, 1 watt thermal, W(*t*) to 1 million watts thermal, MW(*t*). Test reactors operate from about 10 MW(*t*) to 1 GW(*t*). The thermal energy released is usually at low temperature and is not converted to electricity but dissipated in cooling towers.

Another class of reactor that serves a very specific purpose is the so-called critical assembly. It is usually a mock-up of a reactor that is in the design phase. It resembles the core of the reactor under design in size and material composition but usually operates at very low power and does not have a sophisticated heat transfer system. It is possible to check many design calculations on a critical assembly before proceeding to the final specifications of the actual reactor.

Before discussing how to estimate reactor core specifications, we shall enumerate the various objectives that a reactor designer is striving to achieve. In Table 7.1 we are primarily concerned with the power reactor. The ordering is more a matter of logic than priority; usually, all of these objectives must be achieved to some degree.

Table 7.1 The Principal Objectives in the Design of Fission Reactors

1. The reactor must be capable of attaining a chain reaction using materials available.
2. The power level of the reactor must be responsive to the control measures available to the operator.
3. The thermal energy released in the fission reaction must be removed from the core under conditions that do not cause excessive temperatures or excessive temperature differences in core materials or components. In the case of power reactors, however, it is also necessary that the thermal energy be at sufficiently high temperature so that its subsequent conversion to electric energy will be at an acceptably high efficiency.
4. The radioactive fission products formed in the reactions must be contained to a very high degree, even under the severe conditions resulting from the malfunction of the cooling system.
5. The reactor core must be structurally sound. Structural integrity must be assured under a wide range of operating conditions.
6. The design must provide for fuel replenishment, either continuously or intermittently.
7. The fuel cost (especially for power reactors) must be acceptably low.
8. The various materials in the core must exhibit a high degree of chemical inertness in association with each other.

THE ACHIEVEMENT OF A CHAIN REACTION, OR CRITICALITY

We shall deal here with the most fundamental (but not at all the most difficult) design requirement: to achieve a chain reaction or a neutron balance. This is basically a neutron population problem; the neutron "birth" and "death" rates must balance. The births come from the fission reaction. An average fission reaction can be represented

$$^{235}\text{U} + n \longrightarrow \text{Fission Products} + \bar{\nu}n$$

The quantity $\bar{\nu}$ represents the average number of neutrons produced per fission. Remember that there is a variation in the exact way the fissile nucleus splits, so that the number of neutrons produced each time varies from about one to four. The value of $\bar{\nu}$ in the fission of ^{235}U is about 2.5. We can write the rate of fission as in Chapter 4

$$\Sigma_f \phi \text{ fissions/cm}^3 \cdot \text{s}$$

The neutron birth rate is

$$\bar{\nu}\Sigma_f \phi \text{ fission neutrons/cm}^3 \cdot \text{s}$$

For our present purposes we can divide the fate of all neutrons into two kinds: either the neutrons will be absorbed (disappear) into a nucleus in the core, or they will escape (leak) from the core. If the core is large enough, the leakage can be neglected; we will start with this case. It is known as the "infinite medium" case. Because the neutron leakage from many power reactors is small (as a fraction of neutrons produced) this idealization is closer to reality than one might think. The rate of absorption of neutrons is

$$\Sigma_a \phi \text{ neutrons/cm}^3 \cdot \text{s}$$

where Σ_a is the macroscopic cross section for all events involving disappearance of a neutron—even including fission.

The simplest representation of a fission reactor is that of a medium in which neutron birth rate and death rate are in balance, namely

$$\bar{\nu}\Sigma_f \phi = \Sigma_a \phi \tag{4.35}$$

Such a medium is said to be critical. When these two quantities are not in balance, the conventional description is

$$\bar{\nu}\Sigma_f \phi < \Sigma_a \phi \quad \text{(subcritical)}$$

and

$$\bar{\nu}\Sigma_f \phi > \Sigma_a \phi \quad \text{(supercritical)}$$

Example 7.1
In an infinite medium of ^{235}U and water, what weight ratio, $^{235}U/H_2O$, would just achieve critically?

Solution: Without giving a justification for the present, we select cross sections that apply at a neutron energy of 0.025 eV. The neutron production rate is

$$\bar{\nu}\Sigma_f \phi = \bar{\nu} N_{235}\, \sigma_{f,\,235}\, \phi = 2.5 \times N_{235} \times 582 \times 10^{-24} \times \phi$$

Only ^{235}U is involved here, because hydrogen and oxygen nuclei have a zero fission cross section. The neutron absorption rate is

$$\Sigma_a \phi = (N_{235}\, \sigma_{a,\,235} + N_H\, \sigma_{a,\,H} + N_O\, \sigma_{a,\,O})\phi$$

$$\sigma_a = \sigma_f + \sigma_c$$

Also

$$N_H = 2N_{H_2O} = 2N_O$$

Therefore

$$\Sigma_a \phi = [N_{235}\,(582 + 112) + 2N_{H_2O} \times 0.332 + N_{H_2O} \times .002]\; 10^{-24}\, \phi$$

For the criticality condition we set $\bar{\nu}\Sigma_f\phi$ and $\Sigma_a\phi$ equal.

$$2.5 \times 582\, N_{235} = 694\, N_{235} + 0.664\, N_{H_2O}$$

$$\frac{N_{235}}{N_{H_2O}} = \frac{0.664}{761} = 8.73 \times 10^{-4}$$

This is the ratio of atoms of uranium-235 per molecule of water. The weight ratio would therefore be

$$\frac{8.73 \times 10^{-4} \times 235}{18} = 0.0114 \text{ kg} - {}^{235}\text{U/kg} - H_2O$$

We can also express the composition as

$$\frac{0.0114}{1.0114} \times 100 = 1.13 \text{ wt\% (w/o) } {}^{235}\text{U}$$

Note that we have not dealt with the very practical question of how one would get a homogenous mixture of ^{235}U in water or what the composition might be be to achieve criticality for some finite volume of solution. In practice, a class of research reactor called, misleadingly, the water boiler has in fact a core of UO_2SO_4 dissolved in water. We shall discuss the relationship between critical composition and volume later in this chapter. It is interesting to note, however, that in a water boiler reactor the critical composition is about 4.6 w/o ^{235}U. The increase from 1.13 w/o is primarily due to the small volume (about 30 liters) of the core of these reactors, a circumstance that leads to the leakage of a high percentage of the neutrons.

The Dynamic Behavior of a Fission Reactor

We now turn to the second design requirement, namely, that the power level of the reactor must be controllable. If a reactor is just critical, the neutron population is constant; so also is the power, since the power is proportional to $\Sigma_f \phi$ or $\Sigma_f n\bar{v}$. A reactor that always operated at the same power level would not be much use, since the reactor is initially at zero power. Likewise, it must be shut down occasionally. In short, the reactor operator must be able to make the reactor supercritical or subcritical.

When a reactor is supercritical, it means that the production of neutrons exceeds their disappearance. The inevitable result (just as when births exceed deaths in a human population) is that the neutron population will increase. The rate of such an increase is a matter of great importance, so we wish to be quantitative about it. We tentatively define a new term on the basis of the neutron balance, $k = \bar{\nu}\Sigma_f/\Sigma_a$. It is the multiplication constant (the infinite multiplication constant, since we are dealing with an infinite medium) representing the number of neutrons in one generation per neutron in the previous generation. When k is other than one, the neutron population, or flux, changes

$$n(t) = n(0)\ [k]^g$$

$$\phi(t) = \phi(0)\ [k]^g \tag{7.1}$$

where g is the number of neutron generations or lifetimes between time 0 and t. The value of g is simply t/l, where l is the neutron lifetime. The lifetime of a neutron is the mean path for absorption divided by the speed.

$$l = \frac{\lambda_a}{v} = \frac{1}{\Sigma_a v} \tag{7.2}$$

We are usually interested in k values very close to unity. It will help our derivation to define

$$\delta k \equiv k - 1 \tag{7.3}$$

We can now write the rate of change of the neutron population

$$\phi(t) = \phi(0)\ [1 + \delta k]^{t/l}$$

or

$$\ln \frac{\phi(t)}{\phi(0)} = \frac{t}{l} \ln\ [1 + \delta k]$$

$$= \frac{t}{l}\ [\delta k - \frac{\delta k^2}{2} + \frac{\delta k^3}{3} - + \ldots]$$

which, for small δk, can be written

$$\ln \frac{\phi(t)}{\phi(0)} = \frac{\delta k}{l}\ t$$

or

$$\phi(t) = \phi(0) e^{\frac{\delta k}{l} t} \tag{7.4}$$

Example 7.2
How fast would the power rise in a reactor of pure ^{235}U in which $\delta k = 0.01$?

Solution: Consider a neutron with an energy of 1 MeV traveling in metallic ^{235}U. The value of N is 0.048×10^{24} nuclei ^{235}U/cm^3 The value of σ_a is 1.4 barns.

$$\Sigma_a = (.048)\ (1.4) = 0.0672\ \text{cm}^{-1}$$

$$v = \sqrt{2E/m} = [(2)\ (1)\ (1.60 \times 10^{-6})/(1.66 \times 10^{-24})]^{1/2}$$

$$= 1.39 \times 10^9\ \text{cm/s}$$

The lifetime of such a neutron is

$$l = 1/v\Sigma_a = \frac{1}{(1.39 \times 10^9)(0.0672)} = 1.07 \times 10^{-8} \text{ s}$$

Now, if the value of k is 0.01, we see that

$$\phi(t) = \phi(0) \exp\left(\frac{10^{-2}}{1.07 \times 10^{-8}}\right) t \cong \phi(0)e^{10^6 t}$$

which means that the flux (and, therefore, the thermal power) will increase by a factor of e every microsecond. The value of δk might be compared to an interest rate except that it is not on a "per-annum" basis; it is on a "per-neutron-lifetime" basis. The resulting equation is a "compound interest" expression.

The rate of rise of the power indicated by the above equation would be totally unmanageable in terms of a control system's ability to respond. It so happens that the lifetime calculated is shorter than in an actual reactor, because most neutrons will have energies below 1 MeV. Even in a thermal neutron reactor, however, the neutron lifetime is around 10^{-4} seconds, which in the above case would lead to an equation

$$\frac{\phi(t)}{\phi(0)} \simeq e^{100t}$$

This rate of rise in power would still present a formidable control problem.

The result of Example 7.2 indicates that reactor power level should be extremely sensitive to any change in the neutron balance or the multiplication constant, k. Fortunately, there is another phenomenon that enters the picture to reduce drastically the rate of power increase for small positive values of δk. A small fraction (0.7% in the case of ^{235}U) of neutrons produced in fission are not emitted immediately in the fission process but are emitted in the radioactive decay of a number of fission-product nuclides. (One such isotope is ^{88}Br.) These are called "delayed neutrons." In terms of animal populations, such neutrons have a long gestation period. These neutrons, whose emission is delayed anywhere from a few tenths of a second to minutes, have an effective lifetime that is much longer than $1/v\,\Sigma_a$. The result is that as long as the margin of supercriticality is less than the yield of delayed neutrons, the rate of power increase is very modest, permitting ample time for response of control mechanisms.

Although there are six different "parents" of delayed neutrons, the general effect of delayed neutrons can be illustrated by assuming a single parent. In keeping with conventional terminology, we refer to those neutrons that are not delayed but that come directly from the fission reaction and whose lifetime is represented by l as "prompt" neutrons. The fraction of neutrons that are delayed is called β and, therefore the fraction of "prompt" neutrons is $1 - \beta$. The behavior of the population of prompt neutrons is exactly as has been described before, except that instead of k neutrons in generation $g + 1$ for every one in generation g, it is $k - \beta$ or

$$\phi_p(t) = \phi_p(0)e^{\frac{\delta k - \beta}{l} t} \tag{7.5}$$

Here we designate the prompt neutrons with the subscript p and delayed neutrons with d. Differentiating Eq. 7.5 and canceling the neutron velocity

$$\frac{dn_p}{dt} = \frac{\delta k - \beta}{l} n_p \tag{7.6}$$

The delayed neutrons are the product of the radioactive decay of the parent; the rate of production of delayed neutrons is just equal to the rate of decay of the parent

$$\frac{dn_d}{dt} = \lambda N \tag{7.7}$$

where λ is the decay constant and N the concentration of the delayed neutron parent. This parent is a fission product; its concentration is described by the activation differential equation give in Chapter 4.

$$\frac{dN}{dt} = \beta\bar{\nu}\Sigma_f\phi - \lambda N = \beta\bar{\nu}\Sigma_f n v - \lambda N \tag{7.8}$$

But

$$\bar{\nu}\Sigma_f = k\Sigma_a$$

and

$$l = \frac{1}{v\Sigma_a}$$

Therefore

$$\frac{dN}{dt} = k\frac{\beta}{l}n - \lambda N \tag{7.9}$$

We now have two simultaneous differential equations

$$\frac{dn}{dt} = \frac{dn_p}{dt} + \frac{dn_d}{dt} = \frac{\delta k - \beta}{l} n + \lambda N$$

$$\frac{dN}{dt} = k\frac{\beta}{l}n - \lambda N \tag{7.10}$$

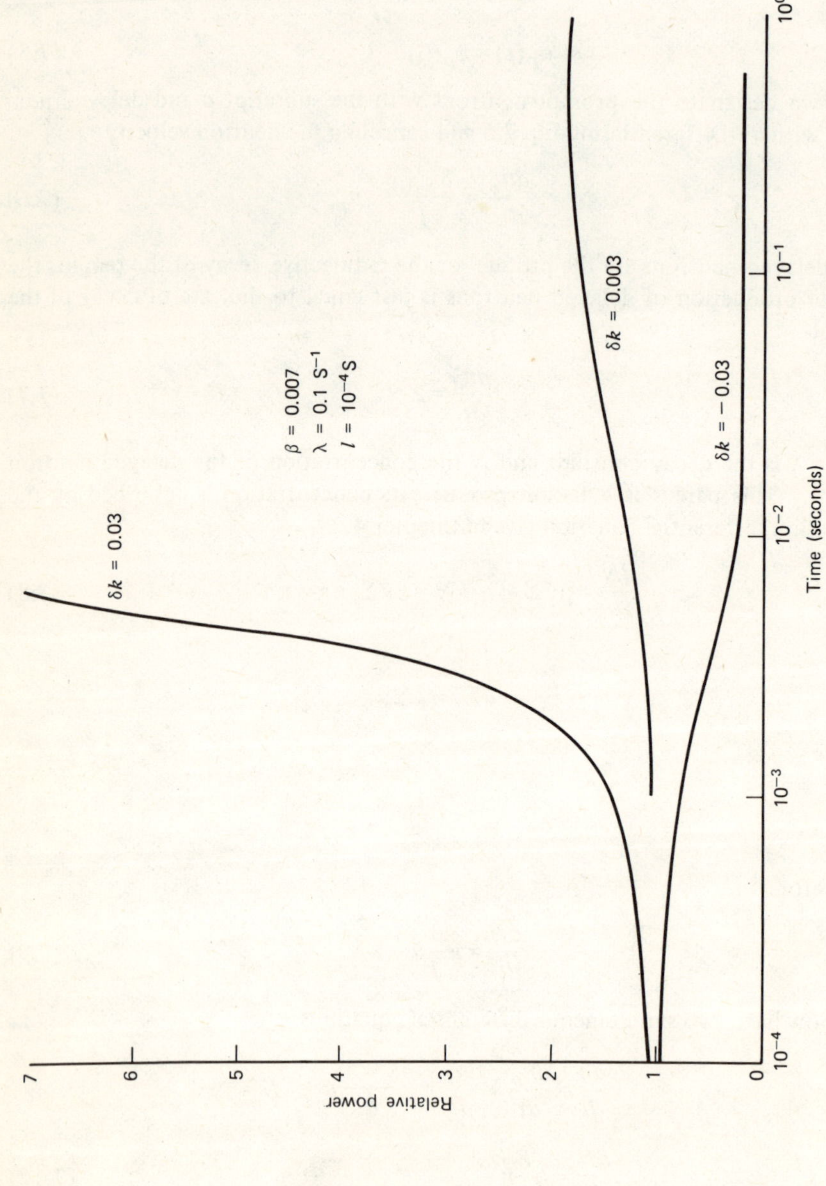

Figure 7.1 Reactor power following step changes in multiplication factor.

The solutions are in the form

$$n(t) = A_{11}e^{\alpha_1 t} + A_{12}e^{\alpha_2 t}$$
$$N(t) = A_{21}e^{\alpha_1 t} + A_{22}e^{\alpha_2 t} \tag{7.11}$$

We shall not go into the details of evaluating the constants. However, the results are important, so they will be presented.

$$\frac{n(t)}{n(0)} = \frac{\beta}{\beta - \delta k} \exp\left(\frac{\lambda \delta k}{(\beta - \delta k)} t\right) - \frac{\delta k}{\beta - \delta k} \exp - \left(\frac{\beta - \delta k}{l} t\right) \tag{7.12}$$

In equation 7.12 some product terms containing λ have been omitted, since they have a negligible influence. Solutions to Eq. 7.12 can provide considerable insight into the dynamic behavior of reactors. Figure 7.1 shows the rise in power level following a step change in k from 1.0 to $1 + \delta k$. The pronounced effect of the δk value relative to β can be seen. When $\delta k = 0.003$, the rate of rise of the neutron population is rather modest. The times involved are well within the response time of control mechanisms. When δk is 0.03, however, the rate of rise is extremely rapid. Eq. 7.12 contains the explanation. When $\delta k < \beta$, the second term on the right hand side dies out as time increases; the first term becomes dominant, resulting in a simple exponential rise

$$\frac{n(t)}{n(0)} = e^{t/T} \tag{7.13}$$

where

$$T = \frac{\beta - \delta k}{\lambda \delta k} \quad (\delta k < \beta) \tag{7.14}$$

T is called the period or e-folding time, the time for power to increase by a factor of e. On the other hand, when $\delta k > \beta$, the second term of Eq. 7.12 is positive and increasing with time. The period for the rate of rise becomes

$$T = \frac{l}{\delta k - \beta} \quad (\delta k > \beta) \tag{7.15}$$

This result is the same as in Eq. 7.4, except that δk is replaced by $\delta k - \beta$.

A more physical explanation of the difference between the situations in which $\delta k < \beta$ and $\delta k > \beta$ is that in the former case the delayed neutrons are a necessary part of the chain reaction. As a result, the rate of rise in neutron population is controlled by the rate at which the delayed neutron production changes. When $\delta k > \beta$, the chain is maintained by prompt neutrons alone. Because of the very short lifetime of prompt neutrons, the rise in population is very rapid. This fundamental difference between the supercriticality condition when $\delta k < \beta$ and one when $\delta k > \beta$ is well recognized in the field of reactor theory. A

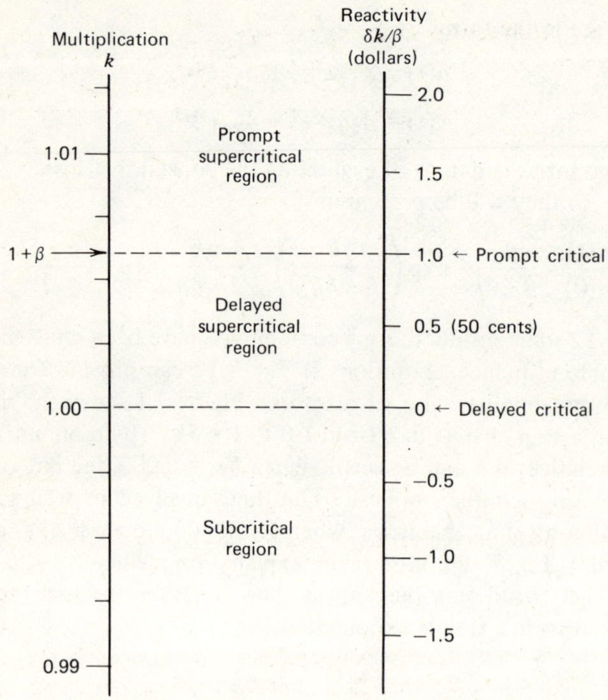

Figure 7.2 The nomenclature of reactor criticality states.

nomenclature has developed around it, as illustrated in the diagram of Figure 7.2. The deviation of k from unity is called reactivity. It may be expressed either as δk, a fraction or percent, or as $\delta k/\beta$. When $\delta k = \beta$ or $\delta k/\beta = 1.0$, the reactivity is said to be 1 dollar (at least it is in the United States). Accordingly, $\delta k/\beta = 0.5$ represents a reactivity of 50 cents, and so on.

Figure 7.1 also shows the results of negative δk. Clearly the neutron population or reactor power drops rapidly. It should be noted, however, that after an initial drop there is a "tail" to the curve, which is controlled by the decay of the parent of the delayed neutrons.

To this point we have illustrated the dynamic behavior of a reactor based on either no delayed neutrons (Eq. 7.4) or one group of delayed neutrons (Eq. 7.12). Actual reactor behavior requires an assumption of six groups of delayed neutrons, distinguished by half-lives varying from a fraction of a second to about one minute. We say an "assumption," because there appear to be many more than six neutron-emitting isotopes among short-lived fission products, but many have half-lives so close together that they behave as single emitters. Although it is not practical to give the complete solution to the time-dependent flux equation with six groups of delayed neutrons, we can give in a simple fashion the asymptotic, or stable, period, T. Following a change in k to some fixed value,

Table 7.2 Period, T, as a Function of Reactivity

	REACTIVITY, $\delta k/\beta$ (DOLLARS)					
	^{233}U (β = 0.0027)		^{235}U (β = 0.0065)		^{239}Pu (β = 0.0021)	
T (s)	$l = 10^{-6}$ s	$l = 10^{-4}$ s	$l = 10^{-6}$ s	$l = 10^{-4}$ s	$l = 10^{-6}$ s	$l = 10^{-4}$ s
1×10^{-4}	4.798	–	2.537	–	5.705	–
4×10^{-4}	1.956	–	1.387	–	2.185	–
1×10^{-3}	1.383	–	1.155	–	1.474	–
4×10^{-3}	1.095	–	1.037	4.785	1.117	–
1×10^{-2}	1.036	4.795	1.012	2.533	1.044	5.701
4×10^{-2}	0.999	1.946	0.989	1.372	0.998	2.172
1×10^{-1}	0.979	1.358	0.966	1.119	0.973	1.443
4×10^{-1}	0.918	1.013	0.888	0.926	0.900	1.018
1×10^{0}	0.839	0.877	0.789	0.805	0.810	0.858
4×10^{0}	0.641	0.650	0.565	0.569	0.602	0.613
1×10^{1}	0.476	0.480	0.399	0.400	0.440	0.444
4×10^{1}	0.244	0.245	0.190	0.190	0.219	0.221
1×10^{2}	0.132	0 133	0.099	0.099	0.116	0.117
4×10^{2}	0.042	0.042	0.030	0.030	0.035	0.036
1×10^{3}	0.018	0.018	0.013	0.013	0.015	0.015

the reactor flux or power will shortly vary as exp (t/T). Values of the reactivity, $\delta k/\beta$, as a function of T are given in Table 7.2. The relationship depends on the particular fissioning isotope, as shown, because the fraction of each group of delayed neutrons varies with the isotope.

Example 7.3

If the uranium-235 in the $^{235}U/H_2O$ critical mixture of Example 7.1 is increased 1%, at what rate would the power level increase?

Solution: From Example 7.1 we have 8.73×10^{-4} atoms of ^{235}U per molecule of H_2O. We adjust this value to 8.8173×10^{-4} atoms of ^{235}U and recalculate $\bar{\nu}\Sigma_f$ and Σ_a:

$$\bar{\nu}\Sigma_f = 2.5 \times 8.8173 \times 10^{-4} \times 582$$
$$= 1.283^{1}$$
$$\Sigma_a = 8.8173 \times 10^{-4} \times 694 + 1 \times 0.664$$
$$= 1.276^{1}$$

[1]Normally the unit of the macroscopic cross section is cm^2/cm^3, or cm^{-1}. In this case we need only deal in ratios, that is, N_{235}/N_{H_2O}, and, arbitrarily, set $N_{H_2O} = 1$. The atom density is not important. Note that whatever absolute density might be used, it would cancel in the calculation of k.

$$k = \frac{\bar{\nu}\Sigma_f}{\Sigma_a} = \frac{1.283}{1.276} = 1.0055$$

$$\delta k \equiv k - 1 = 0.0055$$

$$\beta = 0.0065 \text{ (Table 7.2)}$$

$$\frac{\delta k}{\beta} = \frac{0.0055}{0.0065} = 0.846 \text{ dollar}$$

The neutron lifetime in a light-water reactor is about 10^{-4} seconds. Interpolating from Table 7.2, we obtain a period of about 0.8 seconds, which means the power level would rise

$$P(t) = Ce^{t/0.8}$$

For example, it would double every 0.5 seconds. This rate of rise is by no means slow but is still well within the response capabilities of control mechanisms.

Control of a Fission Reactor

In the preceding section we discussed the response of the power level of a critical reactor to a deviation from criticality. The objectives in the design of a reactor control system are to be able to raise and lower power level at a desired rate, to be able to hold power level steady within desired limits, and to be able to respond to various changes in the multiplication constant, k, that may result either from natural phenomena or from possible accidental occurrences. The usual control mechanisms that are part of the reactor system are directed to changing the neutron capture in the core or, in other words, changing Σ_a in the multiplication constant, k.

$$k = \frac{\bar{\nu}\Sigma_f}{\Sigma_a}$$

One could, of course, change Σ_f, but in practice, movement of fissile material is avoided. Control elements are made of some nonfissile material that has a high value of Σ_a. Such materials as boron, cadmium, hafnium, indium, and silver are examples. If these materials are inserted into the core, k decreases; if they are withdrawn, k increases. The control elements are designed so that they can be moved into or out of the core at the will of the operator or in response to a servomechanism.

It is very important to understand that in addition to mechanical devices by which the operator can affect k, certain natural responses to changes in power level that affect k are almost always operating. These responses are inherent in the physical nature of the reactor. They are particularly important, since they do not depend for their operation on the functioning of electronic or electromechanical devices. Consider, for example, the $^{235}U/H_2O$ reactor of Example 7.3. A 1% increase in ^{235}U concentration results in a reactivity of 85 cents, which in turn results in the power increasing on a period of 0.5 seconds. However, this increase in fission power causes the temperature of the water to increase. As a result, several changes in the physical situation take place. The neutron spectrum changes, causing the neutron microscopic cross sections to change. The density of the water decreases. If the temperature rises high enough, the water can boil. The latter two phenomena reduce the macroscopic cross section, thereby increasing the neutron mean free path and increasing the fraction of neutrons that leak from the core. This will always have a negative effect on k. Often it is within the designer's ability to make the net effect negative. Such negative feedback effects introduce a great deal of self-control into a reactor. The result is that reactors are by and large easily controlled. The operator does not sit at the controls running control rods rapidly in and out in an effort to hold power level steady. These remarks are made in an effort to give some perspective to the situation that is not apparent in dynamics equations such as Eq. 7.12, but that involves physical situations whose analyses are beyond the scope of this text.

THE NEUTRON SPECTRUM: FAST REACTORS AND SLOW (THERMAL) REACTORS

Reactors are classified according to their neutron spectrum into two categories, fast reactors and slow (or thermal) reactors. The meanings become obscured by popular usage and by the jargon of the trade. The distinctions are important to an understanding of the nuclear energy industry, however, so we shall spend some time on an explanation of the differences.

In fission reactors, the principal source of neutrons is the fission reaction. These source neutrons do not have a single energy but, rather, a spectrum (see Figure 7.3). The spectrum is a modified normal distribution with an average energy of 2 MeV. Over 99% of fission neutrons have energies between 0.05 MeV and 10 MeV. This source spectrum represents the highest energy spectrum, or an upper limit in a reactor, because nothing can speed up a neutron in this energy range. There are phenomena, however, that do slow neutrons or degrade their energy, namely, elastic and inelastic scattering. Consequently, a reactor always has a neutron spectrum that is shifted downward in energy from the fission spectrum. The highest energy spectrum attainable would be realized in a pure metallic fissile isotope such as ^{235}U or ^{239}Pu, where a neutron undergoes a minimum number of energy-degrading scatters before being absorbed. Any nonfissile material present will cause additional spectral degradation. We use the expression "spectral degradation" to indicate a shift in the neutron spectrum

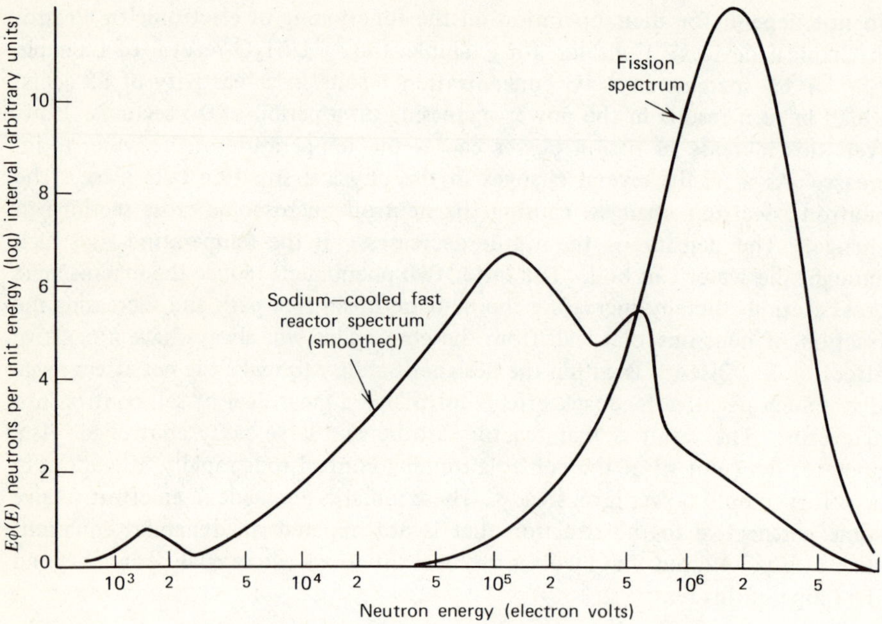

Figure 7.3 Fission neutron spectrum and fast reactor spectrum.

toward lower energies. It is obviously a relative term. Reactor designers also refer to "spectral softening," meaning a shift toward lower energies; "spectral hardening" is the converse.

A fast reactor can be defined as one in which no deliberate attempt is made to slow neutrons or to degrade the spectrum. In the course of trying to achieve design objectives (Table 7.1), however, it is necessary to introduce a coolant, structural materials, and fertile materials into a reactor core. The materials will all have the effect of degrading the spectrum. This effect is incidental, however; the materials are not deliberately included for that purpose. A typical spectrum in a sodium-cooled fast reactor is shown in Figure 7.3.

The second spectral category of reactors is the slow reactor. In such a reactor a material, called the moderator, is included for the purpose of slowing the neutrons to very low energies, that is, in the range of 0.02 to 1 eV. Because there is a minimum energy (threshold) for inelastic scattering around 100 keV, it is necessary to rely on elastic scattering in order to slow neutrons to the eV range. The most effective materials in slowing neutrons down are nuclei of low mass (see Chapter 4). The principal moderator materials, therefore, are hydrogen (mass number, 1), deuterium (mass number, 2) and carbon (mass number, 12). All slow-neutron power reactors are moderated by one of these. The usual forms of hydrogen, deuterium, and carbon are light water, heavy water, and graphite, respectively. Several other elements of low mass number will indeed

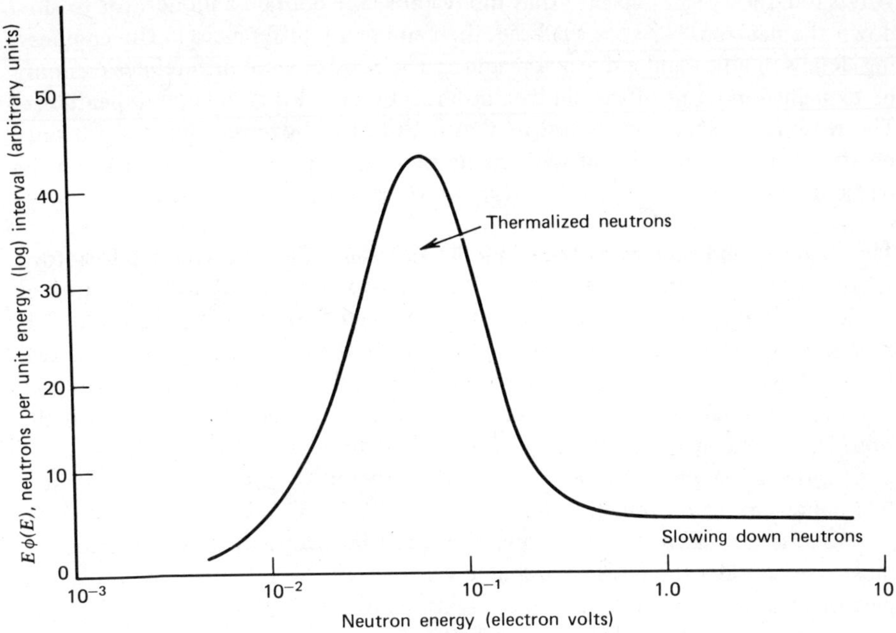

Figure 7.4 Thermal neutron spectrum.

slow down neutrons but are eliminated as moderators for practical reasons: helium because of its low density; lithium and boron because of their high neutron absorption cross sections; and beryllium because of its high cost and toxicity.

A typical slow reactor spectrum is shown in Figure 7.4. It is a modified Maxwellian spectrum. The neutrons lose energy on each succeeding collision until their energy is down to a few tenths of an electron volt. At this energy, the neutron's speed is not much higher than the atoms or nuclei of the moderator. The latter begin to "hit back," the neutrons sometimes have their energy increased in a scattering, and the slowing-down process comes to a halt. The neutrons get into thermal equilibrium with the atomic motion of the moderator and behave rather like a perfect gas. These are called thermal neutrons. A slow reactor is therefore often called a thermal reactor. (The choice of words is probably unfortunate, due to the confusion with thermal energy.) The exact spectrum depends, among other things, on the temperature of the moderator.

The Differences between Fast and Thermal Reactors

Just what is the difference between a fast reactor and a thermal reactor? Certainly neither a fast spectrum nor a thermal spectrum is an end in itself. The spectrum of a reactor is not mentioned in the design objectives of Table 7.1.

Why should a designer specify that the reactor core contain a moderator to slow down the neutrons? As we shall see, there are many differences in the engineering details of a fast and a thermal reactor. The fundamental differences can only be explained in terms of the neutron cross sections and their energy dependence. The reasons for specifying a fast or thermal reactor, however, depend not only on the physical properties of fuel (fissile plus fertile) materials, but also on the particular historic and economic setting in which the decision is made.

The Quantity and Concentration of Fissile Material in Fast and Thermal Reactors. A good portion of the explanation of the relative amount and concentration of fissile material required in fast and thermal reactors is contained in the difference in the neutron cross sections that obtain in the neutron spectra of the two types. Table 7.3 contains some representative cross section data for each type.[2] The most obvious difference is that the microscopic cross sections for the fissile isotopes are much larger at thermal energies than at fast energies. In order to obtain the same *macroscopic* cross section, therefore, the concentration in a thermal reactor can be much lower.

Because our interest is primarily directed at the economic power reactor, we must also consider the fertile material whose conversion to fissile is an important part of reactor economics. The cross sections of ^{238}U relative to the cross sections of the fissile nuclides are much higher in the fast reactor. The effect that these values have on the concentration of fissile material in a fissile-fertile mix can be shown by a simple example. We ask what concentration ratio of ^{239}Pu to ^{238}U would be necessary to give an infinite multiplication, k, of unity, using each set of cross sections from Table 7.3.

$$k = \frac{\bar{\nu}\Sigma_f}{\Sigma_a} = \frac{N_{239}\nu\sigma_{f,239} + N_{238}\nu\sigma_{f,238}}{N_{239}\sigma_{a,239} + N_{238}\sigma_{a,238}}$$

Table 7.3. Physical Properties That Distinguish Fast and Thermal Reactors

	FAST REACTOR			THERMAL REACTOR		
	^{235}U	^{239}Pu	^{238}U	^{235}U	^{239}Pu	^{238}U
Neutrons per fission, $\bar{\nu}$	2.6	3.1	2.6	2.4	2.9	0
Neutrons per absorption, $\bar{\eta}$	2.1	2.6	0.3	2.0	1.9	0
Fission cross section, $\bar{\sigma}_f$	1.9	1.8	0.06	280	790	0
Capture cross section, $\bar{\sigma}_c$	0.45	0.35	0.46	56	416	2.40
Absorption cross section, $\bar{\sigma}_a$	2.35	2.15	0.52	336	1206	2.40

[2]See Reference 1 at end of chapter.

Designate

$$\frac{N_{239}}{N_{238}} = e$$

For a fast reactor

$$k = \frac{3.1 \times 1.8e + 2.6 \times 0.06}{2.15e + 0.52} = 1.0$$

$$e = 0.11$$

For a thermal reactor

$$k = \frac{2.9 \times 790e}{1206e + 2.40} = 1.0$$

$$e = 0.0022$$

While these calculations of k are greatly oversimplified relative to actual reactor analysis, the much higher concentration of fissile material required in the fast reactor is clear. This is an effect that is based on very basic physical properties. Table 7.4 gives more typical values for a fast and thermal reactor. More than any other characteristic of reactors, the required concentration of fissile material has influenced the historic development of power reactors.

The fact that thermal reactors can be made critical at low fissile concentrations means that they can be fueled with natural uranium and, conversely, that natural uranium can serve as fuel only in thermal reactors. In situations where natural uranium is the only fuel available, therefore, there is no alternative. The higher concentration of fissile material required to fuel a fast reactor can be obtained in one of two ways: separation or concentration (enrichment) of the isotope ^{235}U found in natural uranium, or operation of a thermal reactor that serves to convert ^{238}U to ^{239}Pu as a by-product. Both of these methods were developed and used in the United States during World War II. The so-called "Hanford Piles" were natural-uranium-fueled thermal reactors, in which ^{239}Pu was formed for weapons purposes. The plutonium so produced could, of course, be used to fuel a reactor (fast or thermal); the decision is a political or economic one, or both.

In summary, a key inherent difference between fast and thermal reactors is that the fast reactor requires a high concentration of fissile material, much higher than the concentration of ^{235}U in natural uranium. The thermal reactor, therefore, permits easier entry into the power reactor business because it is much less demanding in terms of the fissile material concentration and quantity needed to start. Certain designs of power reactors can even be operated on natural uranium. It is primarily for this reason that thermal reactors dominate the

Table 7.4 Typical Core Composition of Fast and Thermal Reactors

	1000 MW(*e*) FAST REACTOR CORE VOLUME 4.2 m^3 (LMFBR)				1000 MW(*e*) THERMAL REACTOR CORE VOLUME 33 m^3 (PRESSURIZED WATER)			
MATERIAL	FUEL (WT %)	WT IN CORE (kg)	CORE WT (%)	CORE VOLUME (%)	FUEL (WT %)	WT IN CORE (kg)	CORE WT (%)	CORE VOLUME (%)
*Fuel**		15,260	75	41		100,500	63	30
Fissile U	–				2.37			
Fertile U	84				97.63			
Fissile Pu	11				–			
Fertile Pu	5				–			
	100				100			
Coolant/Moderator								
H_2O		–	–	–		14,400	9	50
Sodium		1,020	5	38		–	–	–
Structure		4,070	20	21		44,700	28	20
			100	100			100	100

*Fuel is defined as the fissile and fertile isotopes and chemically associated atoms. For example, the oxygen in UO_2 is part of the fuel.

power reactor field today. The matter would rest there were it not for a second, related, difference between fast and thermal reactors, namely, their relative capability to convert fertile isotopes to fissile ones.

The Conversion Ratio in Fast and Thermal Reactors. In Chapter 3 we tried to make the point that, in the long run, the conversion of nonfissile ^{238}U to fissile ^{239}Pu (or the conversion of nonfissile ^{232}Th to ^{233}U) is crucial to fission-fuel supply. The only naturally occurring fissile isotope, ^{235}U, is orders of magnitude less abundant than ^{238}U or ^{232}Th. In order to convert a substantial fraction (greater than 1 to 10%) of ^{238}U to ^{239}Pu, it is necessary that more than one fissile atom be formed by conversion for every fissile atom destroyed (Figure 3.9). There must be sufficient neutrons produced in the average fission to permit this high conversion to take place. Competing with the ^{238}U for neutrons are all other nuclides present in the reactor. As an absolute minimum we must have the fissile nuclides themselves, and these compete for neutrons not only in the fission reaction but also in the radiative capture reaction. For example, in ^{239}Pu

$$^{239}\mathrm{Pu} + n \xrightarrow{\sigma_c} {}^{240}\mathrm{Pu}$$

Consequently, a fissile atom's *net* production of neutrons, η, is

$$\eta = \nu \frac{\sigma_f}{\sigma_f + \sigma_c} \tag{7.16}$$

The property η gives the number of neutrons produced per absorption in a fissile nuclide. In order to convert one atom of ^{238}U to ^{239}Pu for every atom of ^{239}Pu destroyed (either by fission or by transformation to ^{240}Pu) the minimum value of η is 2.0. For practical design reasons, η must exceed this value in order to cover loss of neutrons to other materials. The value of η for the three fissile nuclides ^{233}U, ^{235}U, and ^{239}Pu is given as a function of neutron energy in Figure 7.5. The reason that a fast reactor favors the conversion process lies in the physical reality displayed there. The value of η is significantly higher in the fast reactor spectrum range (Figure 7.3) than in the thermal reactor spectrum range (Figure 7.4). In an actual reactor, since the neutrons have an energy range, the value of η is spectrum-averaged. Table 7.3 gives some values typical of power reactor neutron spectra.

As discussed in Chapter 3, the usual index of performance relative to the conversion of fertile nuclides is the conversion ratio, defined as the ratio of fertile nuclides converted to fissile nuclides destroyed. In a simple ^{238}U/^{239}Pu system, we can write

$$\text{Conversion ratio } (C) = \frac{N_{238}\,\sigma_{c,238}}{N_{239}\,\sigma_{a,239}}$$

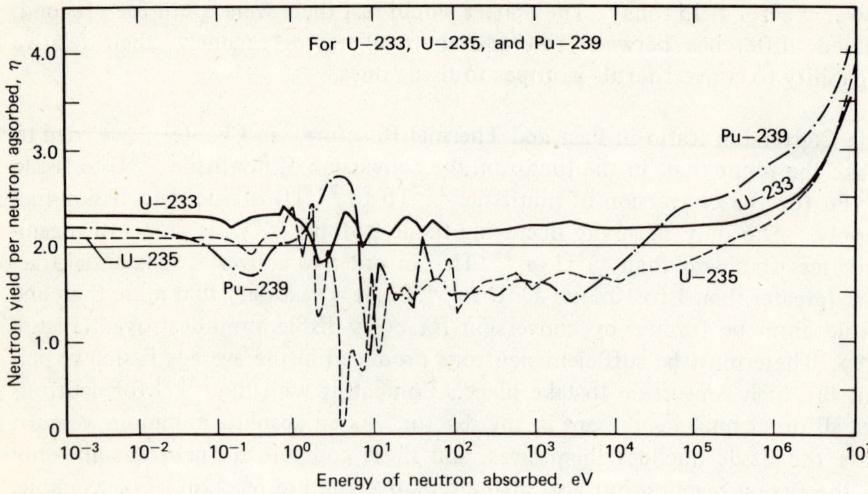

Figure 7.5 Net neutrons produced per absorption in fissile nuclides. From "Final Environmental Statement, Light Water Breeder Reactor Program," ERDA-1541, June 1976.

Using the data from Table 7.3, we can compare this value for fast and thermal reactors. We use N_{239}/N_{238} ratios more typical of actual reactors than those computed in the previous section.

For a fast reactor

$$\frac{N_{239}}{N_{238}} = e = 0.17$$

$$C = \frac{0.46}{0.17 \times 2.15} = 1.26$$

For a thermal reactor

$$e = 0.03$$

$$C = \frac{2.40}{0.03 \times 1206} = 0.07$$

The value of C for a thermal reactor is far too low, because the cross sections of Table 7.3 do not reflect the large capture of slowing-down neutrons by ^{238}U. A value of $C = 0.5$ is closer to reality in a thermal reactor. Nevertheless, the higher conversion in fast reactors is clear. Only in the fast reactor spectrum do the physical properties $\overline{\sigma}_f$, $\overline{\sigma}_c$, $\overline{\nu}$, and so on, have values to make breeding ($C >$ 1) clearly possible.

The dichotomy of the fast and thermal power reactors can now be expressed as follows: it is easier to get into the large-scale fission power business with thermal reactors, because they are less demanding of fissile material quantities and concentration; and in order to stay in the fission power business over the long term, fast reactors appear to be necessary, because only they can use a large fraction of the ^{238}U and ^{232}Th natural resources. Incidentally, Figure 7.5 makes it clear why a spectrum intermediate between thermal and fast spectra is of little interest; the η values are a minimum in this region.

This neat summary is slightly clouded by the $^{232}Th/^{233}U$ cycle. As can be seen from Figure 7.5, the η value of ^{233}U is about 2.3 at thermal neutron energies. This fact does offer the marginal possibility of a thermal reactor breeder. In fact, two reactor development projects in the United States, the light water breeder reactor (LWBR) and the molten salt breeder reactor (MSBR), are based on this possibility. Furthermore, it has been argued that the Canadian design heavy water reactors can achieve a high uranium-thorium resource utilization by using the $^{232}Th/^{233}U$ cycle. The possibility exists, therefore, that a thermal reactor system could act either as an alternative to fast reactors or as an interim system between present-day, low-conversion thermal reactors and fast breeder reactors. The difficulties are: since natural thorium contains no fissile material, any $^{232}Th/^{233}U$ system must start with a natural uranium system; and the conversion ratios in a thermal reactor on the $^{232}Th/^{233}U$ cycle are sufficiently low that much of the potential for high uranium resource utilization could be eroded by rather modest reactor design modifications that would be required in practice by safety or economic considerations.

Other Differences. There are certainly many other differences between fast and thermal reactors besides fissile concentration and conversion ratio. However, most of the important differences flow from these two. For example, the higher concentration of fissile material in fast reactor fuel means that it has a higher monetary value. It is an economic requirement, therefore, that the reactor be operated at a higher thermal power rating (MW(t)/kg). This requirement creates the incentive to use a liquid metal heat transport medium, because it facilitates high heat transfer rates. A liquid metal coolant is not a necessity in a fast reactor; there is also support for a helium-cooled design. A fast reactor cannot use water as a heat transport medium, of course, because water is such an effective neutron moderator. The absence of a moderator plus a higher power density results in a much lower volume for a fast reactor core, as can be seen in Table 7.4.

Neutron Spectrum Calculations

Now that we have discussed the importance of the neutron spectrum to reactor characteristics, we should address the question of how one knows what the spectrum of a reactor will be as a function of the materials that will be in it.

This happens to be one of the more involved topics of nuclear reactor theory, so we shall not probe it very deeply. An overview is very desirable, however.

At the start of this chapter we designated a neutron balance

$$\bar{\nu}\Sigma_f\phi = k\Sigma_a\phi$$

The problem here, which we momentarily ignored in the interest of making our point about criticality, is that the neutrons do not all have a single energy or speed, and the cross sections are energy-dependent. As explained in Chapter 4, the absorption rate must be calculated

$$\int_0^\infty \Sigma_a(E)\phi(E)\,dE \qquad (4.38)$$

and the neutron production rate is

$$\int_0^\infty \bar{\nu}\Sigma_f(E)\phi(E)\,dE$$

The above criticality expression must be written, therefore,

$$\int_0^\infty \bar{\nu}\Sigma_f(E)\phi(E)\,dE = k\int_0^\infty \Sigma_a(\mathrm{E})\phi(\mathrm{E})\,\mathrm{dE} \qquad (7.17)$$

or

$$\overline{\nu\Sigma_f\phi} = k\overline{\Sigma_a\phi}$$

We assume that the functions $\nu\Sigma_f(E)$ and $\Sigma_a(E)$ are known, but the spectrum, $\phi(E)$, is not known. The function, $\phi(E)$, depends on all the materials in the reactor and must somehow be calculated. There is no analytical solution to Eq. 7.16; much of the work in reactor theory is devoted to its various approximate solutions. The most popular approach is the so-called multigroup method, which is a finite difference scheme in the energy dimension. The energy scale is broken into N groups. The reaction rate in the gth group is

$$\Sigma_g\phi_g = \int_{\check{E}_g}^{\hat{E}_g} \Sigma(E)\phi(E)\,dE \qquad (7.18)$$

where $\hat{E}_g$ and $\check{E}_g$ are the upper and lower energy boundaries, respectively, of energy group g. The criticality expression, Eq. 7.16, then becomes

$$\frac{1}{k}\sum_{g=1}^{N} \nu\Sigma_{f,g}\phi_g = \sum_{g=1}^{N} \Sigma_{a,g}\phi_g \qquad (7.19)$$

The multigroup approach simply breaks up the integrals of Eq. 7.16, making the averaging of the cross sections, or $\phi(E)$, somewhat easier. In regions where the functions are smooth, broad energy limits can be used; where the functions are irregular (see Figure 4.4), a fine energy structure can be used. Although the analytical process of solving multigroup equations is well beyond the scope of this text, there is one example that gives a hint of the methodology and also produces a useful result. We shall analyze the slowing-down region (Figure 4.4) in a thermal reactor. We define two groups of neutrons

$$\phi_1 = \int_{\check{E}_1}^{\hat{E}_1} \phi(E)\, dE$$

$$\phi_2 = \int_{0}^{\check{E}_1} \phi(E)\, dE$$

The higher energy group is in the energy region in which the elastic slowing-down process takes place. It extends from about 0.5 eV to 500 keV. The lower energy group contains the thermal neutrons. The rate at which neutrons transfer from group 1 to group 2 is calculated as follows:

$$\Sigma_{s,1}\phi_1 = \text{scatterings/cm}^3 \cdot \text{s}$$

From Eq. 4.50, we know it takes an average of

$$\frac{\ln \hat{E}_1/\check{E}_1}{\xi}$$

scatterings to reduce a neutron's energy from $\hat{E}_1$ to $\check{E}_1$ by means of elastic scattering. Therefore, the rate of transfer from group 1 is

$$\frac{\xi\Sigma_{s,1}\phi_1}{\ln \hat{E}_1/\check{E}_1} = \frac{\xi\Sigma_{s,1}\phi_1}{13.8}$$

The neutrons entering the lower energy group only "leave" by being absorbed (assuming that leakage does not occur). Therefore

$$\frac{\xi\Sigma_{s,1}\phi_1}{13.8} = \Sigma_{a,2}\phi_2$$

or

$$\frac{\phi_2}{\phi_1} = \frac{\xi\Sigma_{s,1}}{13.8\Sigma_{a,2}}$$

This ratio is a measure of how well thermalized the neutron spectrum is. There is no hard and fast rule, but a rough criterion might be a ratio of group 2 to group 1 fissions of 20 : 1.

$$\frac{\sigma_{f,2}\phi_2}{\sigma_{f,1}\phi_1} = 20$$

We can obtain the σ_f values from Table A.3. The neutron slowing-down region happens to coincide with the resolved resonance region (Figure 4.4). The cross section in this region is often expressed as an integral

$$I = \int_{\check{E}_1}^{\hat{E}_1} \sigma \frac{dE}{E} = \bar{\sigma} \ln \frac{\hat{E}_1}{\check{E}_1} \tag{7.20}$$

This so-called resonance integral has this form because it can be shown that in the slowing-down region

$$\phi(E) \propto \frac{1}{E}$$

The fission resonance integral of ^{235}U from Table A.3 is 275 barns. Therefore

$$\sigma_{f,1} = \frac{275}{13.8} = 19.9 \text{ barns}$$

$$\frac{\phi_2}{\phi_1} = \frac{20\sigma_{f,1}}{\sigma_{f,2}} = \frac{20 \times 19.9}{582.2} = 0.68$$

Also

$$\frac{\xi\Sigma_{s,1}}{\Sigma_{a,2}} = \frac{13.8\phi_2}{\phi_1} = 9.4$$

The significance of this analysis is that it can tell us how much moderator is needed in order to achieve thermalization of neutrons.

Example 7.4

Estimate the mole ratio of moderator to ^{235}U required to achieve thermalization in a homogeneous mixture with each of the moderating materials: H_2O, D_2O, and graphite (carbon).

Solution: We shall infer the necessary moderator properties from data given by Duderstadt (see Reference 1).

	ξ	$\xi\Sigma_{s,1}$ (cm^{-1})	ρ (g/cm^3)	$N \times 10^{-24}$ ($mol./cm^3$)	$\xi\sigma_{s,1}$ (barns)	$\xi\sigma_{s,1}/\sigma_{a,2}$	$\sigma_{a,2}$ (barns)
H_2O	0.920	1.35	1.0	0.0335	40.3	71	0.57
D_2O	0.509	0.176	1.1	0.0331	5.32	5670	0.001
C	0.158	0.060	1.60	0.0803	0.75	192	0.039

Note that we assign a microscopic cross section to a molecule. This is a bookkeeping operation in which the recorded cross section is simply the sum of the cross sections of the nuclear constituents of the molecule. From our criterion for thermalization

$$\frac{\xi\Sigma_{s,1}}{\Sigma_{a,2}} = 9.4$$

We designate the moderator by the subscript m and ^{235}U by the subscript 25.

$$\frac{\xi\Sigma_{s,1}}{\Sigma_{a,2}} = \frac{N_m(\xi\sigma_{s,1})_m}{N_m\sigma_{a,2,m} + N_{25}\sigma_{a,2,25}}$$

We neglect the contribution of ^{235}U to $\xi\Sigma_{s,1}$. The thermal absorption cross section of ^{235}U, from Table A.3, is 680.8 barns. Substituting in the above equations for H_2O

$$40.3\frac{N_m}{N_{25}} = 9.4\left(0.57\frac{N_m}{N_{25}} + 680.8\right)$$

$$\frac{N_m}{N_{25}} = 183$$

Similar calculations for D_2O and graphite give

D_2O: $$\frac{N_m}{N_{25}} = 1200$$

C: $$\frac{N_m}{N_{25}} = 9000$$

Example 7.5

The values for the mole ratio of moderator to fissile material can be converted to volume ratios. In an actual power reactor, however, the fuel will contain fertile material, ^{238}U, as well as fissile. Consider that the fuel is UO_2 of density 10 g/cm^3. Compute the volume ratio of moderator to fuel for each moderator in the previous example, assuming that the fuel is 2.5% ^{235}U in the case of H_2O moderator and 0.71% ^{235}U (natural uranium) in the other cases.

Solution: The volume ratio can be written

$$\frac{V_m}{V_{UO_2}} = \frac{N_m \times A_m/\rho_m}{N_{UO_2} \times A_{UO_2}/\rho_{UO_2}} = \frac{(N_m/N_{UO_2}) \times A_m/\rho_m}{A_{UO_2}/\rho_{UO_2}}$$

The value of N_m/N_{UO_2} can be calculated from the values of N_m/N_{25} obtained in the previous example.

$$\frac{N_m}{N_{UO_2}} = \frac{N_m}{N_{25} + N_{28}} = \frac{N_m/N_{25}}{1 + N_{28}/N_{25}}$$

For the case of H_2O moderator

$$\frac{N_m}{N_{UO_2}} = \frac{183}{1 + 97.5/2.5} = 4.58$$

$$\frac{V_{H_2O}}{V_{UO_2}} = \frac{4.58 \times 18/1.0}{270/10} = 3.0$$

Similar calculations give

$$\frac{V_{D_2O}}{V_{UO_2}} = 5.8$$

$$\frac{V_C}{V_{UO_2}} = 17.8$$

More than any other single phenomenon, the volume of moderator required to thermalize neutrons is the distinguishing feature of thermal reactor core designs. We refer to the light-water-moderated reactors (LWRs), the heavy-water-moderated reactors (CANDU), and the graphite-moderated reactors (Magnox, HTGR). Some of the characteristics of these types are listed in Tables 7.7 and 7.8. The larger volume of the D_2O-moderated and graphite-moderated reactor for a given thermal power generation is clear from Table 7.8.

The reactor designer, of course, is not interested in neutron thermalization per se (see Table 7.1). However, power cost is important. In this respect, a moderator is a mixed blessing. On the one hand, it takes up space and reduces the power per unit volume of core, which is an economic disadvantage. On the other hand, the microscopic cross sections of the fissile materials are much higher when a moderator is present (low neutron energies). This effect reduces the quantity and concentration of fissile material required, thereby effecting an economy. Light water has a disadvantage not shared by D_2O or graphite, namely, a relatively high thermal absorption cross section (see the table in Example 7.4). Therefore, there is a special incentive to keep its volume low in LWRs. The optimum moderator-to-fuel volume ratio involves the tradeoff of these and other competing factors. The ideal moderator would be one with the slowing-down power ($\xi\sigma_{s,1}$) of H_2O and the thermal absorption cross section of D_2O, but, unfortunately, nature did not provide it.

Example 7.6
Table 7.4 shows 38 % by volume of a conceptual fast reactor core to be occupied by sodium. Will this much sodium thermalize many neutrons? Suppose H_2O is substituted for the sodium. Would a thermal reactor result?

Solution: Use a value of 800 kg/m^3 for the density of hot liquid sodium. The value of σ_s is obtained from Table A.3.

$$\Sigma_s = N\sigma_s = \frac{0.8}{23} \times 6.023 \times 10^{23} \times 3.2 \times 10^{-24}$$

$$= 0.067 \text{ cm}^{-1}$$

$$\xi = 0.845 \quad \text{(Table 4.1)}$$

$$\xi\Sigma_s = 0.057 \text{ cm}^{-1}$$

This is the value that applies in the pure sodium. Since the sodium occupies only 38 v/o of the core, the volume-averaged value

$$\xi\Sigma_s = 0.38 \times 0.057 = 0.022 \text{ cm}^{-1}$$

We calculate the value of the thermal neutron absorption cross section. From Table 7.4, the density of ^{238}U and ^{239}Pu, averaged over the core volume, is

$$^{238}\text{UO}_2: \frac{15{,}260 \times 0.84}{4.2} = 3050 \text{ kg/m}^3 = 3.05 \text{ g/cm}^3$$

$$^{239}\text{PuO}_2: \frac{15{,}260 \times 0.11}{4.2} = 400 \text{ kg/m}^3 = 0.40 \text{ g/cm}^3$$

We shall only include contributions to $\Sigma_{a,2}$ from

$$^{238}\text{U}: \frac{3.05}{270} \times 6.023 \times 10^{23} \times 2.70 \times 10^{-24} = 0.0184 \text{ cm}^{-1}$$

$$^{239}\text{Pu}: \frac{0.40}{271} \times 6.023 \times 10^{23} \times 1011.3 \times 10^{-24} = 0.899 \text{ cm}^{-1}$$

$$\text{Na}: \frac{0.38 \times 0.8}{23} \times 6.023 \times 10^{23} \times 0.53 \times 10^{-24} = 0.0042 \text{ cm}^{-1}$$

$$\Sigma_{a,2} = 0.0184 + 0.899 + 0.0042 = 0.922 \text{ cm}^{-1}$$

$$\frac{\xi\Sigma_s}{\Sigma_{a,2}} = \frac{0.022}{0.922} = 0.024$$

This value is well below the value of 9.4 developed in the text as a criterion of a well-thermalized reactor. Sodium definitely will not thermalize neutrons. It does have the more subtle effect of lowering the mean energy of the neutron spectrum of a fast reactor.

If H_2O (density 1.0 g/cm^3) is substituted for the sodium

$$\xi\Sigma_s = 0.38 \times 1.35 = 0.51 \text{ cm}^{-1} \quad \text{(see Example 7.4)}$$

The contribution of H_2O to $\Sigma_{a,2}$ is about 0.007 cm^{-1}.

$$\frac{\xi\Sigma_s}{\Sigma_{a,2}} = \frac{0.51}{0.924} = 0.55$$

Such a reactor could not be described as a thermal reactor, but there would be a substantial fraction of thermal fissions. The neutron spectrum would be grossly altered by the substitution of water for sodium.

THE CRITICAL SIZE OF A REACTOR

We have already discussed the infinite model of a reactor—one from which no neutrons escape. Clearly, any real system is finite, so we must deal with neutron escape. The way in which any given neutron escapes from the core or any region is simple enough. The neutron scatters in such a direction that the distance to the edge of the medium along the line of flight is s cm. The probability that it will indeed travel that far without undergoing another interaction

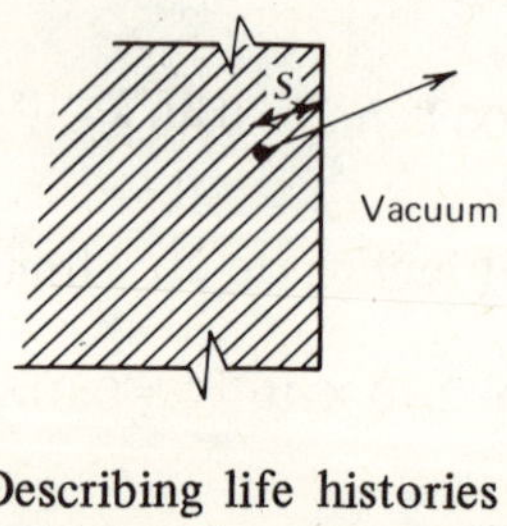

is $e^{-\Sigma s}$ (see Chapter 4). Describing life histories of a statistically significant sample of neutrons one at a time to determine whether they escape or not is a very tedious but not impossible task. Such a technique is the heart of the so-called Monte Carlo calculation. The more common approach, however, is to treat neutrons collectively, or as populations, describing their net flow in a manner not unlike a fluid. This field of analysis is called neutron transport theory. Clearly, if one knows the net flow of neutrons, **J** neutrons/cm^2 · s, everywhere in a region, then integration over the surface

$$\int_s \mathbf{J} \cdot \mathbf{ds}$$

will give the leakage. We now have a neutron leakage term to add to the fission neutron source term, $\bar{\nu}\Sigma_f \phi$, and the absorption term, $\Sigma_a \phi$. Note that these two are on a per-unit-volume basis. We can place the leakage term on a volume basis by application of Gauss's theorem

$$\int_s \mathbf{J} \cdot \mathbf{ds} = \int_V \boldsymbol{\nabla} \cdot \mathbf{J}\, dV$$

The criticality condition becomes $\nabla \cdot J + \Sigma_a \phi(r) = S(r)$

$$\bar{\nu}\Sigma_f \phi - \Sigma_a \phi - \boldsymbol{\nabla} \cdot \mathbf{J} = 0 \tag{7.21}$$

One approximation to neutron transport theory is that neutrons obey a simple diffusion rule, known as Fick's law.

$$\mathbf{J} = -D\boldsymbol{\nabla}\phi = -Dv\boldsymbol{\nabla}\mathbf{n} \tag{7.22}$$

This approximation is very widely used. It says that neutrons exhibit a net flow in the direction of a low neutron concentration. Intuitively, one might expect that the diffusion coefficient is related to the neutron mean free path (Eq. 4.9). Indeed, an approximation gives the simple relationship

$$D \cong \frac{\lambda}{3} = \frac{1}{3\Sigma_t} \tag{7.23}$$

If Eq. 7.22 is substituted in Eq. 7.21

$$\boldsymbol{\nabla} \cdot D\boldsymbol{\nabla}\phi + \bar{\nu}\Sigma_f \phi - \Sigma_a \phi = 0 \tag{7.24}$$

which we can also write

$$D\nabla^2\phi + (k - 1)\,\Sigma_a \phi = 0 \tag{7.25}$$

and

$$\nabla^2\phi + B^2\phi = 0 \tag{7.26}$$

where

$$B^2 = \frac{k - 1}{D/\Sigma_a} \tag{7.27}$$

Equation 7.26 is a second-order partial differential equation that has discrete solutions, depending on the geometrical configuration. Some solutions are given in Table 7.5. The B^2 term is called the buckling. It is sometimes distinguished as materials buckling when represented by Eq. 7.27, and sometimes as geometric buckling when given in terms of dimensions, as in Table 7.5. In Table 7.5, the J_0 term represents a *zeroth*-order Bessel function of the first kind. The value 2.405 is its first zero.

Table 7.5 Geometric Buckling Formulae and Flux Distribution for Simple Reactor Core Configurations

Parallelepiped

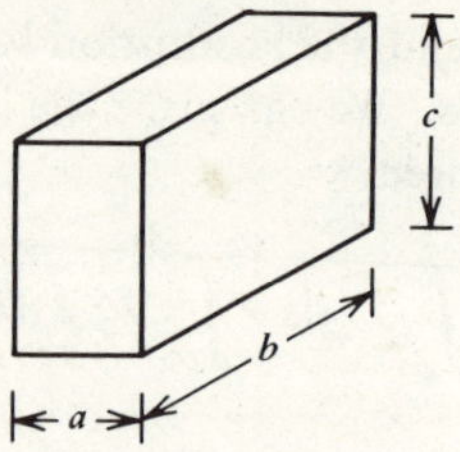

$$B^2 = \frac{\pi^2}{a^2} + \frac{\pi^2}{b^2} + \frac{\pi^2}{c^2}$$

$$\phi(x, y, z) = A \cos\left(\frac{\pi x}{a}\right) \cos\left(\frac{\pi y}{b}\right) \cos\left(\frac{\pi z}{c}\right)$$

Cylinder

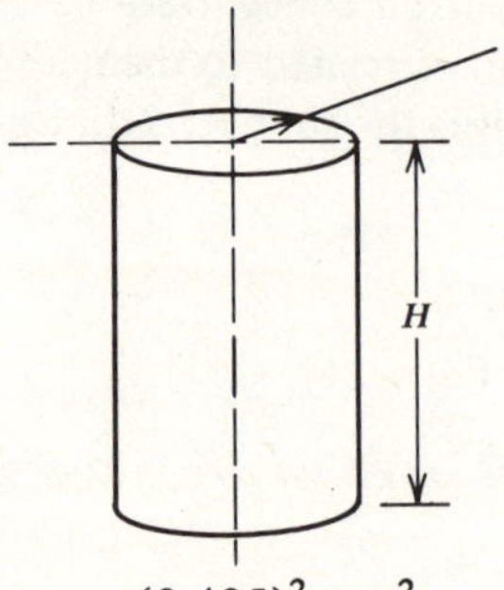

$$B^2 = \frac{(2.405)^2}{R^2} + \frac{\pi^2}{H^2}$$

$$\phi(r, z) = AJ_0\left(\frac{2.405r}{R}\right) \cos\left(\frac{\pi z}{H}\right)$$

Sphere

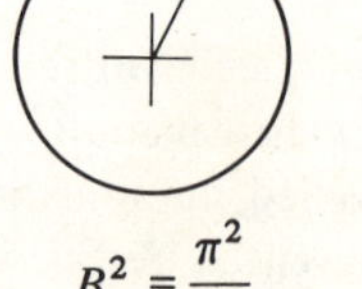

$$B^2 = \frac{\pi^2}{R^2}$$

$$\phi(r) = \frac{A}{r} \sin\left(\frac{\pi r}{R}\right)$$

Equation 7.24 is written without explicit recognition of the energy-dependence of neutron cross sections. It is a one-group or one-speed model of a reactor. In practice, it is necessary to integrate over all neutron energies just as in Eq. 7.17, but with the leakage term of Eq. 7.24 included. The most common practice is to use a multigroup formulation analogous to Eq. 7.19.

In spite of limitations, Eq. 7.24 and its solutions (Table 7.5) contain a great deal of information about critically reacting systems. A criticality expression, with leakage, can be written by rearranging Eq. 7.27

$$k_{\text{eff}} = \frac{k}{1 + (D/\Sigma_a)B^2} \tag{7.28}$$

where B^2 values are configuration-dependent, as shown in Table 7.5. The criticality condition is that k_{eff} equal one. This equation separates conveniently into

$$k_{\text{eff}} = \underbrace{k}_{\substack{\text{infinite}\\ \text{medium}}} \times \underbrace{\frac{1}{1 + (D/\Sigma_a)B^2}}_{\substack{\text{nonleakage}\\ \text{probability,}\\ \text{finite medium}}}$$

In one-group analysis the term D/Σ_a is defined

$$L^2 = \frac{D}{\Sigma_a} \cong \frac{1}{3\Sigma_t\Sigma_a} \tag{7.29}$$

where L is called the diffusion length. It can be shown that the mean crow's-flight distance a neutron travels to extinction (absorption) in an infinite medium is

$$\sqrt{\overline{r^2}} = \sqrt{6L^2} = 2.45\,L \tag{7.30}$$

The gross distance traveled is (Chapter 4) λ_a or $1/\Sigma_a$. Although we do not make explicit use of Eq. 7.30, it is useful in giving some perspective on the likelihood of neutrons leaking from a finite medium.

In thermal reactors, a two-group (slowing-down and thermal) approximation leads to a modification of Eq. 7.28

$$k_{\text{eff}} = \frac{k}{1 + M^2B^2} \tag{7.31}$$

where M^2, called the migration area, substitutes for L^2 and accounts for neutron travel while slowing down, as well as while the neutron is thermal. The moderator is the most important material in determining the value of M^2 in thermal reactors. Some values of diffusion properties for popular moderators are given in Table 7.6. The values of L^2 for D_2O and graphite are misleading as predictors of thermal neutron leakage in a reactor because, inevitably, the fuel and other core materials greatly increase the thermal Σ_a, so that the effective L^2 is greatly

Table 7.6 Some Neutron Diffusion Properties of Moderators

MODERATOR	DENSITY (g/cm^3)	THERMAL NEUTRONS D (cm)	Σ_a (cm^{-1})	L (cm)	$M^2 - L^2$ (cm^2)	M (cm)
H_2O	1.00	0.16	0.019	2.9	26	5.9
D_2O	1.10	0.87	3×10^{-5}	170	130	170
C	1.60	0.84	3.1×10^{-4}	52	370	55

reduced. The major contribution to M^2, therefore, comes from the slowing-down region in which neutron scattering is the dominant phenomenon. It is for this reason that $M^2 - L^2$, the slowing-down contribution, is shown separately in Table 7.6.

As is evident in Table 7.6, a neutron travels much farther in D_2O or graphite than in H_2O. To keep neutron leakage to the same percent, D_2O and graphite reactors must be larger. This effect is closely related to the fact that more of these moderators are needed to thermalize the neutrons. More neutron collisions are needed to thermalize than in H_2O and, in the course of undergoing a larger number of collisions, a neutron will be, on the average, farther from its starting point.

Example 7.7

A solution of uranium salt in H_2O in a cylindrical container 25.4 cm in diameter is found to be critical at a height of 72.4 cm. The uranium is enriched in the isotope ^{235}U and can be treated as pure ^{235}U. The mole ratio $H_2O/^{235}U$ is 120. What is the effective value of M^2? If the uranium is in the form $UO_2(NO_3)_2$ and the solution density is 1.16 g/cm^3, what is the critical mass? What would be the critical depth if this solution were placed in a container of square cross section, 25.4 cm on a side?

Solution: From prior calculations we can conclude that this solution will thermalize neutrons, so that the use of 0.025 eV cross sections is justified. We calculate the infinite multiplication constant.

$$k = \frac{\bar{\nu}\Sigma_f}{\Sigma_a} = \frac{N_{235}(\nu\sigma_f)_{235}}{N_{235}\sigma_{a,235} + N_H\sigma_{a,H}}$$

We can neglect the absorption cross section of oxygen.

$$k = \frac{2.5 \times 582.2}{680.8 + 240 \times 0.332} = 1.91$$

From Eq. 7.31

$$M^2B^2 = k - 1 = 0.91$$

From Table 7.5

$$B^2 = \frac{(2.405)^2}{(25.4/2)^2} + \frac{\pi^2}{(72.4)^2} = 0.0377 \text{ cm}^{-2}$$

$$M^2 = \frac{0.91}{0.0377} = 24.1 \text{ cm}^2$$

Compare with the value of 34 cm^2 for H_2O from Table 7.6. If the uranium is in the chemical form $UO_2(NO_3)_2$, then the mass ratio is

$$120 \times \frac{18}{391} = 5.52 \text{ kg-}H_2O/\text{kg-}UO_2(NO_3)_2$$

The weight fraction of $UO_2(NO_3)_2$ is

$$\frac{1}{1 + 5.52} = 0.153$$

The concentration is

$$0.153 \times 1.16 = 0.178 \text{ g/cm}^3$$

The concentration of ^{235}U is

$$0.178 \times \frac{235}{391} = 0.107 \text{ g/cm}^3$$

The critical volume is

$$\frac{\pi}{4} \times (25.4)^2 \times 72.4 = 36{,}686 \text{ cm}^3$$

The critical mass, defined as the mass of fissile material required for criticality, is

$$0.107 \times 36{,}686 = 3920 \text{ g-}^{235}\text{U}$$

If this same solution is placed in a different container, the infinite multiplication and M^2 would remain the same, since they are functions only of composition. The formula for B^2 will change, however. For a cuboid (Table 7.5)

$$B^2 = \frac{\pi^2}{a^2} + \frac{\pi^2}{b^2} + \frac{\pi^2}{c^2} = 0.0377 \text{ cm}^{-2}$$

$$a = b = 25.4 \text{ cm}$$

$$\frac{\pi^2}{c^2} = 0.0071$$

$$c = 37.2 \text{ cm}$$

The volume and, therefore, the critical mass are considerably reduced in this configuration.

Example 7.8

Consider the solution of Example 7.7, contained in a cylinder 25.4 cm in diameter. The critical height is 72.4 cm. Suppose this solution were reduced in density from 1.16 g/cm^3 to 1.00 g/cm^3. What would be the new critical height and critical mass?

Solution: The value of M^2 varies inversely as the square of the density. For example

$$L^2 \propto \frac{1}{\Sigma_t \Sigma_a} \propto \frac{1}{N^2} \propto \frac{1}{\rho^2}$$

The new value of M^2, therefore, is

$$M^2 = 24.1 \times \left(\frac{1.16}{1.00}\right)^2 = 32.4 \text{ cm}^2$$

The value of k, the infinite multiplication, is unaffected by a density change.

$$M^2 B^2 = \text{k} - 1 = 0.91$$

$$B^2 = \frac{0.91}{32.4} = 0.0281 \text{ cm}^{-2}$$

$$B^2 = \frac{(2.405)^2}{(25.4/2)^2} + \frac{\pi^2}{\text{H}^2} = 0.0281$$

$$\frac{\pi^2}{\text{H}^2} = 0.0281 - 0.0359 = -0.008 \text{ cm}^2$$

This result indicates that there is no positive value of the height that would make such a solution critical. The neutron leakage radially is simply too great. Although the effect of a density change is more pronounced in such a small (high geometric buckling) system, it is also present in actual power reactors. When a power reactor is started and brought to power, there are significant density changes as the temperature rises.

Example 7.9

If the H_2O of the solution in Example 7.7 should be replaced by D_2O such that the $D_2O/^{235}U$ mole ratio remains at 120, would the solution be critical in a cylinder 25.4 cm in diameter and 72.4 cm high?

Solution: The only contribution to the thermal Σ_a would be from ^{235}U, because Σ_a of D_2O is negligible.

$$k = \frac{2.5 \times 282.2}{680.8} = 2.14$$

The value of M^2 would change.

$$L^2 = \frac{D}{\Sigma_a}$$

$$\Sigma_a = \frac{0.107}{235} \times 6.023 \times 10^{23} \times 680.8 \times 10^{-24} = 0.187 \text{ cm}^{-1}$$

Use the D value of D_2O (Table 7.6).

$$L^2 = \frac{0.87}{0.187} = 4.7 \text{ cm}^2$$

The value of $M^2 - L^2 = 130 \text{ cm}^2$ can be used directly since it is unaffected by ^{235}U.

$$M^2 = 130 + 4.7 = 134.7 \text{ cm}^2$$

From Example 7.7

$$B^2 = 0.0377 \text{ cm}^{-2}$$

$$M^2B^2 = 5.1$$

$$k_{\text{eff}} = \frac{k}{1 + M^2B^2} = \frac{2.14}{5.1} < 1$$

This substitution of D_2O for H_2O in this small a container will definitely result in a subcritical system.

Power Distribution in a Reactor

We have already noted several times that the thermal power density at any point in a reactor is proportional to the fission rate. In Example 4.12, the thermal power density, q''', is given

$$q''' = 200 \times 1.602 \times 10^{-13} \Sigma_f \phi$$

$$= 3.204 \times 10^{-11} \Sigma_f \phi \text{ W/cm}^3 \qquad (7.32)$$

allowing 200 MeV per fission. It is understood that the $\Sigma_f \phi$ term is obtained by integration of the energy-dependent fission cross section over the neutron spectrum in a manner adequate for the particular reactor. In multigroup notation, the thermal power density is written

$$q''' = 3.204 \times 10^{-11} \sum_{g=1}^{N} \Sigma_{f,g} \phi_g \text{ W/cm}^3$$

In a homogeneous reactor, Σ_f is everywhere the same. However, it is apparent from the solutions to Eq. 7.26, given in Table 7.5, that the flux varies with position in the reactor. The thermal power density must vary accordingly. The spatial variation of power density is very important in power reactor design. The thermal power that can be extracted from a reactor core is usually limited by the conditions existing at or near the point of maximum power density. These conditions include fuel temperature, fuel clad temperature, heat flux into the coolant, and so on. The variation of power density means, of course, that some portions of the core are not generating thermal energy at a rate that design conditions would otherwise permit.

The variation in power density in actual reactors is indicated in Table 7.8 by the average and maximum values of the heat flux. The heat flux is the rate of thermal energy transfer into the coolant in watts per square centimeter of coolant channel surface area. The usual index of power variation in a core is the maximum-to-average power density, called max-to-average, or peak-to-average, power. In the case of the PWR in Table 7.8, for example, the max-to-average power is 168.6/54.1 or 3.12.

Generally it is desirable to reduce the max-to-average power. This is also referred to as flattening the power. There are a number of design expedients that effect some degree of power flattening. The most common is to surround the core with a reflector region, a nonfissioning material that scatters neutrons, returning some of them to the core, thus raising the flux and power at the core boundary. The reflector is discussed in more detail in the next section. Another expedient is to increase the fissile material composition toward the edge of the core. Since the flux tends to drop there, an increase in Σ_f will make $\Sigma_f \phi$ flatter. In a typical PWR, for example, the uranium might have three ^{235}U enrichments in the initial load, 2.2%, 2.3%, and 2.7%, with the latter at the core edge. The converse of increasing the fissile material near the edge of the core is increasing neutron absorber concentration near the center. This expedient is also practiced to some extent by manipulating control rods in power reactors.

Example 7.10

The data of Table 7.8 show that the Pickering CANDU reactor has a diameter of 7.5 m and a height (actually length, because the CANDU's axis is horizontal) of 6.0 m. Calculate the max-to-average power for a bare reactor of this dimension and compare with the data of Table 7.8. From the information given, calculate the maximum thermal neutron flux in the core.

Solution: The flux distribution over the core is given (Table 7.5).

$$\phi(r, z) = AJ_0\left(\frac{2.405\, r}{R}\right) \cos\left(\frac{\pi z}{H}\right)$$

In order to find the volume-averaged flux, we must integrate over the volume. The differential volume at any point (r, z) is

$$dV = 2\pi r\, dr\, dz$$

and

$$\overline{\phi} = \frac{\int \phi(r, z)\, dV}{\pi R^2 H}$$

$$\int \phi(r, z)\, dV = 2\pi A \int_0^R rJ_0\left(\frac{2.405\, r}{R}\right) dr \int_{-H/2}^{H/2} \cos\left(\frac{\pi z}{H}\right) dz$$

$$= \frac{4AR^2 H J_1(2.405)}{2.405}$$

$$\overline{\phi} = \frac{4AJ_1(2.405)}{2.405\,\pi}$$

$$\phi_{\text{max}} = \phi(0, 0) = A$$

$$\text{Maximum-to-average} = \frac{\phi_{\text{max}}}{\overline{\phi}} = \frac{2.405\,\pi}{4J_1(2.405)}$$

$$J_1(2.405) = 0.5191$$

$$\text{Maximum-to-average} = 3.64$$

We see that the max-to-average flux and, therefore, the max-to-average power is not dependent on the dimensions (radius and height) of the right cylinder.

From Table 7.8, for the CANDU

$$\text{Maximum heat flux} = 111\ \text{W/cm}^2$$

$$\text{Average heat flux} = 59.5\ \text{W/cm}^2$$

Therefore, the max-to-average power is 111/59.6, or 1.87. This value is a considerable improvement over the 3.64 value calculated, primarily because of the power flattening of the reflector.

The thermal power of the reactor is related to the flux.

$$P = 3.204 \times 10^{-11} \int_V \Sigma_f \phi(r, z)\, dV = 3.204 \times 10^{-11}\, \Sigma_f \overline{\phi}\, V\ \text{W}$$

We can compute the Σ_f from the data of Table 7.8 as follows:

$$\text{Core volume} = 265\ \text{m}^3 = 2.65 \times 10^8\ \text{cm}^3$$

There is 92,600 kg of natural uranium in the core. The mass of ^{235}U is

$$0.0071 \times 92{,}600 = 657 \text{ kg} = 6.57 \times 10^5 \text{ g}$$

The average number density of ^{235}U is

$$N_{25} = \frac{6.57 \times 10^5 \times 6.023 \times 10^{23}}{2.65 \times 10^8 \times 235} = 6.35 \times 10^{18} \text{ nuclei/cm}^3$$

$$\sigma_f = 582.2 \times 10^{-24} \text{ cm}^2$$

$$\Sigma_f = 3.70 \times 10^{-3} \text{ cm}^{-1}$$

The thermal power is 1742 MW(t) = 1.742×10^9 W(t). The average thermal flux, assuming all fissions caused by thermal neutrons, is

$$\bar{\phi} = \frac{1.742 \times 10^9}{3.204 \times 10^{-11} \times 3.70 \times 10^{-3} \times 2.65 \times 10^8}$$

$$= 5.5 \times 10^{13} \text{ neutrons/cm}^2 \cdot \text{s}$$

To obtain the maximum thermal neutron flux, multiply by the max-to-average already inferred from Table 7.8.

$$\phi_{max} = 1.87 \times 5.5 \times 10^{13} = 1.04 \times 10^{14} \text{ neutrons/cm}^2 \cdot \text{s}$$

The Effects of Heterogeneities in Reactor Cores

The models of the reactor core that we have been using to this point have all been homogeneous. The nuclear properties are everywhere the same. However, while it is possible to design a power reactor core that is homogeneous–uranium salts dissolved in water, for example–most designs that have evolved are heterogeneous, for a variety of reasons. The heterogeneities to which we refer are the different solid materials as well as liquid or gaseous materials, or both, in the core. The nuclear properties change abruptly and sometimes drastically from one region to another. These heterogeneities have a marked effect on core behavior.

In Chapter 4, we treated the calculation of number densities of various nuclides in a heterogeneous medium. The approach, however, was to treat the mixture of phases as if they were blended into a homogeneous mixture without a change in total volume. When macroscopic cross sections are calculated from such number densities and reaction rates then calculated as $\Sigma\phi$, we implicitly assume that the flux, ϕ, is the same in each phase or material making up the heterogeneous medium. This may not correspond to the physical reality.

The existence of heterogeneities–regions of different nuclear properties–can cause wide variation in the neutron flux and the neutron spectrum over short distances (centimeters and less). These flux variations are not to be confused with the flux variation over an entire core as given by the equations in Table 7.5 and calculated in Example 7.10. Flux variations due to heterogeneities occur over much shorter distances and are superimposed on the overall core variations. The basic reason for these spatial variations of flux is that adjacent materials can have widely differing absorption cross sections. The flux is necessarily depressed in a region of higher Σ_a. For example, in a control rod of ^{10}B, where the thermal-neutron Σ_a might be as high as 490 cm^{-1}, a thermal neutron cannot penetrate, and the flux will drop to zero within a rod of less than one cm diameter.

Another very important heterogeneous effect occurs in materials with high resonance peak cross sections (Figure 4.4). Uranium-238 and thorium-232 (the fertile materials) are the most important cases in point. The neutron flux is greatly reduced within these materials at the energy of the resonances. The neutron capture rate is therefore much lower in ^{238}U when this material is concentrated in fuel rods than if the same amount were uniformly distributed throughout the moderator. In the case of ^{238}U, this effect is of such magnitude that a natural uranium reactor, whether with D_2O or graphite moderator, can be made critical only if it is heterogeneous. The fuel and moderator must be separated. Otherwise, neutron capture in the ^{238}U resonances will be so great that the multiplication will always be less than one. The perception of this heterogeneous effect was very important to the design of the first reactor.

Naturally, not all heterogeneities involve adjacent materials with widely varying neutron cross sections. Some heterogeneous systems behave very much as though they were homogeneous. Fast reactors, for example, follow a homogeneous model much more closely than do thermal reactors, because in a fast-neutron spectrum all cross sections are much lower than in a thermal spectrum.

We can get a semiquantitative idea of whether a material will act neutronically as a heterogeneous region by examining its diffusion length, L, as given by Eq. 7.29, for neutrons of the energy of interest. Since the root mean distance a neutron will travel is 2.45 L (Eq. 7.30) we can make a judgment whether the dimensions of the material are large or small compared with 2.45 L. If the dimensions are relatively large, the material is said to be "black" to neutrons; heterogeneity effects will be significant. If the dimensions are small relative to 2.45 L, the region will be "transparent" to neutrons, and there will be little depression of the flux. It should be noted that heterogenity effects are most pronounced in reactors with a separate moderating medium (mostly thermal reactors), because in such cases the contrast in nuclear properties between regions is greatest. The local spatial flux variations caused by heterogeneity are very important in reactor design. A large portion of the calculational effort in core design goes into the estimation of these variations.

Core Reflectors and Blankets

In the treatment of the finite core we assumed that the core was surrounded by a vacuum—once a neutron leaves, it never returns. (Given its low density, atmospheric air is not significantly different than a vacuum.) Since a neutron's return to the core can be effected simply by surrounding the core with material of high scattering cross section, it is very reasonable to do so. Such a surrounding region is called a reflector. It should have a high Σ_s and a low Σ_c (if it captures the neutrons they obviously will not return). In the case of a fast reactor, it is important not to slow down the neutrons. If they were thermalized, the fission cross section for such neutrons at the core-reflector interface would be high and the local power density, $\Sigma_f \phi$, would be excessive. Consequently, high atomic mass materials with a low capture cross section would make good fast reactor reflectors. Iron and lead are examples. In a thermal-neutron reactor, on the other hand, it is desirable that the reflector also slow down neutrons efficiently. Otherwise the neutrons will get so far from the core edge that they will not get back. This means that the desirable properties, high $\xi\Sigma_s$ and low Σ_c, are the same as for the moderator. Therefore the reflector is usually pure moderator material. In an oversimplified way of looking at it, the designer simply continues the core but leaves out the fuel.

Figure 7.6 shows the effects of a reflector on the flux and power distribution in a thermal and fast-neutron core. The effect is somewhat analogous to placing a thermal insulator around a thermal energy source. Instead of the temperature profile being flattened, however, the neutron flux is flattened. In the case of a thermal-neutron reactor, there is also the effect of an abrupt decrease in thermal absorption cross section together with continued neutron slowing down, which creates a bulge in the thermal neutron population. The flattening in flux and power profiles achieved by a reflector is significant. (See Example 7.10.) For this reason alone, power reactor designs always call for a reflector.

A reflector is used in fast-neutron power reactors, but it also plays an important additional role. We have already mentioned that the core of a fast reactor must contain fissile material in higher concentration than in a thermal reactor. For economic reasons, therefore, it is desirable to keep the core small in order to keep the monetary investment in fuel low. The result, expressed in terms of Eq. 7.31

$$k_{\text{eff}} = \frac{k}{1 + M^2 B^2}$$

is that B^2 is high and the neutron leakage is high. From a criticality standpoint this situation does not present a problem, because the infinite multiplication, k, is high. However, those leaking neutrons represent potential ^{239}Pu atoms that can be produced in the conversion reaction, ^{238}U(n, γ). To achieve breeding,

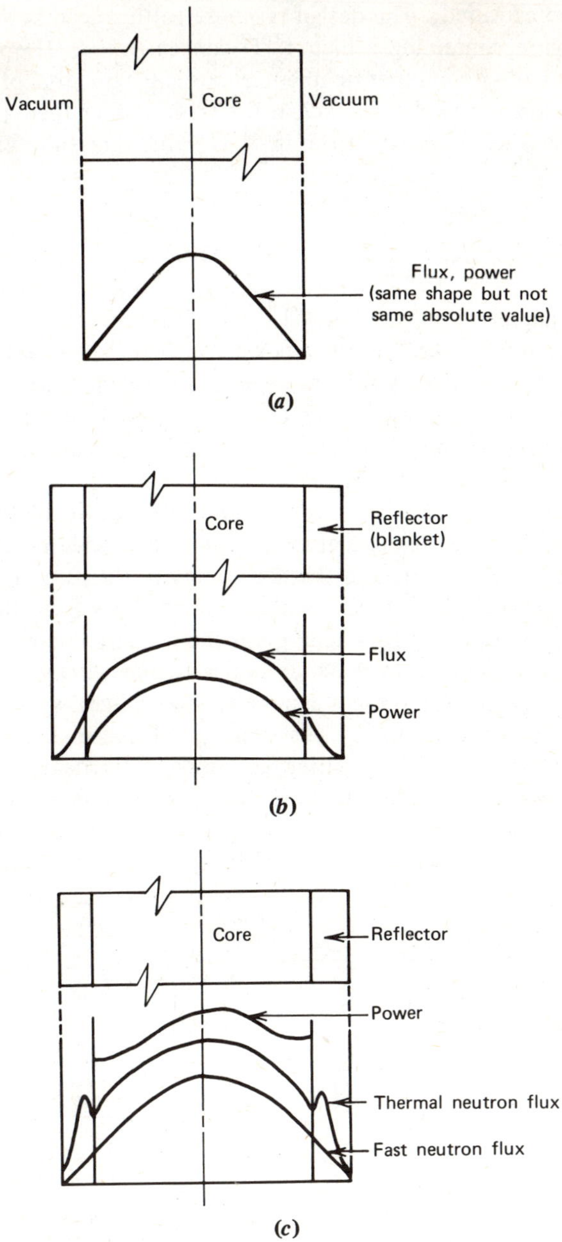

Figure 7.6 The effects of a reflector. (*a*) Unreflected core fast or thermal reactor. (*b*) Reflected fast-neutron reactor. (*c*) Reflected thermal-neutron reactor.

they must be so captured. The design response to this need is to surround the core with a region containing a high ^{238}U concentration. (Depleted uranium from the tails of the uranium enrichment plants is the material of choice.) This surrounding region is called a *blanket* in the trade, rather than a reflector. Its primary purpose is to capture neutrons in ^{238}U rather than to return neutrons to the core.

NUCLEAR POWER SYSTEMS

So far in this chapter we have discussed the reactor core and its various characteristics. The function of the core in a power system is to convert fission energy to thermal energy, preparatory to conversion of the thermal to electrical energy. In this role, the reactor represents one component or process in a power cycle. The design of the reactor is greatly dependent on the requirements of this system.

Figure 7.7 is a schematic of a nuclear power system. It is termed "generic" because it contains the principal features common to a wide variety of systems. The fission energy is converted to thermal energy in the reactor. The thermal energy is transferred to the primary coolant, which flows from the reactor to a heat exchanger. Here the thermal energy in the primary coolant is transferred to a secondary fluid, which may be, as indicated here, the working fluid in a power cycle. This fluid at present is almost always light water (H_2O), but it could be a gas or another vapor. The steam produced in the heat exchanger flows to the turbines, to the condenser, and back to the heat exchanger. This part of the system is similar to a fossil-fired system, except that steam temperatures and pressures are sometimes different. Some reactor systems differ in a significant way from the generic system of Figure 7.7; these differences will be discussed.

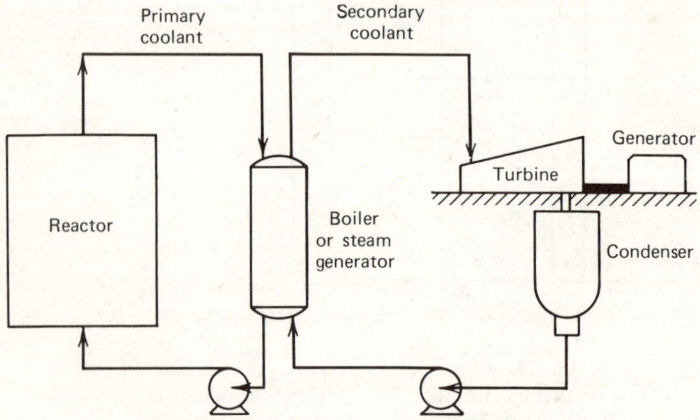

Figure 7.7 Generic nuclear power system.

A person whose knowledge of a power plant was limited to Figure 7.7 would be shocked on visiting an actual plant to see the veritable jungle of pipes and equipment. There are high-pressure turbines, low-pressure turbines, intermediate turbines, and feedwater heaters and reheaters. For the most part these ancillary items are there because they contribute to increased efficiency of the conversion process. Nevertheless the essential features of a plant appear in Figure 7.7.

The Reactor as Heat Exchanger

Earlier in this chapter we dealt with the critical size of a reactor, or the size as determined by maintenance of the neutron chain reaction. It is important to understand that power reactors also function as heat exchangers. It is the heat-exchanger requirements that determine the size of a reactor core, because it turns out that a larger core volume is required to meet these requirements than to meet those of criticality. A typical reactor, for example, is made up of fuel bundles or assemblies containing a matrix of cylindrical rods. The rods are formed by inserting uranium dioxide pellets into tubes of cladding metal. A manufacturer will produce a standard fuel assembly and then market reactors of different power ratings simply by varying the number of assemblies. The General Electric Co.,[3] for example, offers the following boiling water reactor (BWR) thermal power ratings:

THERMAL POWER MW(t)	NUMBER OF FUEL ASSEMBLIES	MW(t)/ASSEMBLY
1956	400	4.89
2444	500	4.89
2894	592	4.89
3579	732	4.89
3833	784	4.89

The constancy of the value of megawatts per assembly makes it clear that the size of the core is determined strictly by the power rating. If more assemblies are added to a core that is already critical, wouldn't that make it supercritical? Indeed it would, but there is a simple design answer to this problem. As fuel assemblies are added, control elements are also added so that the effective multiplication (Eq. 7.31) is kept constant. In a BWR, for example, there is one control element for every four fuel assemblies.

[3] See Reference 2 at end of chapter.

Example 7.11
In a BWR of recent design, the maximum heat flux is specified as 354,000 Btu/h · ft² (1117 kW/m²). The overall peak-to-average power is 2.22. A fuel assembly contains 64 rods (8 × 8); an assembly occupies 0.02323 square meters of core area in the horizontal plane. The fuel rods have an outside diameter of 12.52 mm. The active (fueled) length of the rods is 3.76 m. How many fuel assemblies must be in a core that is to generate 3579 MW(t)? What is the effective core diameter?

Solution: The average heat flux is

$$q'' = \frac{1117}{2.22} = 503.2 \text{ kW/m}^2$$

The heat transfer area required is

$$\frac{3579 \times 1000}{503.2} = 7113 \text{ m}^2$$

The heat transfer area per single fuel rod is

$$\pi \times 12.52 \times 10^{-3} \times 3.76 = 0.1479 \text{ m}^2/\text{rod}$$

The number of rods required is

$$\frac{7113}{0.1479} = 48{,}089$$

There are 64 rods to a fuel assembly, which means 751 fuel assemblies in the core. Since each of these represent 0.02323 m² of core horizontal cross section, the effective diameter is

$$\frac{\pi}{4} D^2 = 751 \times 0.02323$$

$$D = 4.71 \text{ m}$$

Manufacturers tend to standardize on fuel assembly length. The rated power of a reactor can be adjusted by specifying the number of assemblies in the core.

We must take a closer look at the nature of a reactor as a heat exchanger. In the more common heat exchanger, thermal energy in one fluid is transferred to another (such as in the steam generator in Figure 7.7). The reactor, however, usually involves creation of thermal energy in a solid, followed by transfer of this thermal energy to the primary coolant through a series of conductive and convective resistances.

The heat-transfer situation in a fuel rod is depicted in Figure 7.8. We assume that the neutron flux, ϕ, is not a function of radial position within the rod. The volumetric thermal power density is

$$q''' = 3.204 \times 10^{-11}\ \Sigma_f \phi \quad \text{W/cm}^3 \tag{7.32}$$

This rate of thermal energy is created out of nuclear energy by the fissions within the UO_2 (fuel) pellets. This generation of thermal energy initially raises the fuel temperature, but if the fuel is not to melt and if a steady state is to be achieved, then thermal energy must be conducted out of the fuel at the same rate it is being generated. We can describe the conduction process with a dif-

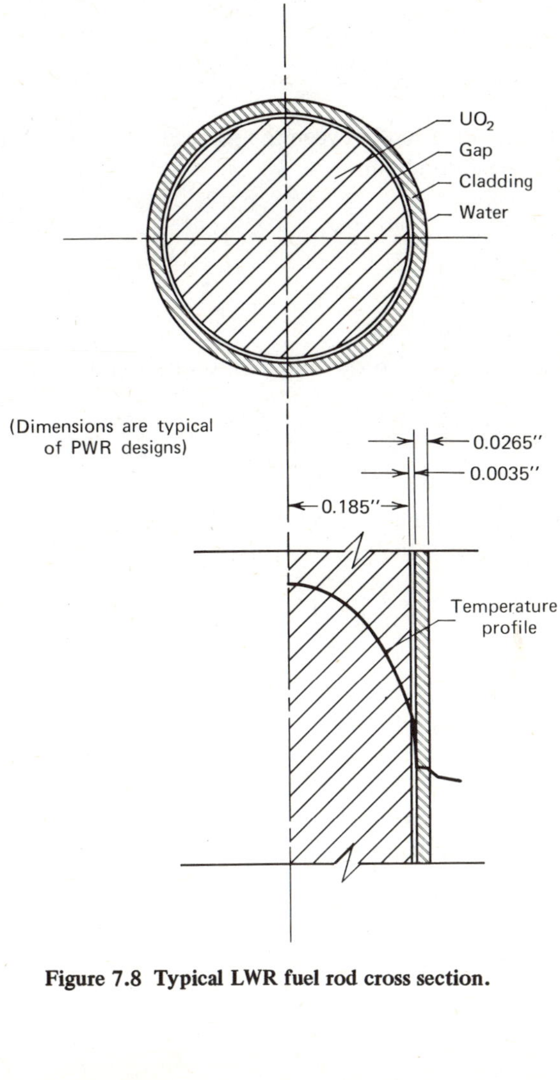

Figure 7.8 Typical LWR fuel rod cross section.

ferential equation. We assume that all the thermal energy moves radially; none is conducted in the axial direction of the fuel rod. Then at any value of the rod radius, r, for a unit length

$$\pi r^2 q''' = \text{thermal power generated inside } r$$

must be conducted radially past r, or

$$\pi r^2 q''' = -2\pi r\, k \frac{dt}{dr} \tag{7.33}$$

by the conduction equation. We can rearrange this equation

$$\frac{q'''}{2} r\, dr = -k\, dt$$

We can integrate this equation between the rod centerline ($r = 0$) and the outside of the UO_2 rod ($r = r_0$) to obtain

$$\frac{q'''}{2} \int_0^{r_0} r\, dr = -\int_{t_℄}^{t_0} k\, dt \tag{7.34}$$

The thermal conductivity of UO_2, k, is too sensitive a function of temperature to simply treat as a constant, so this equation is often integrated to the form

$$\frac{q''' r_0^2}{4} = \int_{t_0}^{t_℄} kdt = \int_0^{t_℄} kdt - \int_0^{t_0} kdt \tag{7.35}$$

Values of $\int_0^t kdt'$ for UO_2 are available. The value of $t_℄$ in this case must be calculated by trial and error. One can also write

$$\frac{q''' r_0^2}{4\bar{k}} = t_℄ - t_0 \tag{7.36}$$

where $\bar{k}$ is the effective average thermal conductivity. It should be noted that t_0 is the temperature at the surface of the UO_2, not at the surface of the cladding. There are temperature drops across the gap and through the cladding that can be calculated but that will not be calculated here.

One performance parameter often quoted in reactor descriptions is the thermal power per unit length of rod, q'/l. This can be calculated

$$\frac{q'}{l} = \pi r_0^2 q''' \tag{7.37}$$

Note that if we substitute Eq. 7.35 for q''', we obtain

$$\frac{q'}{l} = 4\pi \int_{t_0}^{t_{\text{℄}}} k\,dt \tag{7.38}$$

This value is a function only of the fuel temperature and not of the rod diameter. In other words, for given temperature limits one obtains the same thermal power per unit length, regardless of rod size. One might say, therefore, that the rod radius should be kept low because the thermal power per volume of fuel, q''', goes up inversely as the square radius. There are limits and tradeoffs, of course. As fuel rods become smaller the fabrication costs rise, more cladding is required, and the heat flux at the surface goes up. Heat flux refers to the thermal energy flow rate normal to the cladding per unit area of cladding surface. This is a very important parameter in liquid-cooled reactors, because it can reach a value so high that the liquid at the surface is all vaporized, thus insulating the rod and leading to possible melting of the rod. The heat flux, q'', is calculated

$$q'' = \frac{\pi r_0^2 q'''}{2\pi r_c} = \frac{r_0^2 q'''}{2r_c} \tag{7.39}$$

where r_c is the outside radius of the cladding. It should be noted that we are talking about conditions at a particular point in a reactor core. These values, q''', q'', and so on, will vary over the entire core, because the neutron flux varies (see Figure 7.6).

Example 7.12
The Final Safety Analysis Report (FSAR) for the Rancho Seco PWR of the Sacramento Municipal Utility District indicates that the thermal power per unit length of rod is 19.03 kW/ft. The rod dimensions are shown in Figure 7.8. Calculate the heat flux at the cladding surface and the thermal power density. Estimate the centerline temperature at this thermal power.

Solution: We can calculate the thermal power density from Eq. 7.37.

$$q''' = \frac{q'}{l\pi r_0^2}$$

$$r_0 = 0.185 \text{ in} = 0.4699 \text{ cm}$$

$$q'/l = 19.03 \text{ kW/ft} = 0.624 \text{ kW/cm}$$

$$q''' = \frac{0.624}{\pi \times 0.4699^2} = 0.90 \text{ kW/cm}^3 = 900 \text{ W/cm}^3$$

The heat flux at the clad surface can be calculated from Eq. 7.39.

$$q'' = \frac{r_0^2 q'''}{2r_c}$$

$$r_c = 0.215 \text{ in} = 0.546 \text{ cm}$$

$$q'' = \frac{0.4699^2 \times 900}{2 \times 0.546} = 18.19 \text{ W/cm}^2$$

The temperature drop in the fuel rod can be calculated from Eq. 7.36

$$t_{℄} - t_0 = \frac{q''' r_0^2}{4\bar{k}}$$

The rod outside temperature will be taken as 800°C. We know the water temperature is close to 315°C. There is a large drop across the gap (Figure 7.8), but we shall not attempt to compute it here. We shall assume that the centerline temperature is 2200°C. Between 800°C and 2200°C, the average thermal conductivity of UO_2 is 0.025 W/cm °C.

$$t_{℄} - t_0 = \frac{900 \times 0.4699^2}{4 \times 0.025} = 1987°\text{C}$$

The centerline temperature is estimated to be 800 + 1987 or 2787°C. This value does not agree with the assumed value of 2200°C, so a new value of $\bar{k}$ must be used. The value of $\bar{k}$ between 800°C and 2700°C is 0.027 W/cm °C. The adjusted value of the centerline temperature becomes

$$t_{℄} = \frac{900 \times 0.4699^2}{4 \times 0.027} + 800 = 2640°\text{C } (4784°\text{F})$$

A value of 4400°F is cited in the FSAR.

Specific Reactor Power Systems

If a power reactor designer were starting from scratch, he could in principle meet the design objectives of Table 7.1 with many different reactor types. A basic reactor type is defined by specifying a material for each of the generic components of the reactor: fissile material, fertile material, coolant material, cladding material, and moderator material. (Not all reactors need all of these components, of course.) There are several materials and material forms that can fill each of these roles. They may be assembled in different configurations. We

Table 7.7 Qualitative Characteristics of Principal Power Reactor Types

REACTOR CLASS	FISSILE MATERIAL	FERTILE MATERIAL	PRIMARY COOLANT	CLADDING MATERIAL	MODERATOR MATERIAL	PRINCIPAL NATIONAL DEVELOPER
U.S. Naval	Uranium-235	None	H_2O	Zirconium	H_2O	United States
Light water (LWR)						
PWR & BWR	Uranium-235/ Plutonium	Uranium-238	H_2O	Zirconium/ stainless steel	H_2O	United States
Gas-cooled						
Magnox	Uranium-235/ Plutonium	Uranium-238	CO_2	Mg alloy (Magnox)	Graphite	United Kingdom
Advanced gas (AGR)	Uranium-235/ Plutonium	Uranium-238	CO_2	Stainless steel	Graphite	United Kingdom
High temp. gas (HTGR)	Uranium-235/ Uranium-233	Thorium	Helium	Graphite	Graphite	United States, Germany
Heavy water						
CANDU	Uranium-235/ Plutonium	Uranium-238	D_2O	Zirconium	D_2O	Canada
SGHWR	Uranium-235/ Plutonium	Uranium-238	H_2O	Zirconium	D_2O	United Kingdom
Thermal breeders						
Light water (LWBR)	Uranium-235/ Uranium-233	Thorium	H_2O	Zirconium	H_2O	United States
Molten salt (MSBR)	Uranium-235/ Uranium-233	Thorium	Fluoride Salts	None	Graphite	United States
Fast reactors						
Sodium-cooled (LMFBR)	Plutonium	Uranium-238	Sodium	Stainless steel	None	France, Germany, United Kingdom, United States, USSR
Gas-cooled (GCFR)	Plutonium	Uranium-238	Helium	Stainless steel	None	United States

see that the number of possible reactor designs is enormous. In practice, however, the development of a single reactor type is a very expensive undertaking, so that the number of different types actually brought to commercial status is rather small. Table 7.7 summarizes the qualitative features of most important reactor types either in use or under development. More quantitative details of the designs already built are given in Table 7.8.[4]

The Pressurized Water Reactor (PWR). The pressurized water reactor (PWR) and boiling water reactor (BWR) both belong to a class known as light water reactors (LWRs). They have so many features in common that it is often convenient to lump them together. Commercial prototypes of both reactors appeared at about the same time. The PWR reactors developed by the U.S. Navy for submarines and, later, surface ships were the forerunners of this class (see Table 7.7). Then, in 1957, the Shippingport PWR, designed and built under the direction of the Naval Reactors Branch of the AEC, began power operation, producing electricity for a utility. Other early PWRs for electric utilities were the Yankee at Rowe, Massachusetts (1961) and the Indian Point at Buchanan, New York (1962).

The pressurized water reactor is a reactor moderated and cooled by light water. The volume of water required for moderating and that required for cooling are so nearly the same that all the water in the core acts in both roles. The choice of light water as a moderator has many pronounced effects on the nature of the core. We can see that a much lower mass of H_2O is required to thermalize neutrons than is true of D_2O or graphite, its two competitors. The result is that the core can be relatively low volume if the water density is high, namely, that of liquid water. Water can be in a liquid state only if its temperature is low or its pressure high. Remember, however, that the water is also serving as the transport medium for the thermal energy that results from the fission process. In order for this energy to be converted to electric energy with reasonable efficiency, the temperature must be high. The situation is further complicated by the fact that the critical temperature, above which the liquid state does not exist, is 374°C (705°F). So there are major design tradeoffs between the need for a high temperature, the need for density, and the penalties associated with high pressure. In the case of the PWR, the design choice is a water temperature around 315°C (600°F) and a pressure of 138 bar (2000 lb/in^2). The water remains in the liquid state because at this pressure the boiling point is 336°C

[4]The engineering details of specific reactors are available from a number of sources. The most comprehensive world wide compilation is published by the International Atomic Energy Agency (IAEA) in Vienna, Austria. In addition to tabulations of specifications, these volumes also contain a small bibliography on each reactor that permits the interested person to seek more detail if he desires. For U.S. reactors, the Preliminary and Final Safety Analysis Report (PSAR and FSAR) can be a good source. These documents are available at public reading rooms maintained by the AEC. They can also be obtained by school libraries.

Table 7.8 Characteristics of Different Power Reactor Systems

	UNITS	OCONEE 1 (PWR)	QUAD CITIES 1 (BWR)	BERKELEY (MAGNOX)	DUNGENESS B (AGR)	FORT ST. VRAIN (HTGR)	PICKERING (CANDU)	PHENIX (LMFBR)
Energy Conversion								
Thermal power	MWt	2,584	2,511	556	1,458	842	1,742	563
Net electric power	MWe	886	809	138	606	330	508	233
Net efficiency	%	34.3	32.2	24.7	41.6	39.2	29.1	41.4
Core Dimensions								
Diameter (cyl)	m	3.3	4.6	13.4	9.6	5.9	7.5	1.4
Height	m	3.7	3.7	7.3	8.3	4.8	6.0	0.85
Volume	m^3	30.7	61.6	1,032	598	132	265	1.23
Fuel								
Chemical form	–	UO_2	UO_2	U-metal	UO_2	UC_2/ThC_2	UO_2	UO_2/PuO_2
Weight	kg-M[a]	83,000	123,000	231,500	152,000	20,440	92,600	4,160
Isotopic comp.	–	2.1% ^{235}U	2% ^{235}U	Nat. U	2.2% ^{235}U	4.3% ^{235}U	Nat. U	20% ^{239}Pu
Primary Coolant System								
Chemical form	–	$H_2O(\ell)$	$H_2O(\ell, \nu)$	CO_2	CO_2	He	$D_2O(\ell)$	$Na(\ell)$
Pressure (core in.)	kg/cm^2	156	71.4	9.8	34.3	49.2	96.6	7.6
Temperature								
Reactor inlet	°C	290	167	160	320	400	249	400
Reactor outlet	°C	318	278	345	675	770	293	560
Fuel, maximum	°C	2,343	2,416	596	1,631	1,260	2,000	N.A.
Core flow rate	kg/s	15,600	12,350	2,870	3,380	400	17,170	2,760
Thermal Performance								
Power density	kWt/kg-M[a]	31.1	20.4	2.4	10.0	41.2	18.8	135
	kWt/ℓ-core	84.2	40.8	0.54	2.4	6.3	6.6	460
Heat flux								
average	W/cm^2	54.1	41.6	N.A.	17.2	14.2	59.5	143
maximum	W/cm^2	168.6	124.9	N.A.	39.5	37.8	111.0	208
Steam to Turbine								
Pressure	kg/cm^2	64	69	22.6	172	169	N.A.	166
Temperature	°C	299	284	322	571	539	N.A.	510
Superheat	°C	20	0	105	220	190	N.A.	160

[a]The letter M (for metal) here designates the heavy metal element constituents of fuel as charged; for example, oxygen in oxide is excluded.

Figure 7.9 Typical UO_2 pellets used to fuel LWRs. Courtesy of General Electric Co.

(636°F). Under these conditions a thermal efficiency (percentage of thermal energy converted to electric) of about 33% is achievable. Compared with modern fossil power plants and some other power reactor designs, this value is low and represents an inherent weakness of the LWR class of reactors. There are some obvious compensations, one being the relatively small size or high power density.

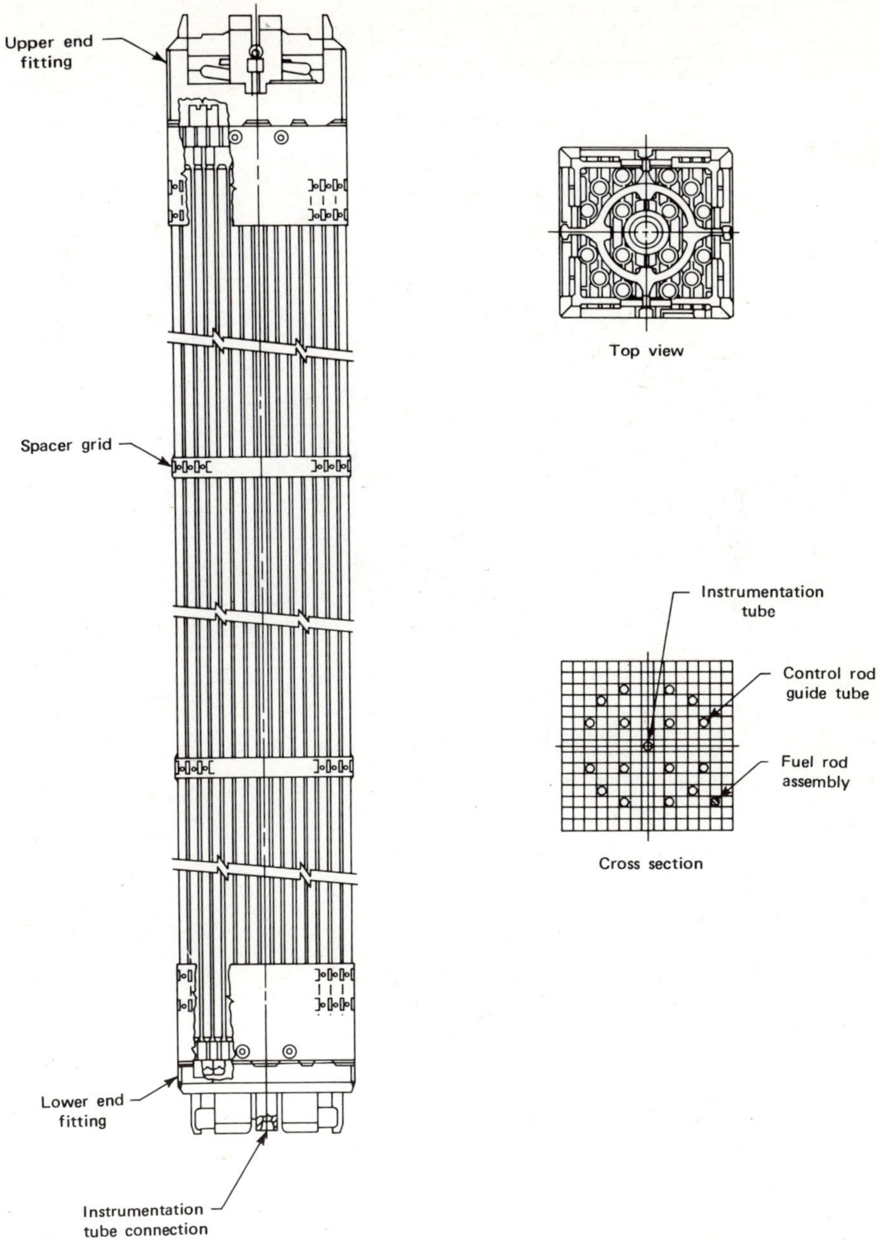

Figure 7.10 Fuel assembly of a PWR. Courtesy of Westinghouse Electric Corporation.

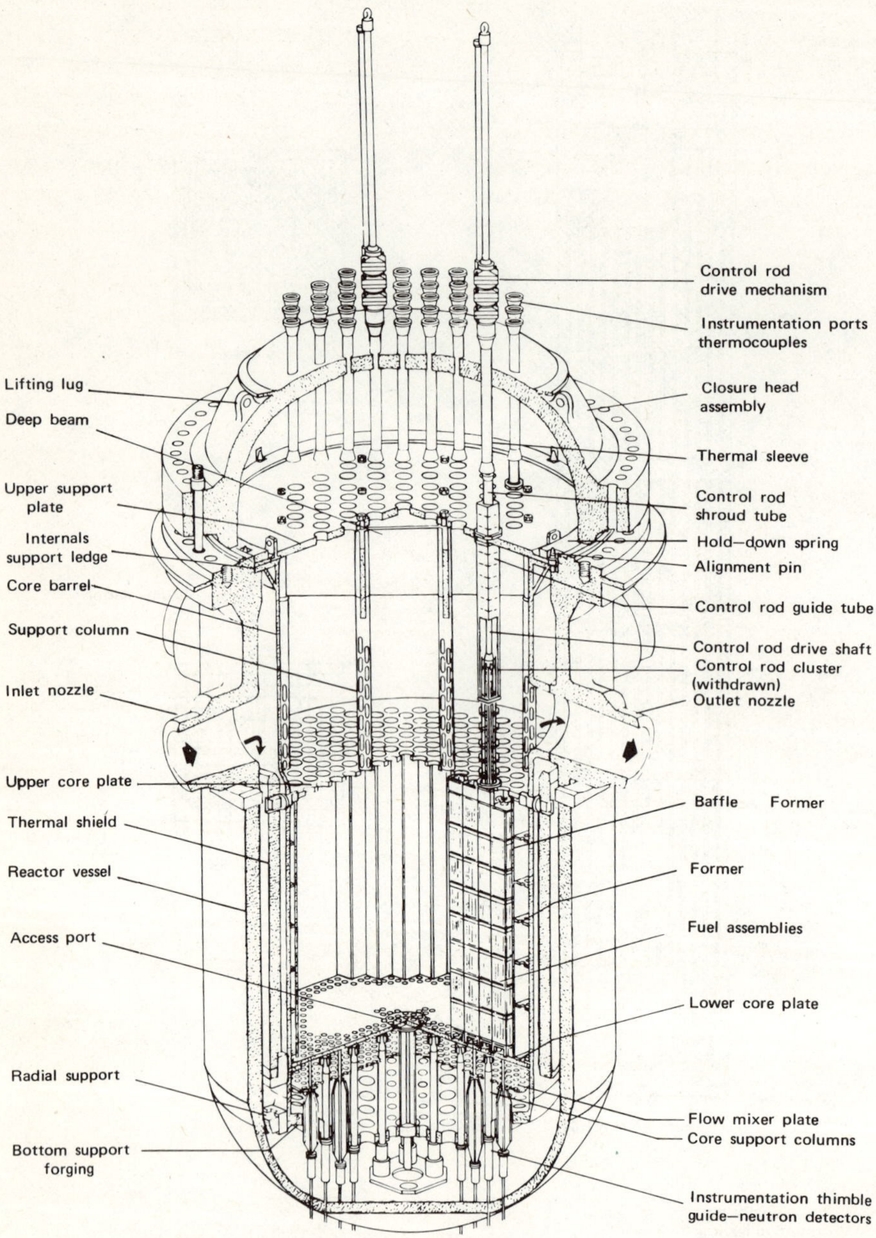

Figure 7.11 Cross section of a typical PWR pressure vessel. Courtesy of Westinghouse Electric Corporation.

Another characteristic of the LWR class reactor is that it requires enriched uranium as fuel. The cross section of hydrogen(H) for neutron capture is relatively large, so that criticality can not be achieved using natural uranium as a fuel. A typical enrichment is 2.5 w/o uranium-235. This characteristic can be an important consideration in a small country's selection of a reactor type, because selection of an LWR usually will imply the need to import enriched uranium; at the present time an enrichment plant represents a large capital expenditure. Reactors moderated with either heavy water or graphite can be fueled with natural uranium.

The basic fuel of PWRs is UO_2 compacted (density about 10 g/cm^3) in the form of pellets about one cm in diameter and 1.5 cm long (Figure 7.9). These pellets are stacked inside a sealed cladding tube of zirconium alloy to make a fuel

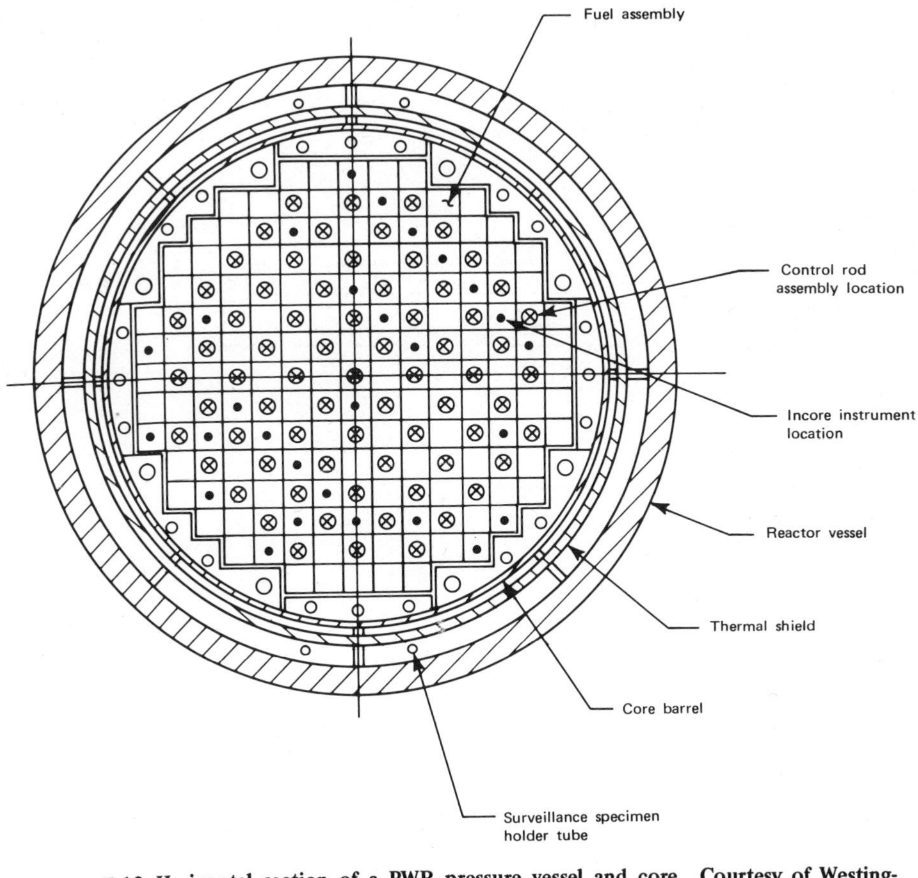

Figure 7.12 Horizontal section of a PWR pressure vessel and core. Courtesy of Westinghouse Electric Corporation.

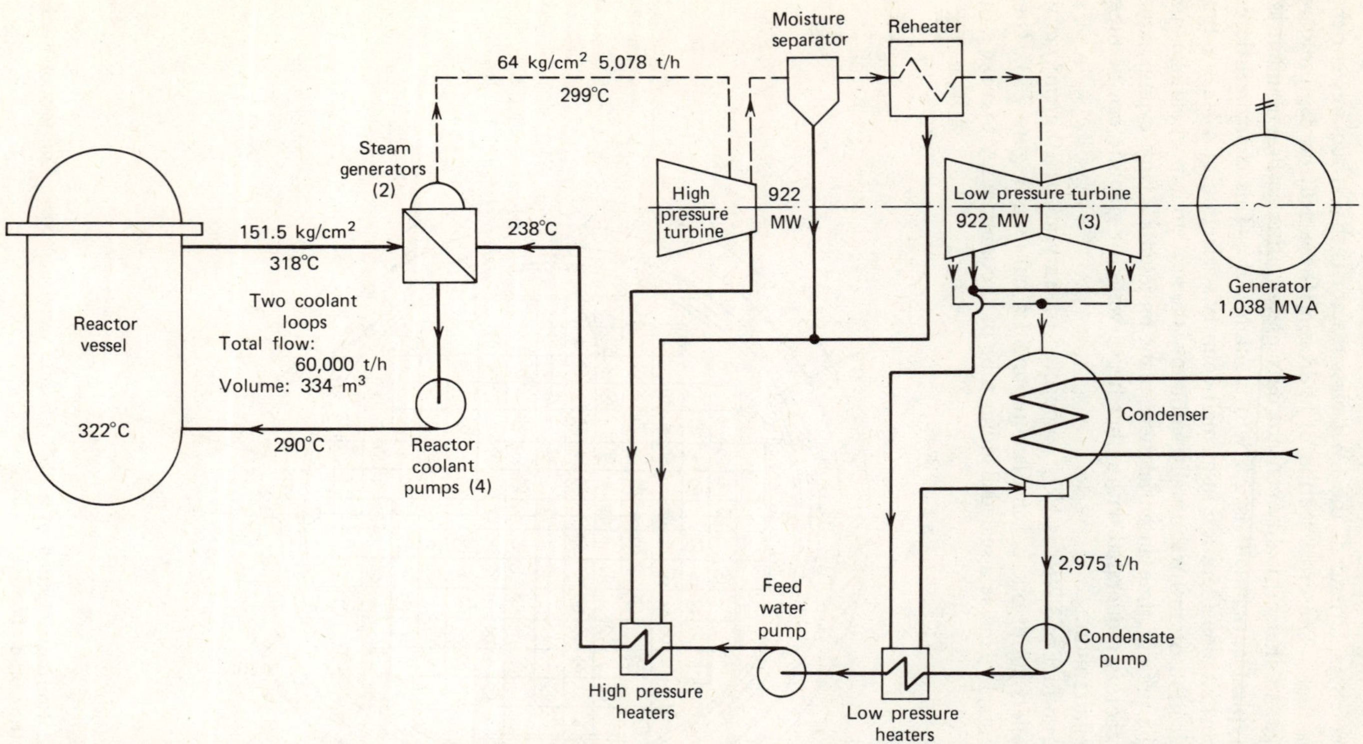

Figure 7.13 Flow diagram of a PWR power cycle. Adapted from "Directory of Nuclear Reactors," Vol. IX, International Atomic Energy Agency, 1971.

rod. These rods in turn are assembled in a square matrix, typically 15 × 15 or 17 × 17 with suitable end fittings and spacers to form a fuel assembly (Figure 7.10). The rods are spaced about three mm apart at their closest approach, so there is sufficient volume for water to cool and moderate. The fuel assembly is the basic fuel unit that is shipped from a fuel fabricator to a reactor-operating utility. At the reactor these fuel assemblies are arranged side-by-side within a pressure vessel (Figure 7.11) on a grid designed to accept and lock the individual end fittings. The fuel assemblies are arranged in a more or less circular plan (Figure 7.12) within the reactor vessel. When the open space is filled with water, these fuel assemblies constitute the reactor core. It will be noted in Figure 7.12 that the assemblies are stacked snugly together; in fact the outer row of rods of one assembly is almost as close to the outer row of the adjacent assembly as to the next row of its own. This close-packed arrangement is the principal characteristic that separates the light water from the heavy water reactors.

Figures 7.10, 7.11, and 7.12 also indicate that control rods are a part of the core. These rods are made of strong neutron absorbing materials such as cadmium, indium, silver, clad in stainless steel to form rods of the same size as fuel rods. They can be moved up and down within guide tubes located in some assemblies of the core matrix. PWRs also use boron dissolved in the coolant/moderator as a control expedient. The boron is added or withdrawn from the water in a separate chemical control system. This kind of control has the advantage of being mechanically simpler and of causing less flux and power distortions than do solid control elements in the core (see section on heterogeneities).

A sketch of a PWR power cycle is shown in Figure 7.13, with typical values of pressure, temperatures, and flow rates. It can be seen to resemble closely the generic system of Figure 7.7.

Example 7.13

Compute the thermal efficiency of the PWR system of Figure 7.13, based on the 922 MW output of the turbine and the reactor flow conditions.

Solution: The diagram shows that 60,000 t/h (metric tons per hour) of water enter the reactor at 290°C and leaves at 318°C. The pressure out is 151.5 kg/cm^2. There is some pressure change within the core, but not enough to affect our calculation. The heat capacity of water at these temperatures is approximately 1.35 cal/g · °C, or 5.63 kJ/kg · °C. The rate of thermal energy transfer to the primary coolant in the reactor is

$$
\begin{aligned}
q' &= \dot{m}\, c_p (t_{\text{out}} - t_{\text{in}}) \\
&= 60{,}000{,}000 \times 5.63 \times (318 - 290) \\
&= 9.46 \times 10^9 \text{ kJ/h} \\
&= 2.63 \times 10^6 \text{ kW} = 2630 \text{ MW } (t)
\end{aligned}
$$

We assume this thermal energy output is equal to the fission energy converted, although some fraction (gamma ray energy, for example) may be deposited outside the core and escape the power cycle. Based on 922 MW(*e*) output, the thermal efficiency is

$$\eta = \frac{922}{2630} \times 100 = 35.1\%$$

It is not clear from the diagram whether the 922 MW is mechanical or electrical energy, so whether it includes bearing friction and windage losses in the turbine is not clear. In any event, the efficiency is a gross efficiency because the figure includes some electricity supplied inside the system to drive pumps, and so on.

Example 7.14

In Figure 7.13 the transfer of thermal energy between primary and secondary coolants takes place in the steam generators. From the data in Figure 7.13 compute the heat transfer as measured from the primary coolant side and the secondary coolant side. They should be almost equal.

Solution: The measure of heat transfer in a steady-flow system like this is the change in enthalpy of the fluid as it passes through the heat exchanger–steam generator in this case. On the primary side, the enthalpy change ($\dot{m}\, c_p\, \Delta t$) in the reactor was calculated in Example 7.13. According to Figure 7.13, the enthalpy change is the same (with opposite sign) in the steam generators. (Obviously, any losses through pipe walls and so on are considered negligible). From Example 7.13, therefore, the heat transfer on the primary side is

$$q' = 9.46 \times 10^9 \text{ kJ/h}$$

On the secondary side the water is converted to steam. We cannot use $c_p \Delta t$ for enthalpy change, therefore, but must obtain enthalpy values directly from the Steam Tables.[5]

$$q' = \dot{m}(h_{\text{out}} - h_{\text{in}})$$

$$\dot{m} = 5.078 \times 10^6 \text{ kg/h}$$

A check of the Steam Tables shows that the water vapor emerging from the steam generators at 64 kg/cm^2 (910 lb/in^2) and 299°C (570°F) is above the saturation temperature for that pressure, that is, it is superheated. The saturation temperature (boiling point) for water at 64 kg/cm^2 is about 279°C, so the steam is 20°C superheated. The specific enthalpy is

$$h_{\text{out}} = 2868 \text{ kJ/kg}$$

[5] See Reference 3 at end of chapter.

Liquid water comes into the steam generator at 238°C and, presumably, about 65 kg/cm^2. The enthalpy is

$$h_{in} = 1028 \text{ kJ/kg}$$

On the secondary side, therefore, the heat transfer is

$$q' = 5.078 \times 10^6 (2868 - 1028) = 9.34 \times 10^9 \text{ kJ/h}$$

We see that the values of q' computed from the primary and secondary sides are in good agreement. We might interpret the discrepancy as heat losses (that is, thermal energy that escapes to the environment) on the primary side, amounting to

$$\frac{9.46 \times 10^9 - 9.34 \times 10^9}{9.46 \times 10^9} \times 100 = 1.3\%$$

This value should not be taken seriously as representative of practice; it looks high, and it is probably an artifact of rounding errors.

Light water reactor cores are loaded in batches; typically one third of a core is replaced at a time. Following the loading operation, the head of the pressure vessel (Figure 7.11) is bolted in place. A reactor typically might operate for a year on a given loading. During this period the consumption of the fissile material is compensated for in part by conversion of the fertile material and in part by withdrawing boron or another strong neutron absorber from the core. When it is necessary to replace some of the fuel, the reactor is shut down, the vessel head removed, and the refueling operation begun.

The Boiling Water Reactor (BWR). The BWR was developed commercially in the same time frame as the PWR. The first commercial prototype was the Dresden 1 Reactor, which achieved full power in 1960. The General Electric Co. is the sole manufacturer of BWRs in the United States. The BWR differs from the PWR primarily in that the water coolant/moderator is allowed to boil in the core. There are, of course, many differences in engineering features. The BWR power system, Figure 7.14, differs from the generic system, Figure 7.7, in that there is no secondary coolant or steam generator, since the steam is actually generated in the reactor core. The temperature and pressure conditions of steam to the turbine are roughly the same in both the BWR and PWR systems, but these conditions are achieved at lower reactor vessel pressure in a BWR (about 70 bar versus 140 bar), because the primary coolant is used directly in the turbine.

A typical BWR fuel rod and fuel assembly are shown in Figure 7.15. They are similar to fuel forms of the PWR, with some important exceptions. In the

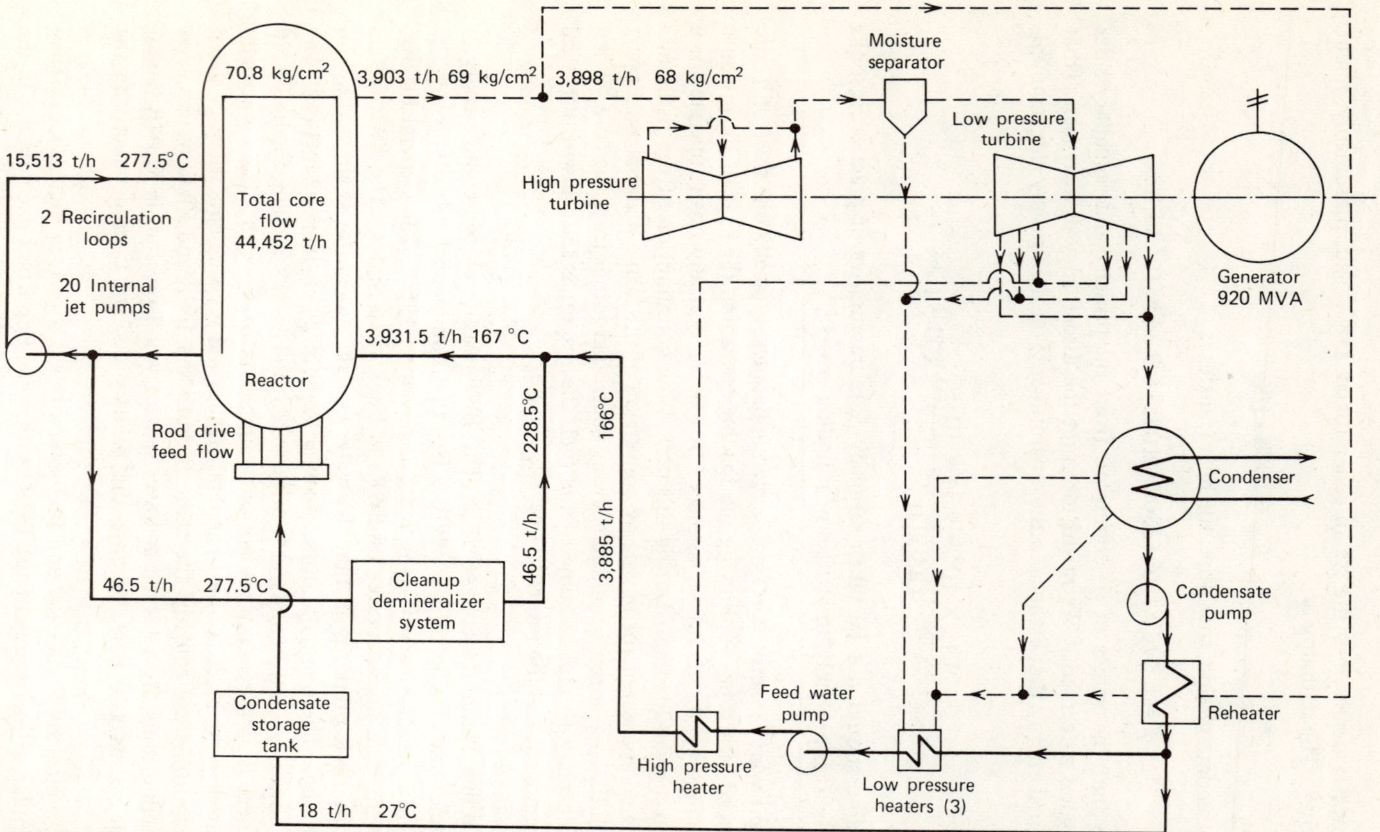

Figure 7.14 Flow diagram of Quad Cities BWR power system. Adapted from "Directory of Nuclear Reactors," Vol. IX, International Atomic Energy Agency, 1971.

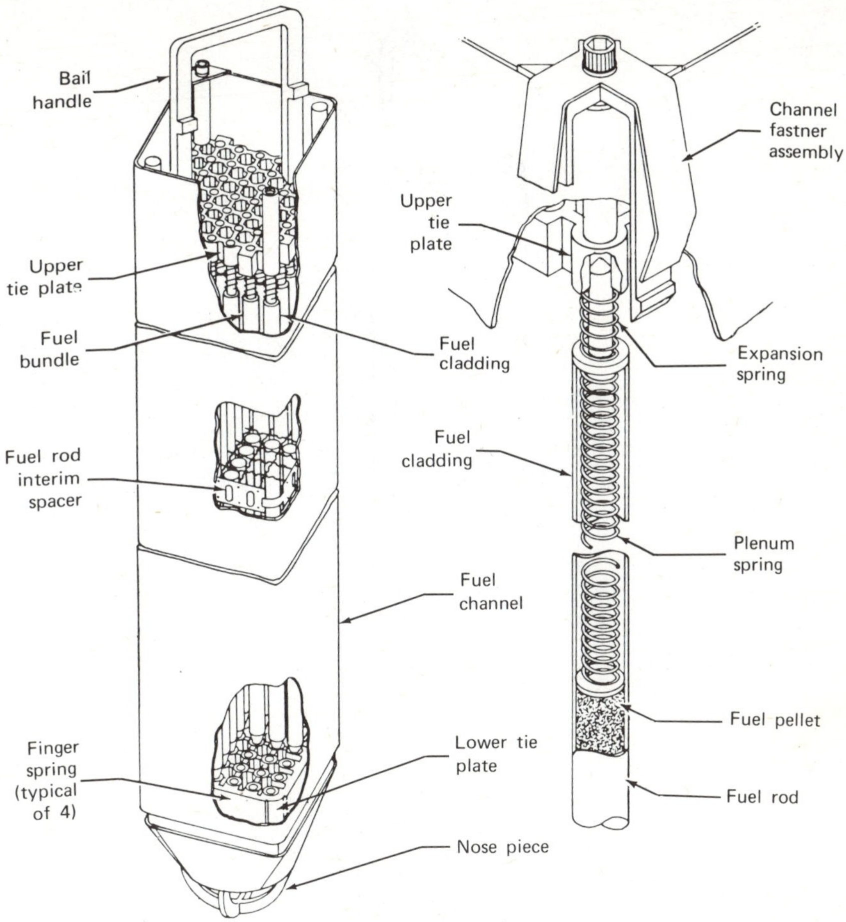

Figure 7.15 Typical BWR fuel rod and fuel assembly. Courtesy of General Electric Co.

BWR the whole assembly is enclosed in the sides so that the coolant cannot move horizontally in the core. In the BWR, the control elements (Figure 7.16) are cruciform blades that run between sets of four fuel assemblies. Figure 7.17 shows the BWR pressure vessel and the essential features of the reactor. In the PWR, the control elements are rods that run in positions within the fuel-rod matrix of an assembly. The PWR relies on boron dissolved in the water for a major part of its control. Dissolving boron in the water of a BWR is not feasible, since the water boils. In a BWR, the control elements are driven from the bottom, exactly the opposite from the PWR. The space above the core is needed for the steam separation.

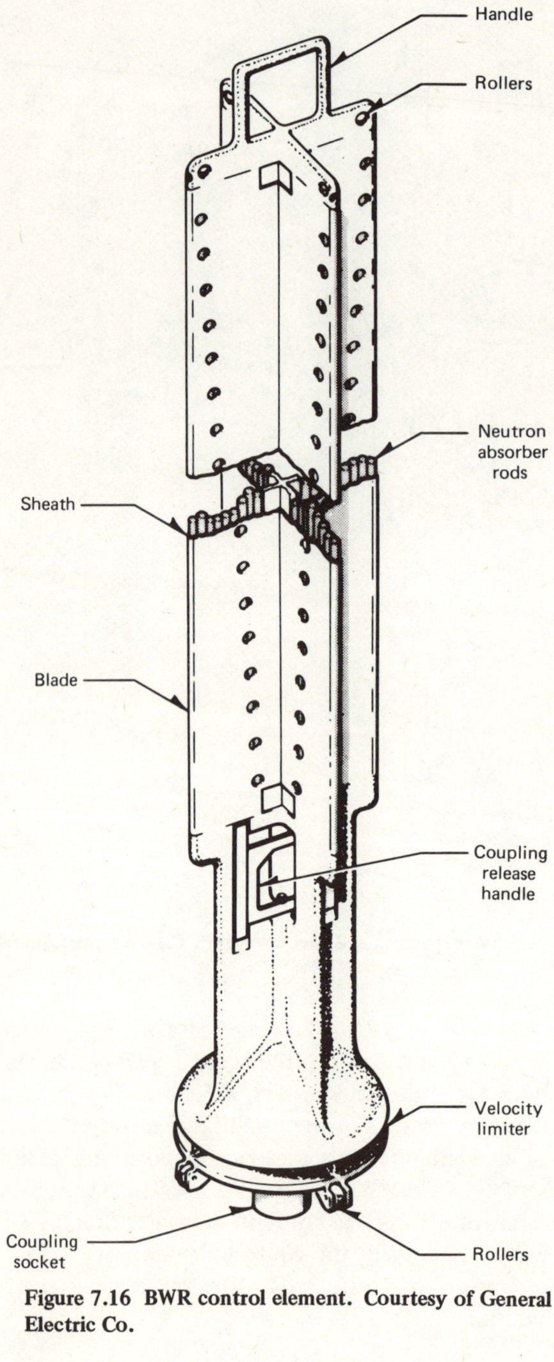

Figure 7.16 BWR control element. Courtesy of General Electric Co.

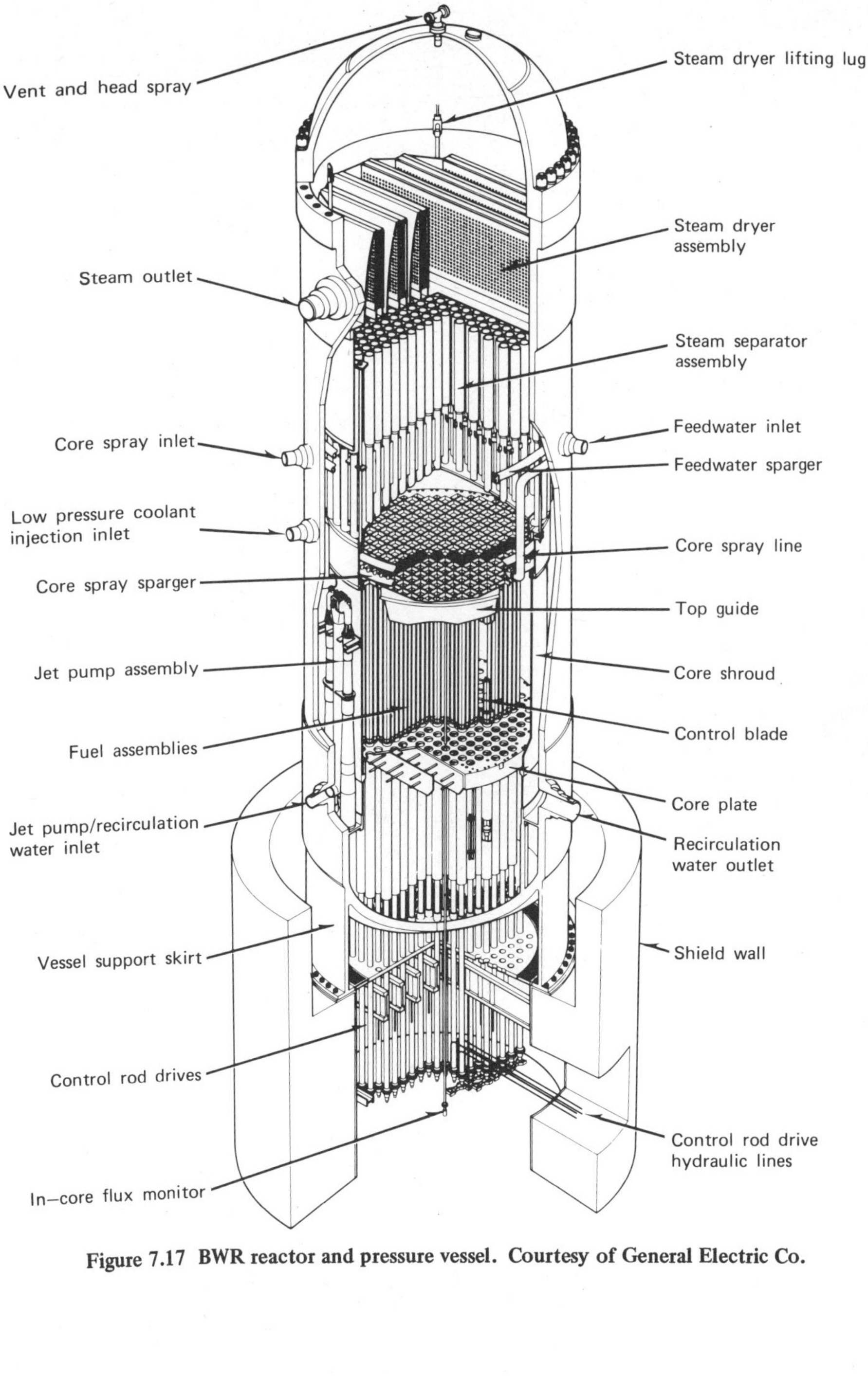

Figure 7.17 BWR reactor and pressure vessel. Courtesy of General Electric Co.

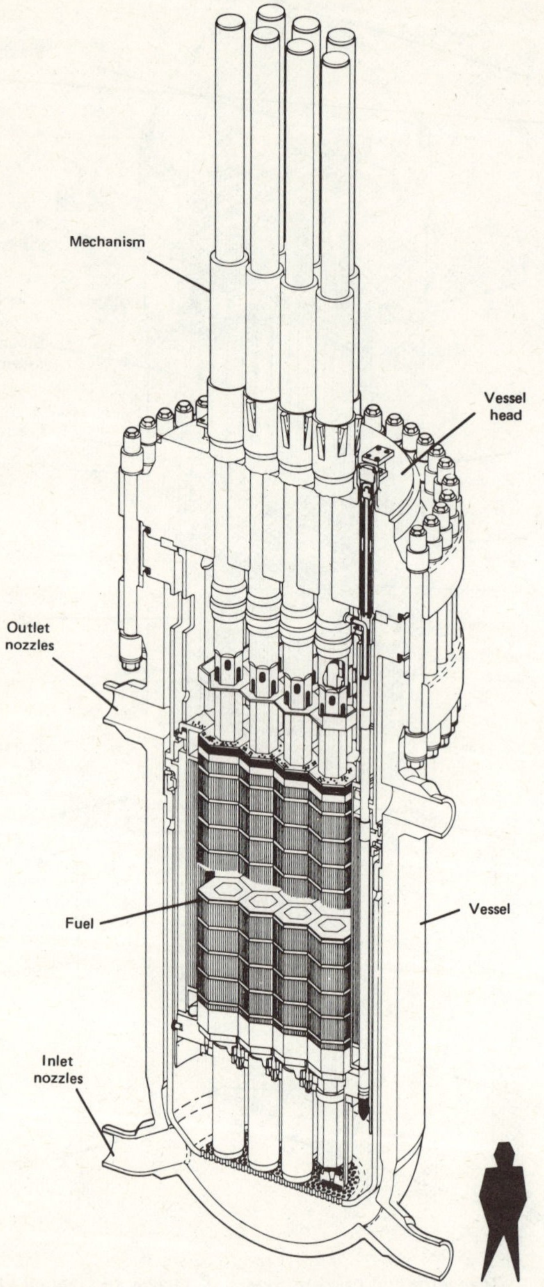

Figure 7.18 A developmental LWBR core in the Shippingport pressure vessel. From "Final Environmental Statement, Light Water Breeder Reactor Program," ERDA-1541, June 1976.

Because the water vapor in the core is low in density relative to liquid, the BWR requires a higher water-to-fuel volume ratio. Not all of the water rising up through the core is vaporized. Rather, it emerges from the core as a liquid-vapor mixture, about 15 w/o vapor. The liquid is separated in the steam separator and recirculated to the bottom of the core by jet pumps inside the reactor vessel. Essentially saturated vapor leaves the top of the reactor vessel.

The Light-Water Breeder Reactor (LWBR). The LWBR is being developed under the direction of the Division of Naval Reactors of ERDA. Its title may be presumptuous, inasmuch as achievement of breeding has yet to be demonstrated. Basically, the LWBR is a PWR designed to operate on the thorium/uranium-233 fertile/fissile system. The pressure vessel and power system of the Shippingport reactor, the first commercial PWR, will be used to demonstrate LWBR characteristics (Figure 7.18). The neutron cross sections of thorium and uranium-233 are such that, contrary to the uranium-238/plutonium fertile/fissile system, breeding can take place in a thermal neutron spectrum. As much uranium-233 is formed via the (n, γ) reaction in thorium as is consumed in fission and capture reactions in uranium-233. The balance is a very close one, however, so that it will take a long time to determine whether this reactor does in fact breed. Nevertheless, this effort may prove very important to the development of nuclear energy in the United States. It promises greater uranium resource utilization than the conventional LWRs, without requiring the same degree of technological development represented by the fast breeder reactor. At a time when the LMFBR development in the United States is going rather badly, intermediate approaches to fuel economy take on greater importance.

As we have said, the LWBR is basically a PWR, so its power cycle would be similar to that of Figure 7.13. The fuel arrangement in the LWBR, however, is very different from the conventional PWR. These differences have as their common purpose the maximization of conversion of ^{232}Th to ^{233}U. Reactor control is effected by movement of fuel assemblies rather than neutron poison materials. The result is that neutrons not needed to sustain the chain reaction are absorbed usefully in thorium rather than in materials such as boron. It will be noted that Figure 7.18 shows no control elements in the core. These substantial changes in engineering design mean that it would be expensive to convert "conventional" LWRs (Figures 7.11 and 7.17) to LWBRs.

Gas-Cooled Power Reactors. Although two of the first reactors built in the world were cooled by air, gas-cooled power reactor designs were not adopted or developed in the early part of the U.S. program. A case can probably be made that gas-cooled designs were not suitable for use as naval reactors and, consequently, did not receive the developmental boost from the naval development program that LWRs and even liquid-metal-cooled reactors did.

The British, on the other hand, did develop gas-cooled designs with graphite as the moderator. The earliest of these was the "Magnox" class (Table 7.7), named after the magnesium alloy used in the cladding. Calder Hall was the first of this class built. It started power operation in 1956. The Berkeley reactor of this class is described in Table 7.8. The gas coolant in these reactors is carbon dioxide. The thermal energy in the CO_2 is transferred to water in the boiler or steam generator. The CO_2 temperature out of the reactor is limited to about 350°C by the cladding properties. As a result, the thermal efficiency is only 25%. Unlike most power reactors, the fuel is metallic uranium. This form of fuel has a disadvantage: it cannot stand the rigors of the reactor environment as long as uranium dioxide can. UO_2 can stay in a reactor 5 to 10 times as long as the metal. These reactors are fueled with natural uranium, so no isotopic enrichment is required.

In a graphite-moderated reactor, the moderator-to-fissile atom ratio is so high that the fuel elements are widely spaced in the moderator. For this and other reasons, it is possible to refuel the reactor (withdraw spent fuel and insert new fuel) while the reactor is at power. This operation is not feasible in the LWRs.

The "Magnox" design was succeeded in Britain by the Advanced Gas Reactor (AGR) design (see Tables 7.7 and 7.8). The cladding was changed to stainless steel, which permitted a higher CO_2 outlet temperature, resulting in a 42% thermal efficiency. The fuel form was changed to UO_2. Since the constituents of stainless steel (iron, chromium, nickel) have much higher neutron capture cross sections than magnox, the uranium has to be enriched in order to achieve criticality.

The U.S. effort in gas-cooled power reactor designs has, with the exception of one abortive effort (the Experimental Gas-Cooled Reactor), been concentrated on the High Temperature Gas Reactor (HTGR). This reactor, interestingly enough, has been developed largely in the private sector by the General Atomic Corp. and its various predecessors. A first version of the HTGR (Peach Bottom 1) began generating power in 1967. A second and much more prototypical HTGR has been built at Fort St. Vrain, Colorado and is scheduled to be on the line in 1976. Its characteristics are given in Table 7.8.

The HTGR is graphite-moderated and helium-cooled (Table 7.7). It is designed to use thorium as fertile material and, therefore, ^{233}U as the fissile material. However, it will have to be fueled initially with ^{235}U and supplied regularly with some make-up ^{235}U (or plutonium), since it will not convert sufficient thorium to ^{233}U to maintain criticality. That is, it is not a breeder. The reason for the choice of the thorium- ^{233}U fertile-fissile system instead of the uranium-Pu system is that the moderator and, therefore, the thermal neutrons are at a very high temperature compared to the water-moderated reactors. At this temperature, plutonium cross sections are highly unfavorable (capture-to-fission ratio is too high) for conversion and, thus, for maintaining criticality over periods of economical length.

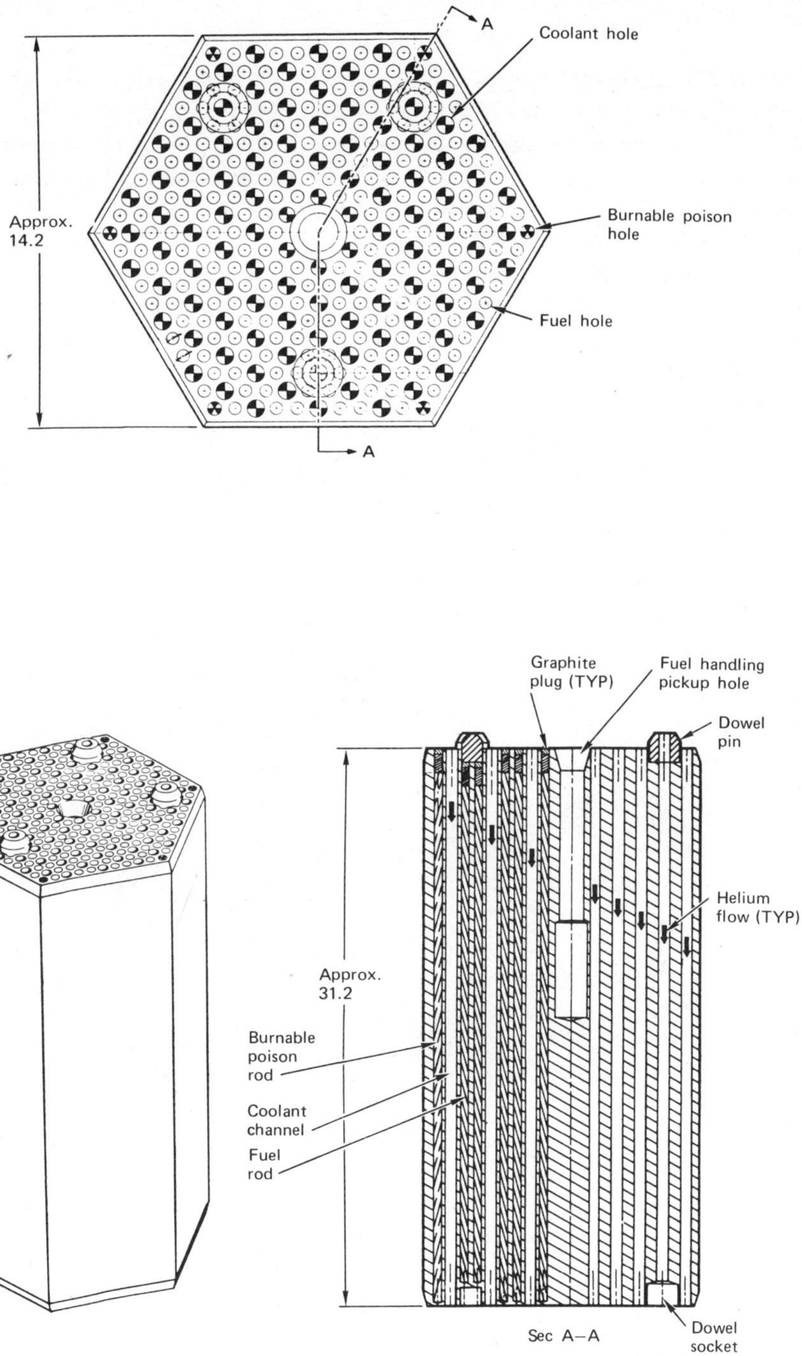

Figure 7.19 HTGR moderator/fuel assembly. Courtesy of General Atomic Company.

The HTGR design involves many innovations that are worth some attention. The fuel assembly is shown in Figure 7.19. A core plan is shown in Figure 7.20 and a vertical section of the reactor core and the reactor vessel is shown in Figure 7.21. The fuel assembly (Figure 7.19) is actually a fuel/moderator assembly, since the entire core is made up of these fuel-moderator modules. They are hexagonal blocks of graphite with some holes drilled to accept fuel and other holes to allow passage of the helium coolant. The HTGR core (Figure 7.21) is formed by stacking these hexagonal assemblies (some 1500 of them in the case of Fort St. Vrain) to the design core dimensions. There are also pure graphite hexagonal blocks from which the reflector is assembled.

The basic fuel forms are small beads or spheres of fissile and fertile material in carbide form (UC_2 and ThC_2). The fissile beads are 0.38 and 0.48 mm in diameter, the fertile beads 0.6 and 0.7 mm. Each carbide fuel sphere is coated or clad with layers of carbon and silicon carbide, which serve to retain fission products. These coated fuel beads are closely packed in 1-cm-diameter holes

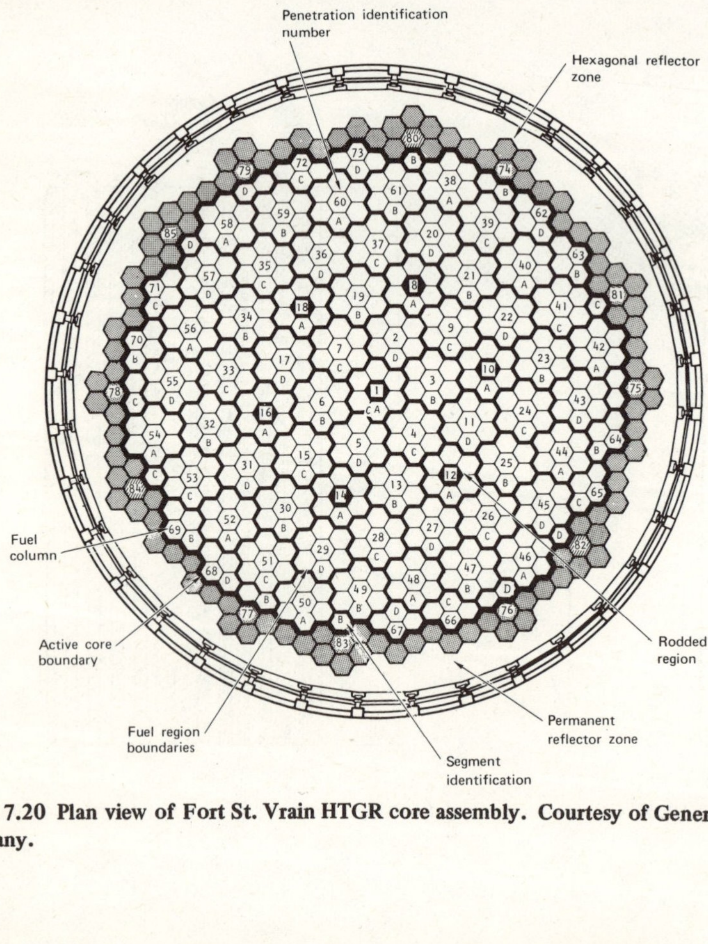

Figure 7.20 Plan view of Fort St. Vrain HTGR core assembly. Courtesy of General Atomic Company.

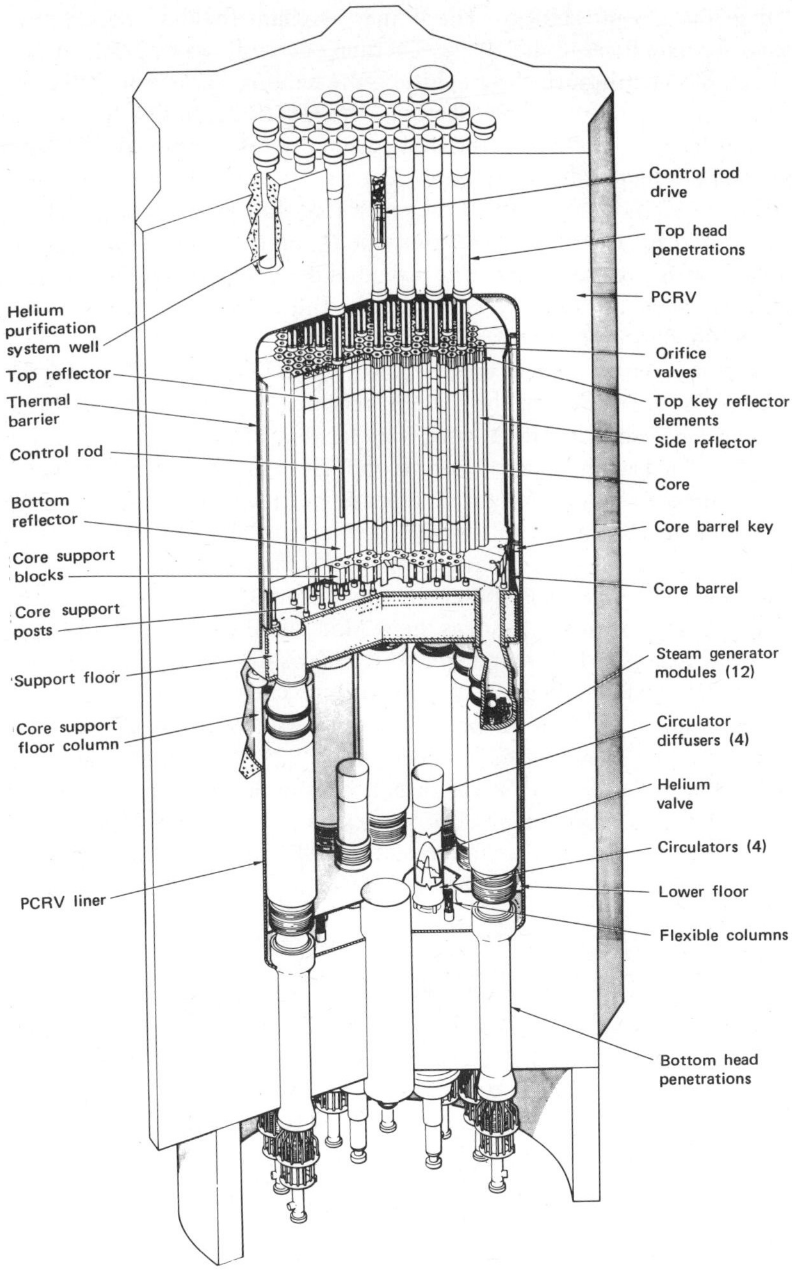

Figure 7.21 Elevation section of Fort St. Vrain HTGR reactor and pressure vessel. Courtesy of General Atomic Company.

drilled in the graphite block. The primary coolant (helium) passes downward through separate holes in the block. Thermal energy flows by conduction out of the fuel beads and through the graphite to the helium.

Unlike the British gas-cooled designs, the HTGR cannot be fueled at power. When the reactor is refueled, the desired number of hexagonal fuel-moderator assemblied are removed as a unit and sent to reprocessing. Another interesting feature of the HTGR design is that the reactor shielding also serves as the pressure vessel (or perhaps it is the other way around). Since concrete has little strength in tension, it is necessary that it be placed under compression by prestressing it. Thus, even when the primary coolant is at the design pressure of 50 bar, the concrete is under compression.

The flow diagram for the HTGR power cycle is shown in Figure 7.22. The helium flows down through the reactor, reaching a temperature of 770°C, and then into the steam generators, which are located within the reactor vessel (Figure 7.21). The gas temperature is high enough to make worthwhile such power cycle features as superheating and reheating, which are standard in modern fossil power plants. Steam to the high pressure turbine is 538°C and 166 bar—conditions close to those of modern fossil power plants.

Although the concept is still at the preliminary design stage, it is possible to use the primary coolant, helium, as the working fluid in a gas turbine cycle, thus eliminating several major equipment items such as the steam generator. In order to achieve reasonable efficiency in a gas turbine cycle, it would be necessary to raise the helium temperature out of the reactor by about 200°C. The gas turbines could be located inside the pressure vessel just as the steam generators are in Figure 7.21. An HTGR gas turbine system could either be designed as a single-fluid cycle, with a thermal efficiency in the range of 35 to 40%, or be operated as a two-fluid cycle in conjunction with a vapor cycle (water or other fluid), achieving a thermal efficiency over 50%.

Example 7.15

The flow diagram of Figure 7.22 shows 1542 t/h helium entering the core at 400°C and leaving at 770°C. Calculate the rate of thermal energy transfer to the helium and compare with the thermal power rating (Table 7.8) of 842 MW(t).

Solution: Few gases are better represented by the perfect gas model than helium. Consequently, we can ignore pressure effects on the heat capacity. The heat transfer in this steady-flow process is equal to

$$q' = \dot{m}c_p(t_{\text{out}} - t_{\text{in}})$$

$$\dot{m} = 1.542 \times 10^6 \text{ kg/h}$$

$$c_p = 1.25 \text{ cal/g °C} = 5.23 \text{ k J/kg °C}$$

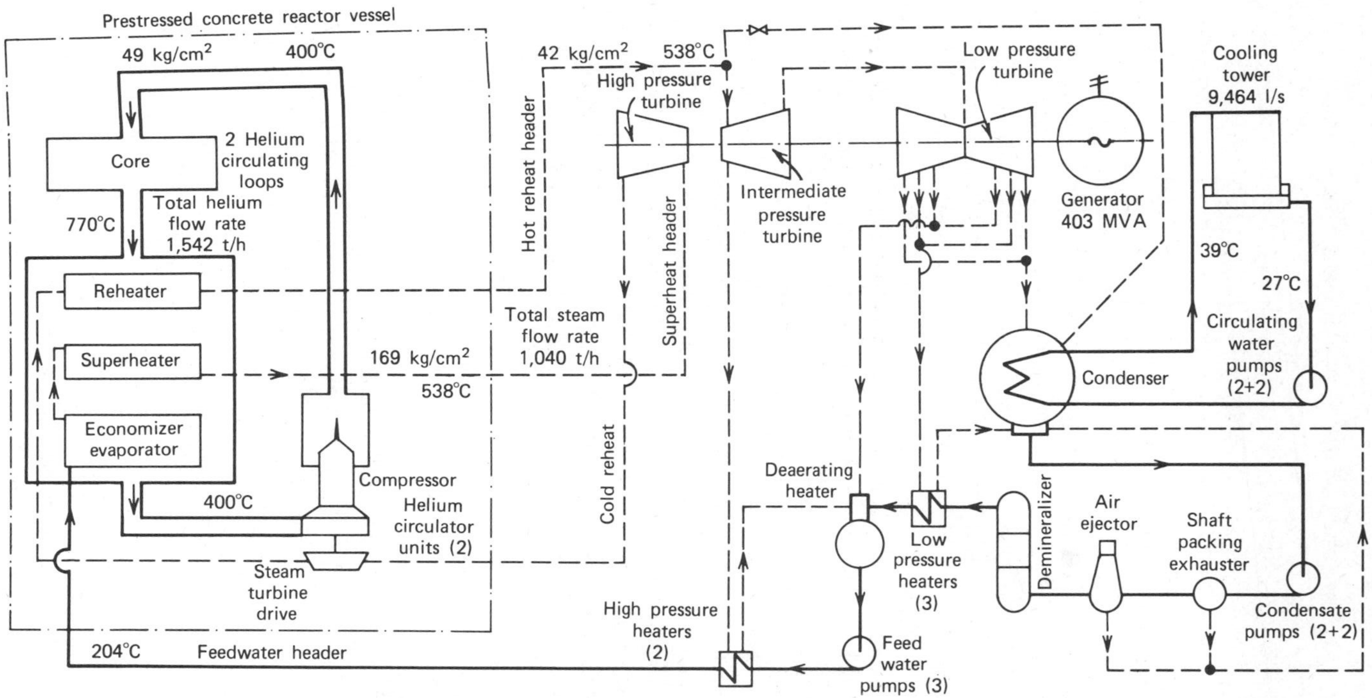

Figure 7.22 Flow diagram of Fort St. Vrain HTGR power system. Adapted from "Directory of Nuclear Reactors," Vol. IX, International Atomic Energy Agency, 1971.

$$q' = 1.542 \times 10^6 \times 5.23\,(770 - 400)$$
$$= 2.984 \times 10^9 \text{ k J/h}$$
$$= 2.984 \times 10^9 \times \frac{10^{-3}}{3600} = 829 \text{ MW}(t)$$

This value is in good agreement with the quoted value of 842 MW(t).

Heavy Water Reactors (HWR). Deuterium, the hydrogen isotope ^2H, has two characteristics that make D_2O reactors very different from an LWR. First, it has almost no tendency to capture neutrons, a fact that makes it possible to fuel a D_2O-moderated reactor with natural uranium. Since D_2O does not compete for neutrons as does H_2O, a higher fraction of neutrons not required in the chain reaction are captured in ^{238}U, resulting in higher conversion to plutonium. Second, deuterium is not as good a moderator as hydrogen. As a result it takes a much larger volume of D_2O to moderate neutrons. Consequently, a heavy water reactor can be fueled with natural uranium and will have a 5 to 10 times larger core than an LWR for a given thermal power output. Because the volume of moderator is so large in an HWR, it is not practicable to use it all as coolant or to keep all the D_2O under the pressure that would be required to maintain it as a liquid at desirable coolant temperatures. Consequently, the choice of coolant in a D_2O reactor can be made independently of the moderator. The most obvious choice is to use a small fraction of the D_2O moderator, but many other coolants have also been investigated, including gases, liquid sodium, and light water.

The development of HWRs to commercial power reactor status has been largely due to the Canadian program, although some work with HWRs has taken place in a number of other countries. The first commercial prototype, the Nuclear Power Demonstration Reactor, was started in 1962. Its electrical output was only 19 MW. The first full-scale CANDU reactor, 508 MW(e), was started at the Pickering Station in 1971. The characteristics of a Pickering CANDU unit are given in Table 7.8. A so-called fuel bundle is shown in Figure 7.23, a view of the reactor in Figure 7.24, and the power system flow diagram in Figure 7.25. D_2O is used as the primary coolant. This D_2O is confined to pressure tubes (calandria tubes in Figure 7.24) that contain the fuel rod bundles (Figure 7.23). The bundles are made up of clad UO_2 rods not unlike those of the BWRs and PWRs. The CANDU fuel is natural uranium, however. A single pressure tube contains 12 fuel bundles, each about one half meter in length; thus the active fuel length of the core is about six meters. The fuel is divided into bundles in order to facilitate on-power refueling.

The bulk of the D_2O in the core serves only as moderator—not coolant— and is contained at low pressure outside of the pressure tubes in the calandria vessel. The shape of the core is a cylinder lying on its side. Making the fuel channels

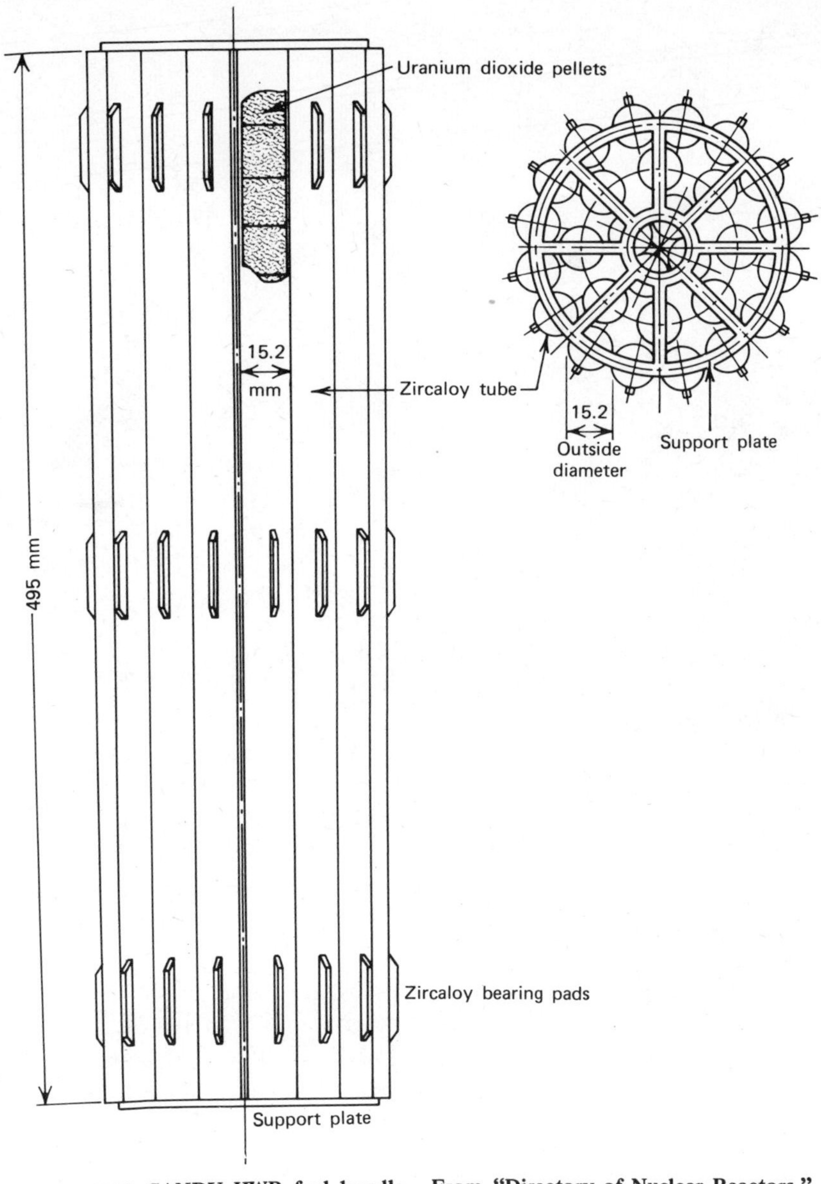

Figure 7.23 CANDU HWR fuel bundle. From "Directory of Nuclear Reactors," Vol. VII, International Atomic Energy Agency, 1968.

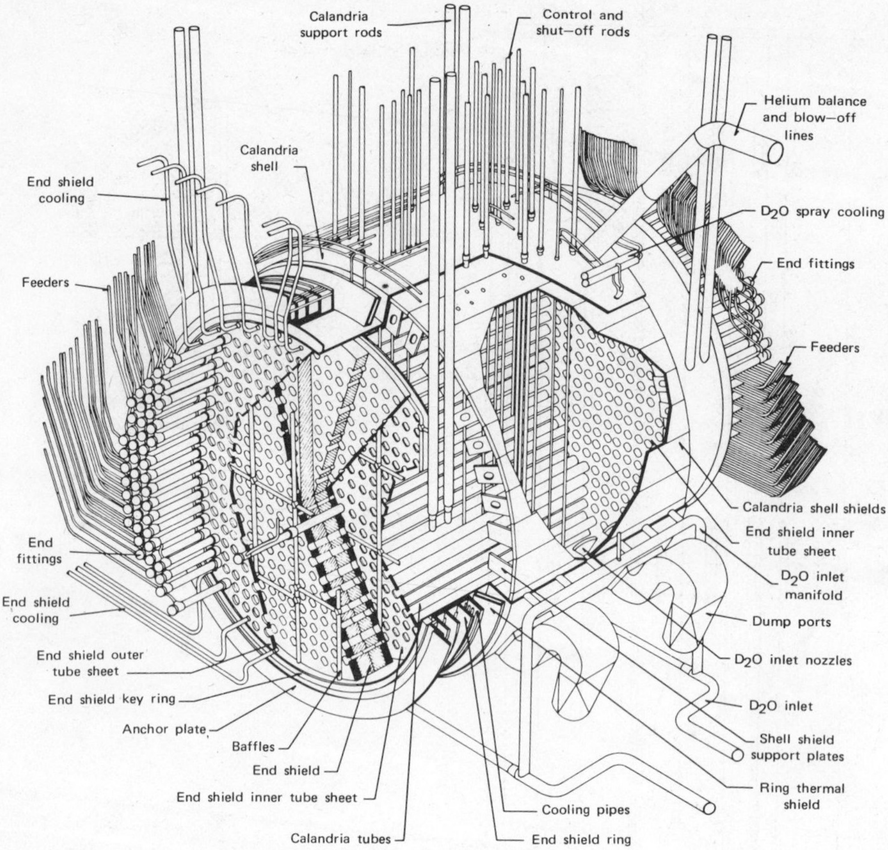

Figure 7.24 Isometric view of Pickering CANDU reactor. From "Directory of Nuclear Reactors," Vol. IX, International Atomic Energy Agency, 1971.

horizontal facilitates charging and discharging fuel from both ends of the core. Thermodynamically, the CANDU reactor is similar to a PWR. The thermal energy in the primary coolant D_2O is transferred to light water in steam generators to produce steam for the turbines. The D_2O flowing to the steam generators is at 85 bar pressure and 293°C temperature.

The Steam Generating Heavy Water Reactor (SGHWR). An interesting variation of the HWR is the SGHWR, in which light water serves as the primary coolant, with D_2O filling most of the moderating function. The light water is allowed to boil (as in a BWR) and, thus, steam flows directly from the reactor to the turbine. A 94 MW(*e*) prototype SGHWR was started at Winfrith, United Kingdom, in 1967. This design takes on renewed interest in the light of a recent (1974)

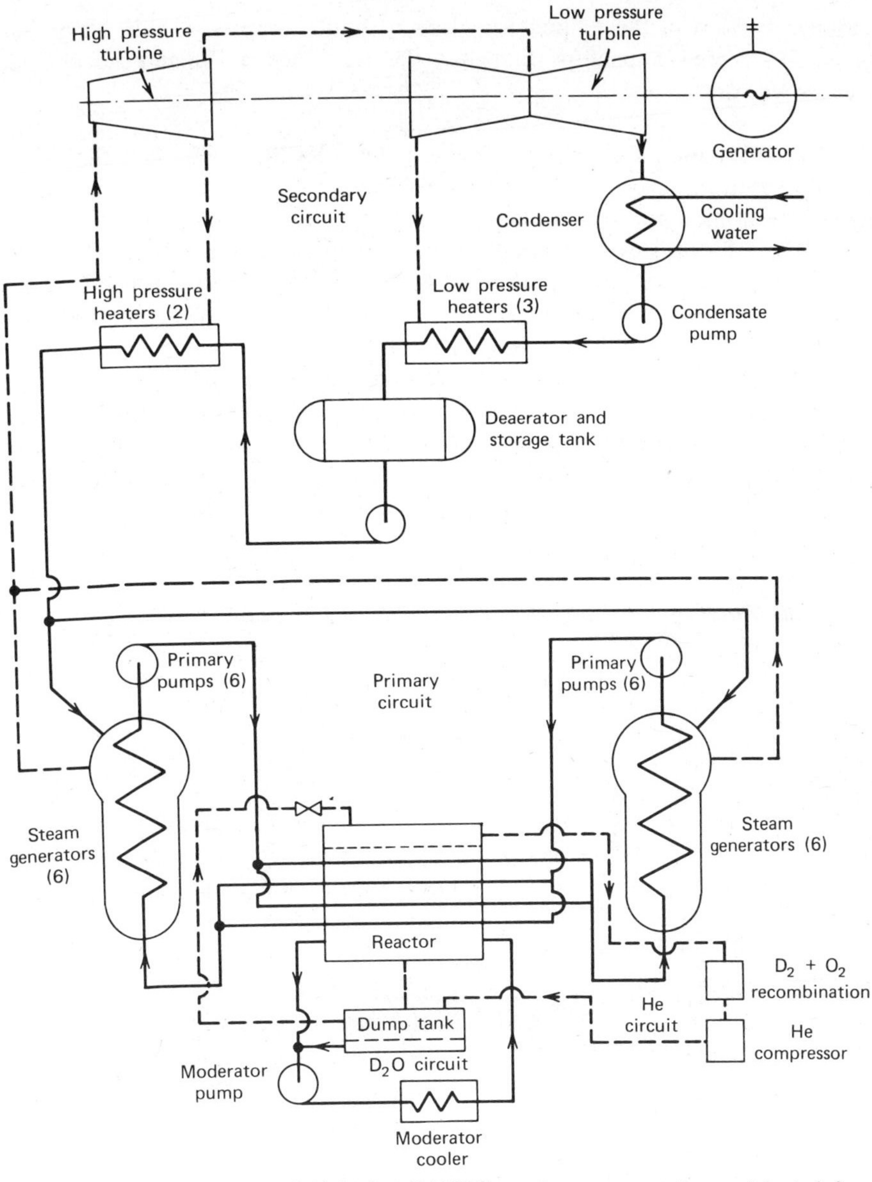

Figure 7.25 Flow diagram of Pickering CANDU reactor power system. Adapted from "Directory of Nuclear Reactors," Vol. IX, International Atomic Energy Agency, 1971.

national decision in the United Kingdom to build reactors of this design as a generation of power reactors to replace the gas-cooled AGRs and precede fast breeder reactors.

The Liquid-Metal-Cooled Fast Breeder Reactor (LMFBR). The fact that a fast neutron spectrum favors the conversion reactions and thus permits breeding has already been discussed in this chapter. By far the favorite design of fast reactor, as evidenced by the development program of many nations, is the LMFBR. The French reactor Phenix is the prototype LMFBR that is furthest advanced. It was brought to full power in 1974. The British Dounreay Prototype Fast Reactor is very close to the Phenix in operating schedule.

A vertical section of the Phenix reactor vessel is shown in Figure 7.26. The core is literally submerged in a large tank of liquid sodium. Because there is no moderator and because sodium is a very efficient heat transfer medium, the core has a high power density (Table 7.8). A flow diagram of the Phenix reactor

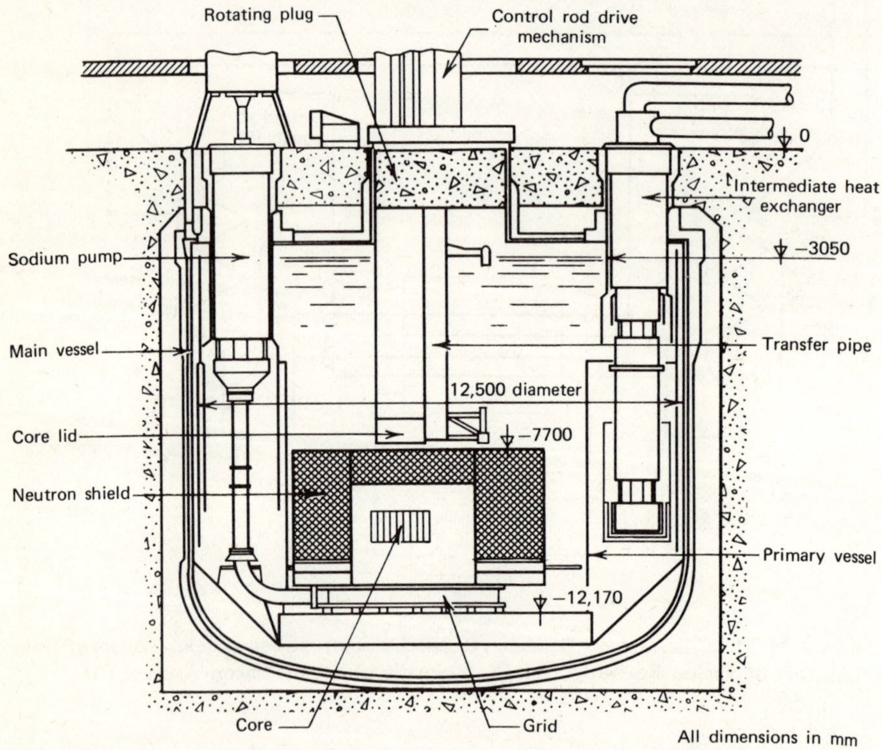

Figure 7.26 Vertical section of Phenix LMFBR and vessel. From "Directory of Nuclear Reactors, Vol. IX, International Atomic Energy Agency, 1971.

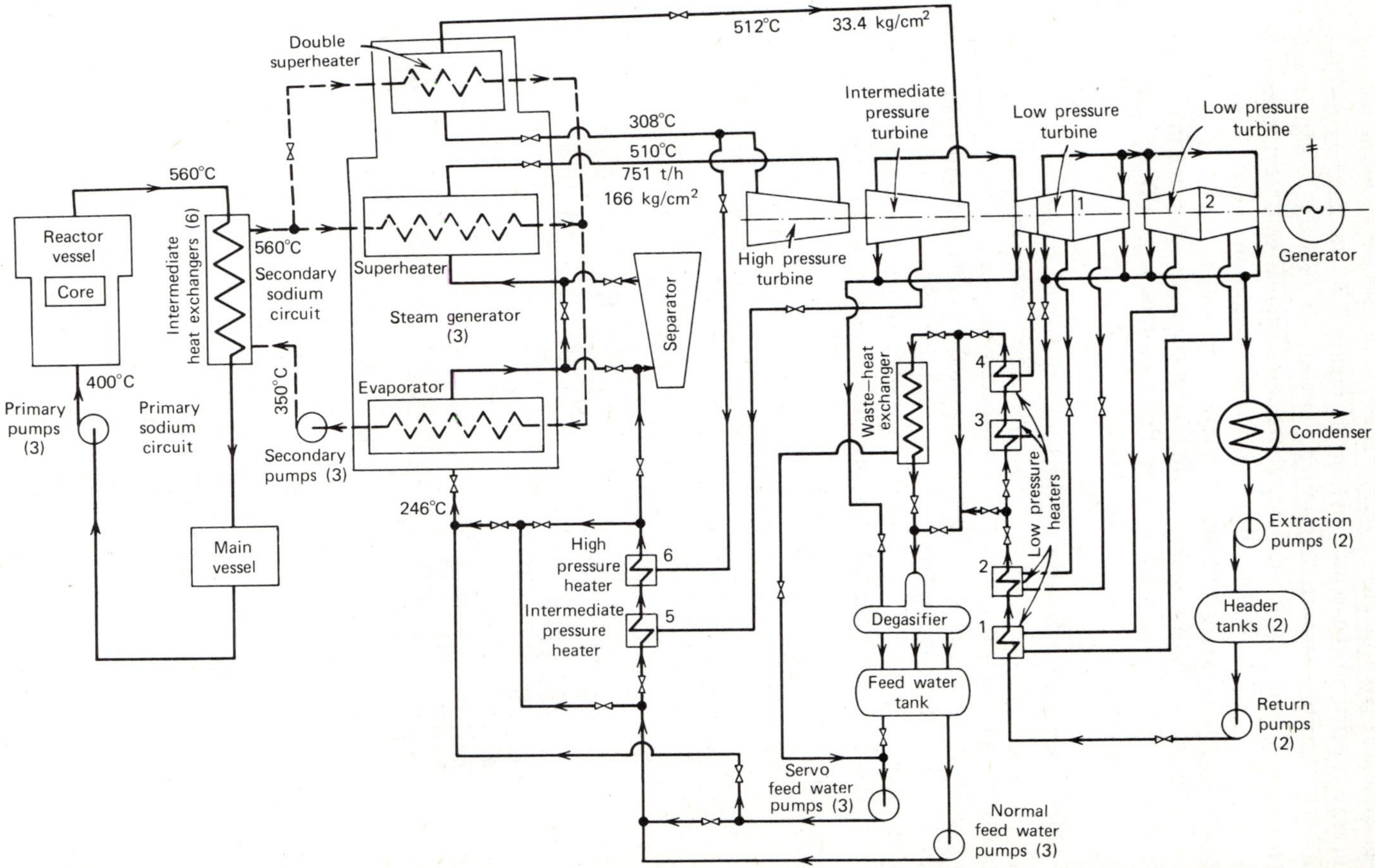

Figure 7.27 Flow diagram of Phenix LMFBR power system. From "Directory of Nuclear Reactors," Vol. IX, International Atomic Energy Agency, 1971.

power system is given in Figure 7.27. These reactors have an extra heat exchanger, compared to the generic system of Figure 7.7. In this intermediate exchanger, thermal energy is transferred between the primary coolant liquid metal and a nonradioactive liquid metal stream. This latter stream goes to a steam generator, where its thermal energy is transferred to the water. This additional exchanger serves to eliminate the possibility that the primary coolant, which is radioactive due to neutron activation in the core, will come in contact with water. The steam conditions to the turbogenerators (Table 7.8) are comparable to modern fossil station practice.

The Molten Salt Breeder Reactor (MSBR). A commercial version of an MSBR has not been built, nor is one planned at this writing. Considerable development work has been performed, however, almost exclusively at the Oak Ridge National Laboratory. A reactor has been built and operated for two and a half years, for example. The MSBR is a thermal-neutron reactor intended to operate on the ^{233}U/thorium fissile/fertile fuel system. In these respects it is like the LWBR and the HTGR already described. The key difference from all other reactors de-

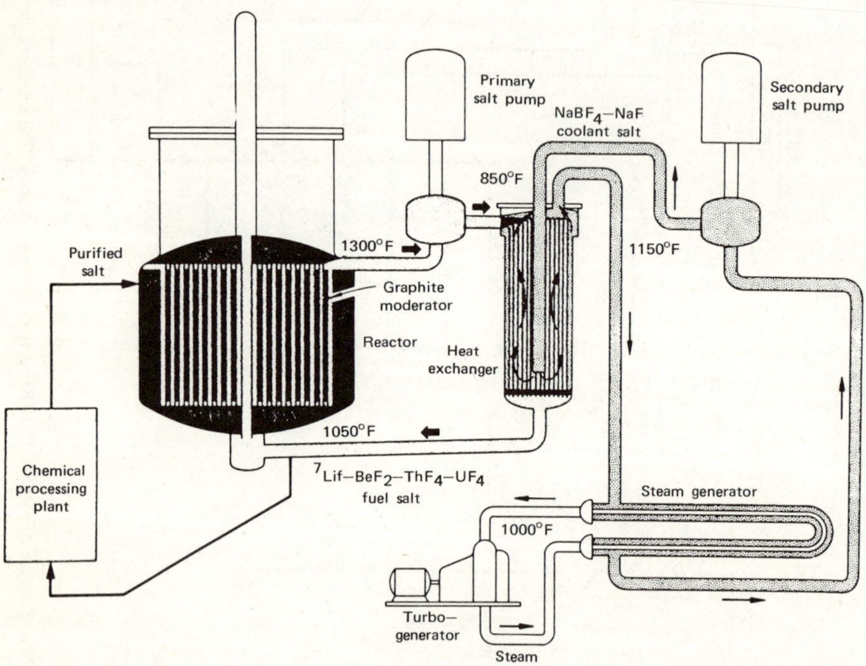

Figure 7.28 Schematic flow diagram of MSBR power system. Adapted from "Proposed Final Environmental Statement, LMFBR Program," WASH 1535, U.S. AEC, December 1974.

scribed in this chapter is that the fuel is in the form of a liquid that also serves as the primary coolant. The liquid is a molten salt complex consisting of fluoride salts of lithium, beryllium, thorium, and uranium (melting point 500°C). Figure 7.28 shows a conceptual flow diagram. The power system is rather like that of the LMFBR, except that the molten salt replaces the molten sodium. The moderator is graphite through which the molten salt flows. The fact that fuel is in liquid form obviates many of the problems of the solid fuel reactors but also introduces other problems. For example, highly radioactive fuel circulates outside the reactor. On the other hand, a by-pass fuel stream can be processed continuously (in the "chemical processing plant" of Figure 7.28), thus keeping the fission product inventory very low. If the fast breeder reactors are not developed in a timely manner, it may be that the MSBR would have a role to play because of its high fuel efficiency. This remark also applies to the HTGR and LWBR.

NUCLEAR FISSION FUEL CYCLE

We have made frequent reference in this and other chapters to the fuel in a nuclear reactor. In Chapter 3 we described very briefly the fuel cycle (see Figure 3.11), that sequence of operations that prepares fuel for insertion in the reactor and that processes the spent fuel removed from the reactor. Because much of the fissile and fertile material in spent fuel is potentially reusable, the sequence of operations is referred to as a cycle. It is not necessarily a cycle, however. Spent fuel could be disposed of without additional processing.

Clearly, the details of the fuel cycle depend on the nature of the fuel in the reactor, its chemical and physical form, its isotopic composition, and so on. We shall not attempt to analyze or describe any of the many possible fuel cycles. We shall discuss two aspects of the fuel cycle that are common to many different cycles. The first is isotopic enrichment of uranium; the second is the isotopic composition of fresh and spent reactor fuel.

Uranium Enrichment

A reactor can be made critical with natural uranium as fuel only under very limited conditions. Those conditions include: a thermal reactor moderated with heavy water or graphite, and a relatively low burnup or energy yield. Enrichment or uranium in the isotope ^{235}U greatly broadens the range of conditions under which the uranium can be used as a reactor fuel. Enrichment carries a cost, however. The evaluation of that cost is the subject of the following analysis.

Without regard to the mechanism of the enrichment process, we can represent the flows of uranium in and out of an enrichment plant.

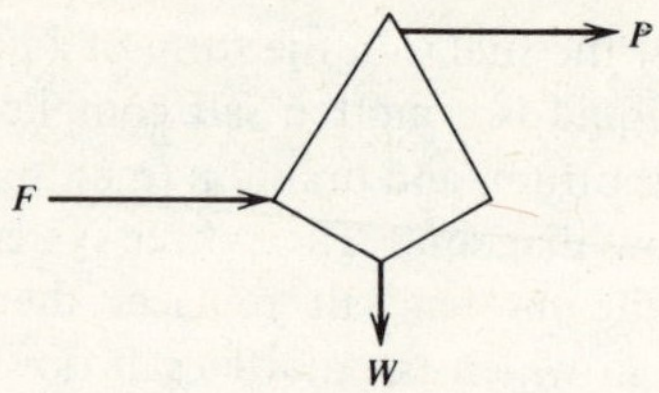

A simple steady-state material balance gives

$$F = P + W \tag{7.40}$$

where F, P, and W are in mass units. We can also make a balance of ^{235}U, representing its weight fraction by x.

$$Fx_f = Px_p + Wx_w \tag{7.41}$$

We shall regard uranium as a binary mixture of ^{235}U and ^{238}U, neglecting any ^{234}U or ^{236}U that might be present. Equation 7.41 can be rearranged

$$\frac{F}{P} = \frac{x_p - x_w}{x_f - x_w} \tag{7.42}$$

For example, if a 3% enriched uranium is to be produced from natural uranium (0.71% ^{235}U) while the enrichment plant is operating with a tails composition of 0.25%, then

$$\frac{F}{P} = \frac{0.03 - 0.0025}{0.0071 - 0.0025} = 5.98$$

which means that 5.98 kg of natural uranium must be supplied for each kg of enriched product produced.

The energy that must be expended in any given separation is proportional to the separative work, SW. This value is given by the expression

$$SW = PV(x_p) + WV(x_w) - FV(x_f) \tag{7.43}$$

where V, called the separative potential, is a unique function of x.

$$V(x) = (2x - 1) \ln \frac{x}{1 - x} \tag{7.44}$$

Separative work is often expressed in separative work units (SWU). However, since $V(x)$ is dimensionless, the units of separative work must be in the same units as F, P, and W It is usually either kilograms or metric tons, but it should be stated explicitly.

Since separative work is proportional to the energy expended in a separation process, and since the enrichment process is energy intensive, the operating cost is often expressed in terms of \$/kg-SWU, which we shall denote C_s. The other component of the cost of enriched uranium is the cost of the feed, C_f, also in \$/kg of feed. The cost of enriched product is

$$C_p = \frac{F \times C_f}{P} + \frac{SW \times C_s}{P} \tag{7.45}$$

The value of tails is taken as zero. Combining Eqs. 7.42, 7.43, and 7.45 gives

$$C_p = \frac{C_f(x_p - x_w)}{x_f - x_w} + C_s \left[V(x_p) + \frac{x_p - x_f}{x_f - x_w} V(x_w) - \frac{x_p - x_w}{x_f - x_w} V(x_f) \right] \tag{7.46}$$

Assuming that the tails have no value, for any fixed ratio of C_f/C_s, a value of tails compositions, x_w, which minimizes C_p can be found by differentiating Eq. 7.46 and setting the derivative, dC_p/dx_w, to zero. The results of this operation give

$$C_f/C_s = (2x_f - 1) \ln \frac{x_f(1 - x_0)}{x_0(1 - x_f)} + \frac{(x_f - x_0)(1 - 2x_0)}{x_0(1 - x_0)} \tag{7.47}$$

and

$$C_p/C_s = (2x_p - 1) \ln \frac{x_p(1 - x_0)}{x_0(1 - x_p)} + \frac{(x_p - x_0)(1 - 2x_0)}{x_0(1 - x_0)} \tag{7.48}$$

where x_0 is that value of x_w that minimizes C_p for the given C_f/C_s value.

It should be noted that at the present time, with government-owned and operated enrichment plants, the value of tails composition is set by the government. It is unlikely that it will be just that value of x_w that minimizes C_p for a given customer. The value of C_s is set by the government. The value of C_f depends on what a given utility or reactor manufacturer has paid for the uranium feed and the processing necessary (for example, chemical conversion to UF_6; see Figure 3.11) to prepare it for the enrichment step.

The foregoing analysis is independent of the particular process used for enrichment. The value of C_s, of course, is dependent on the process. In the United States at present, the only operative process involves the gaseous diffusion mechanism in which UF_6 gas passes through a barrier with very fine pores. The ^{235}U, being slightly lighter than ^{238}U, passes through this barrier a bit faster, and a slight enrichment is effected for each stage. There are alternative processes under development, however. In the United States, the gas centrifuge process is given some chance of replacing gaseous diffusion in time, because the centrifuge process requires far less energy per separative work unit.

Example 7.16

If natural uranium feed to an enrichment plant costs \$50/kg-U, separative work cost \$75/kg-SWU, and the plant is operating at a tails composition of 0.2%, what would be the cost of 3% enriched uranium in \$/kg-U? If the plant were operated at optimum tails composition for these conditions, what would be the product cost and tails composition?

Solution: We solve Eq. 7.46 in steps.

$$V(x_p) = V(0.03) = 3.268$$

$$V(x_w) = V(0.002) = 6.188$$

$$V(x_f) = V(0.0071) = 4.870$$

$$\frac{x_p - x_w}{x_f - x_w} = \frac{0.03 - 0.002}{0.0071 - 0.002} = 5.490 \text{ kg-feed/kg-prod.}$$

$$\frac{x_p - x_f}{x_f - x_w} = \frac{0.03 - 0.0071}{0.0071 - 0.002} = 4.490 \text{ kg-tails/kg-prod.}$$

$$C_p = 5.490 \times 50 + 75[3.268 + 4.490 \times 6.188 - 5.490 \times 4.870]$$

$$C_p = \$598/\text{kg-prod. } (3\% \ ^{235}\text{U})$$

In order to find x_0, the optimum value of x_w, it is necessary to solve Eq. 7.47 by trial and error for the value of C_f/C_s of 50/75 = 0.6667. The value so obtained is $x_0 = 0.00271$. Now solve Eq. 7.48 for C_p.

$$C_p = 75 \times 7.757 = \$582/\text{kg-prod.}$$

The Isotopic Composition of Reactor Fuel

Often the discussion of a reactor fuel focuses on the fuel's composition as it is charged to the reactor. One hears, for example, that the fuel is 3% enriched uranium. We know from the discussions of Chapter 6, however, that as the reactor operates, a large number of isotopic changes are continually taking place. These changes are typically represented by Eq. 6.7 and 6.13. We shall not pursue the analysis further, but the results of such analyses are of interest. Figure 7.29[6] is a representation of LWR fuel isotopic composition as charged and as discharged. We see that at discharge the fuel contains 3.3% fission products. The ^{235}U content has been reduced from 3.3% to 0.84%. Some of the ^{235}U that has dis-

[6]Figures 7.29 to 7.32 are adapted from the work of T. H. Pigford and K. P. Ang, "The Plutonium Fuel Cycles" (presented at symposium on plutonium health implications for people), May 1974.

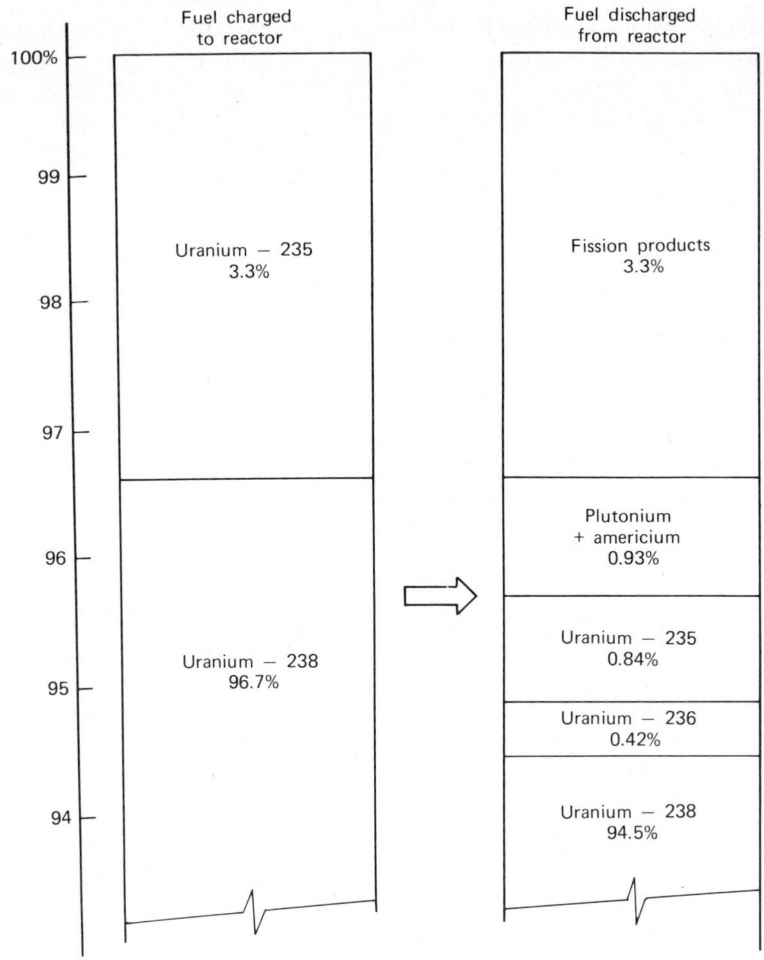

Figure 7.29 Typical isotopic composition of uranium fuel as charged to and discharged from an LWR.

appeared has become ^{236}U, and some of the ^{236}U in turn has been converted to ^{237}Np. Most of the ^{235}U that has disappeared has fissioned. We know this from the relative fission and capture cross sections. It is obvious from mass balance considerations, however, that the 3.3% fission products cannot all come from ^{235}U fissions; some must come from ^{239}Pu or other plutonium isotopes.

Example 7.17

What is the burnup (MW$d(t)$/MTU) of the spent fuel in Figure 7.29?

Solution: The measure of the fission energy which has been released in any spent fuel is the quantity of fission products. Figure 7.29 indicates that for every kg of uranium originally charged to the reactor, there are 33 g of fission products in the spent fuel. (It is convenient to use a base of uranium originally charged, because that does not change.) The 33 g of fission products was originally 33 g of uranium. We assume that the mass of fission products is the same as the mass of the fissioning nucleus. (How much mass is associated with the release and transfer of the thermal energy?) Therefore the number of fissions that must have taken place is

$$\frac{33}{237} \times 6.023 \times 10^{23} = 8.39 \times 10^{22}$$

Here, we use an estimated average mass number of 237 for the fissioning nuclides. The energy per fission is 200 MeV or 3.204×10^{-11} J. Therefore the energy release is

$$q = 3.204 \times 10^{-11} \times 8.39 \times 10^{22} = 2.69 \times 10^{12}\ \mathrm{J}(t)/\mathrm{kg\text{-}U}$$

$$= \frac{2.69 \times 10^{12} \times 10^{-6} \times 10^{3}}{3600 \times 24} = 31{,}100\ \mathrm{MW}d(t)/\mathrm{MTU}$$

A good deal of discussion has been devoted in this chapter and in Chapter 3 to the importance of the conversion of fertile materials to fissile material. A corollary is that the fissile material so converted must be used as fuel. As can be inferred from the values of Figure 7.29, some of the plutonium formed by conversion of ^{238}U is fissioned before the fuel is discharged ("fissioned in situ" is the expression used). However, it is also evident that there is considerable plutonium in the discharged fuel. This plutonium must be utilized if the benefits of conversion are to be realized. The plan is to separate the plutonium from the spent fuel at the reprocessing plant and to incorporate it into fuel material for the same or a different reactor (see Figure 3.11). This procedure is called plutonium recycle. The fuel is often called mixed oxide, reflecting the fact that it would be a mixture of UO_2 and PuO_2. It should be emphasized that at this writing the Nuclear Regulatory Commission had not issued any license necessary for the conduct of plutonium recycle on a regular basis. Experimental mixed oxide fuel assemblies have been tested in a number of power reactors, however.

Figure 7.30 shows one case of the isotopic composition of plutonium recycle fuel in an LWR. In this case, plutonium is blended with natural uranium. Such fuel requires about 5% plutonium in order to attain criticality and to provide for burnup. There is a net disappearance of plutonium between charged and discharged fuel. It is obvious from Figure 7.29 that the 0.9% plutonium in the discharged fuel will not provide this amount of plutonium. The case represented in Figure 7.30 would require the plutonium output from about two uranium-fueled reactors, operating as described in Figure 7.29, to make up the net plutonium consumption. This case represents one of several possible

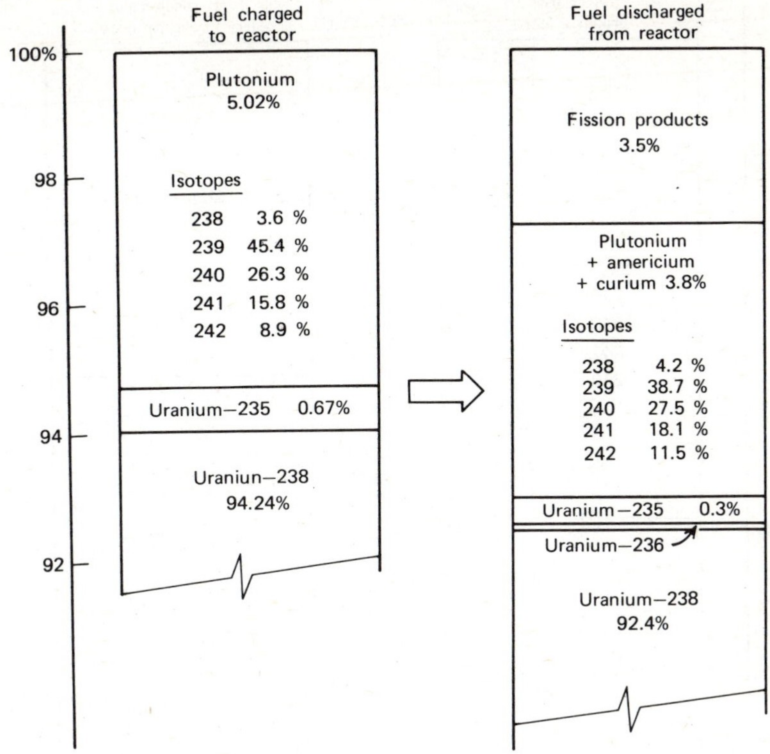

Figure 7.30 Example of isotopic composition of recycle plutonium fuel as charged to and discharged from an LWR.

modes of plutonium recycle operation. Another possibility is simply to recycle to a reactor only that plutonium produced in that reactor, making up net fissile consumption with enriched uranium. This mode is sometimes called self-generation plutonium recycle.

Figures 7.31 and 7.32 show calculated isotopic compositions of fast breeder reactor fuel. Some of the distinguishing features of a fast reactor are evident in these figures. There is the higher fissile concentration in the fuel. Much higher burnups or energy yields are expected from fast reactors because, with their higher conversion ratios, they replace the fissile material consumed and therefore can maintain criticality longer than a thermal reactor. Evidence of the higher burnup is to be found in the higher concentration of fission products in the spent fuel (Figure 7.31). The importance of the conversion in the blanket to the overall breeding process is evident. In Figure 7.31, we can see that there is very little net gain of plutonium, if any, in the core. In the blanket, initially fueled with depleted uranium from enrichment plant tails, there is a large net

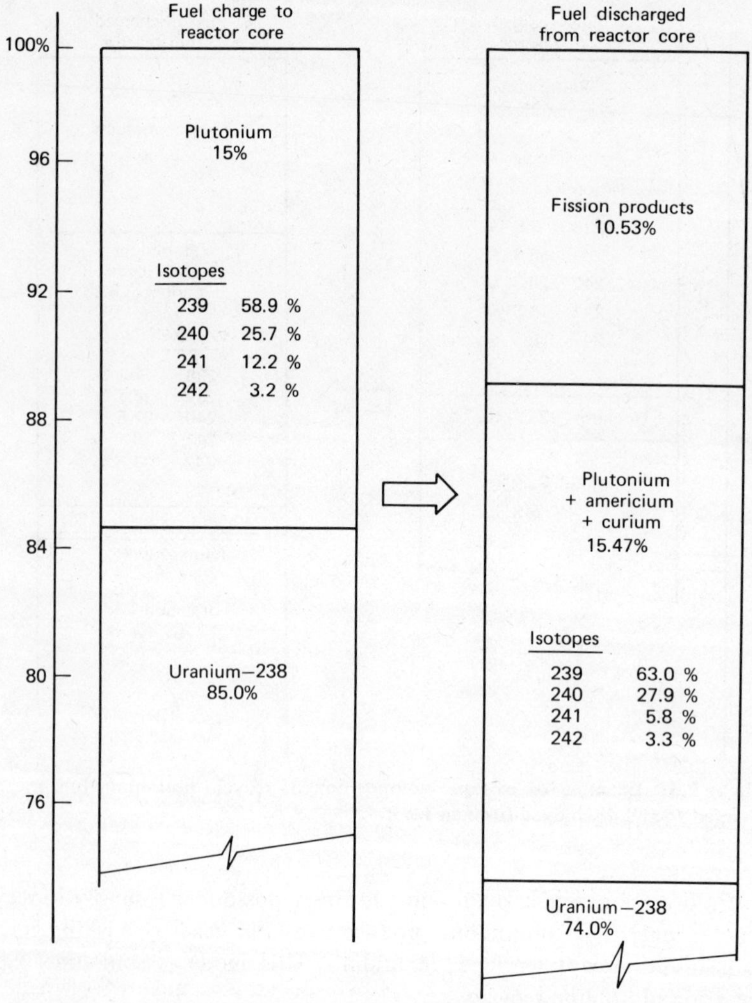

Figure 7.31 Isotopic composition of fuel charged to and discharged from a fast breeder reactor core.

gain of plutonium. It is this plutonium that could be used to provide the starting fissile inventory for another reactor.

Nuclear Fuel Logistics

The isotopic details of nuclear fuel cycles can sometimes serve to obscure the main characteristics. Golan and Salmon[7] have done a good job of consolidating

[7]See Reference 5 at end of chapter.

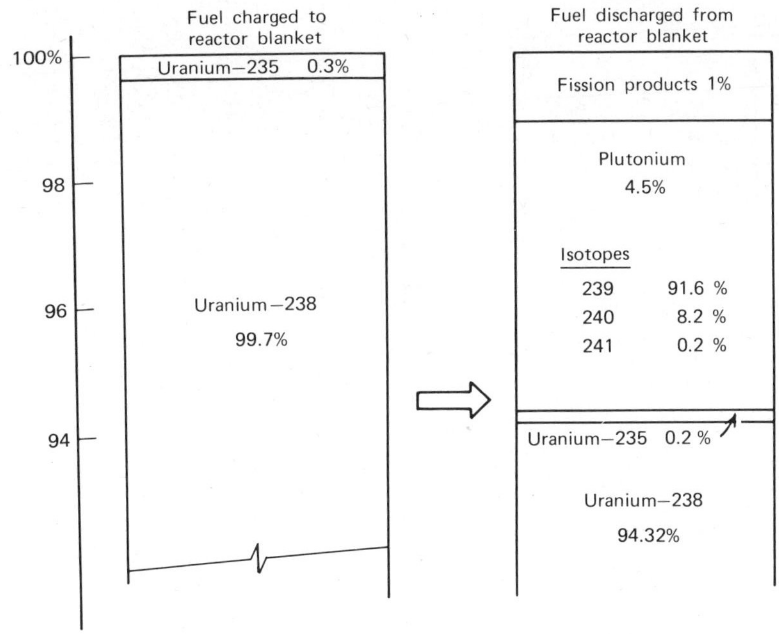

Figure 7.32 Isotopic composition of fuel charged to and discharged from a fast breeder reactor blanket.

a large quantity of detail into a relatively few characteristics of several nuclear fuel cycles. These are given in Table 7.9. The requirements of each fuel cycle are separated into an initial loading, or inventory, requirement and an annual replacement requirement. This distinction is an important one because, if a given reactor type is to be installed at a rapid rate, its inventory requirements can become rate-controlling even though its annual requirements may be modest.

The requirements of Table 7.9 include the quantity of fuel (expressed only in terms of the heavy isotopes U, Pu, and so on) and the quantity of fissile material. In the case of plutonium, for example, only ^{239}Pu and ^{241}Pu are classified as fissile (Check Table A.3). The tons of U_3O_8 required refers to U_3O_8 produced from mined natural uranium. A ton of U_3O_8, as used here, is 2000 lbs, which has an equivalent heavy metal (uranium) content of 769 kg. The tons U_3O_8 requirement is based on the enrichment requirements and is computed by using Eq. 7.42 and the enrichment values shown. Similarly, the separative-work requirements can be computed using Eq. 7.43 and 7.44 and the appropriate values in Table 7.9. (The reader is invited to check the calculations.) The information supplied in Table 7.9 is very useful in assessing the resource requirements (natural uranium and separative work, primarily) of the various fuel cycles.

Table 7.9 Typical Fuel Cycle Logistical Requirements For 1,000-MW(*e*) Nuclear Plants

REQUIREMENTS	LWR(PWR) OPEN CYCLE	LWR(PWR) Pu RECYCLE	HTGR THORIUM WITH U-235 MAKEUP AND U-233 RECYCLE	HTGR THORIUM WITH Pu MAKEUP AND U-233 RECYCLE	LMFBR EARLY/ ADVANCED DEPLETED URANIUM WITH Pu RECYCLE
Initial loading requirement					
Metric ton fuel (heavy isotopes)	74		35		22/17 (core) 30/33 (blanket)
Metric ton fissile	1.9 (2.6 w/o U-235)		1.5 (93 w/o U-235)		2.5/1.8 (Pu)
Tons U_3O_8 [a]	500		400		0
Metric ton separative work units[a]	230		330		0
Annual replacement requirements					
Metric ton fuel (heavy isotopes)	27	27	8	8	10/6 (core) 10/12 (blanket)
Metric ton fissile purchased	0.64 (U-235)	0.50 (U-235)	0.32 (U-235)	0.38 (Pu)	0
Metric ton fissile recycled	0	0.17 (Pu)	0.19 (U-233)	0.25 (U-233)	1.30/0.80 (Pu)
Metric ton fissile sold	0.17 (Pu)	0	0	0	0.19/0.30 (Pu)
Tons U_3O_8 [a]	180	140	92	0	0
Metric ton separative work units[a]	110	86	76	0	0

[a]Based on 0.25% tails.

REFERENCES

1. J. J. Duderstadt and L. J. Hamilton, *Nuclear Reactor Analysis*, Wiley, New York, 1976.
2. E. D. Fuller et al., "General Electric BWR/6 Nuclear System," NEDO-10755, January 1973.
3. J. H. Keenan et al., *Steam Tables*, Wiley, New York, 1969.
4. T. H. Pigford and K. P. Ang, "The Plutonium Fuel Cycles" (presented at symposium on plutonium health implications for people), May 1974.
5. S. Golan and R. Salmon, "Nuclear Fuel Logistics," *Nuclear News* 16, No. 2 (February 1973), 47–53.

PROBLEMS

7.1 An aqueous solution of UO_2SO_4 in light water has a density of 1.10 g/cm^3. The solution contains 5 w/o (wt %) uranium. The uranium is 90 a/o (atomic %) ^{235}U and 10 a/o ^{238}U. Compute the macroscopic fission and aborption cross sections of this mixture for 2200 m/s neutrons. Compute the infinite multiplication factor for this mixture.

7.2 A certain homogeneous reactor has a thermal neutron spectrum and may be considered to be simply a mixture of ^{235}U and H_2O containing 500 moles of H_2O per mole of ^{235}U.

(a) What is the infinite multiplication factor, k, of this mixture?

(b) The reactor in question has a volume of one cubic meter and contains 25,000 grams of ^{235}U. What weight of elemental boron dissolved in this solution would be required to reduce k by 10%?

7.3 Consider a reactor in which the neutron spectrum is thermal. What would be the required ratio of plutonium-239 to uranium-238 atoms such that that the rate of disappearance of ^{239}Pu is just equal to its formation by conversion? If this same mixture was in a fast neutron spectrum, what would be the ratio of ^{239}Pu disappearance to formation?

7.4 A certain homogeneous reactor is just critical when shaped as a cube one meter on a side. If the identical material is shaped as a sphere, what would be the critical size (sphere diameter)?

7.5 The following ingredients are mixed homogeneously to form a medium of density 2.00 g/cm^3.

	wt %
uranium-233 (note 233)	20.00
beryllium	79.90
boron (natural)	0.10

(a) What would be the mean free path of a thermal neutron in this medium for any interaction, for absorption?

(b) What would be the lifetime of a neutron as a thermal neutron, assuming the medium is infinite?

(c) Assuming a thermal neutron spectrum, what would be the infinite multiplication factor of such a medium?

(d) Would such a medium have a thermal spectrum?

(e) The non-leakage probability in this material may be represented as $1/(1 + M^2 B^2)$, where $M^2 = 100\ \text{cm}^2$. Estimate the size of a cube of this material to achieve criticality.

(f) Assuming the neutron spectrum to be thermal, what would be the neutron density (n/cm^3) if the heat power level is 1000 W/cm^3?

(g) On the average, how many collisions would it take to thermalize a 1-MeV neutron in this medium?

7.6 A bare cube of a certain uranium-containing material has critical dimensions 50 cm on a side. What is the thinnest slab of such material that could be critical?

7.7 A typical power reactor in the United States uses uranium in the form of UO_2 rods, about 3/8″ in diameter, as fuel. The uranium is about 3% ^{235}U and 97% ^{238}U. Light water (H_2O) serves as a neutron moderator and as coolant. The volume of water is twice the volume of UO_2, UO_2 density is 9 g/cm^3, and H_2O density is 0.86 g/cm^3.

(a) If the average thermal-neutron flux in the reactor is 10^{13} $n/\text{cm}^2 \cdot \text{s}$, how much ^{235}U will be gone in one year?

(b) Estimate the relative number of thermal neutrons absorbed in ^{235}U, ^{238}U, and H_2O.

(c) If the average heat generation rate in the reactor is 2 MW/ft^3 of reactor volume, what is the average thermal-neutron flux?

7.8 A fission core that is large enough to have negligible neutron leakage is composed of only two nuclides having the following cross sections:

Nuclide	σ_c	σ_f	$\nu\sigma_f$
1	100	500	1250
2	100	0	0

Find the ratio of the two nuclei N_1/N_2 that will just achieve criticality.

7.9 In the neutron spectrum of a particular fission reactor, the following values give the macroscopic cross sections of materials in their pure state:

	Σ_c (cm^{-1})	Σ_f (cm^{-1})	$\nu\Sigma_f$ (cm^{-1})
H_2O	0.01	0.0	0.0
Uranium	0.10	0.25	0.62
Steel	0.20	0.0	0.0

The volume ratio of H_2O to uranium must be 2. What is the maximum volume fraction of steel that can be in the reactor core if the infinite multiplication factor (k) is to be 1.1? (It is greater than 1.0 in order to account for leakage.) Use a homogeneous model here.

7.10 A typical fuel form in a light-water reactor is a rod (really, a column of pellets) of UO_2, 3/8″ in diameter. The UO_2 density is 9 g/cm^3 and the uranium is 3 *a*/o enriched (that is, 3 *a*/o ^{235}U). The microscopic fission cross section of ^{235}U in the reactor neutron spectrum is 350 barns. The neutron flux is 10^{14} neutrons/cm^2 · s.

(a) What is the fission thermal power density in W/cm^3 and Btu/h · ft^3 in the UO_2?

(b) If the average thermal conductivity of UO_2 is 2 Btu/h · ft · °F and the surface temperature is 1000°F, what is the center-line temperature?

7.11 Consider a light-water-moderated power reactor in which the neutrons are well thermalized and the 0.025-eV cross sections apply.

(a) Very roughly, if one added a gram of boron-10 to the reactor core, how much ^{235}U would have to be added to maintain criticality? Be sure to explain the logic of your calculations.

(b) The fuel is UO_2 of density 10 g/cm^3. The uranium is enriched to 3%. A power density of 40 kW(t)/kg-U is achieved. What must be the thermal neutron flux?

(c) If the water suddenly disappeared from the reactor, leaving simply a void where it had been, would the reactor still be critical? Explain the basis of your answer.

7.12 A nuclear fuel is in the chemical form UO_2 and PuO_2. The density is 10 g/cm^3. The isotopic composition is

^{239}Pu	1 a/o
^{235}U	2 a/o
^{238}U	97 a/o

In a thermal neutron flux of 10^{13} neutrons/cm^2 · s, what is the thermal power density (kW(t)/kg − U + Pu) in that fuel?

7.13 A reactor that contains enough carbon to thermalize the neutrons also contains 10 w/o iron and 10 w/o sodium. Assuming a homogeneous model, what w/o ^{235}U would be required to achieve criticality? The absorption cross section of the carbon can be neglected.

Chapter 8

Fusion reactors

It was shown in Chapter 3 that there are fusion reactions that meet all of the conditions of a large-scale energy source save one: up to the present time there is no way to conduct the fusion reaction in a controlled manner with a net energy yield (that is, more fusion energy released than energy expended in obtaining the release). We do not have a fusion reactor. However, there has been a great deal of work devoted to the experimental and theoretical basis for a fusion reactor. The purpose of this chapter is to introduce some of the simpler portions of the theory and to describe present conceptual fusion power plant designs. Most of the presentation is devoted to the approach that has been worked on the longest, namely, conducting the reactions in a high-temperature, low-density plasma confined within a magnetic field. A brief treatment is also given a more recent alternative approach, laser-induced fusion.

MAGNETIC CONFINEMENT

It was recognized almost immediately that if the threshold energy necessary to overcome coulombic repulsion of reactants was to be obtained from the thermal motion of the reactants themselves, the reactants would be in the state of a totally ionized plasma. Temperatures of the order of 100 million °K would be required. However, the fact that the reactants would exist as free ions mingling with free electrons offered the possibility that the plasma could be confined within a magnetic field. This very simple concept opened the door to a large number of questions: What configuration of magnetic field? What strength and spatial variation of strength? Should it be steady-state or pulsating? Experimental programs over some 25 years have addressed these and other questions; final answers have yet to be reached.

The fusion plasma magnetic confinement systems that have been tried can be divided into two broad classes: open and closed. These two are depicted in Figure 8.1. In the open system the magnetic flux lines do not close. This creates the obvious problem that the charged plasma particles will leave the field, moving parallel to the flux lines. One answer to this leakage is to create a

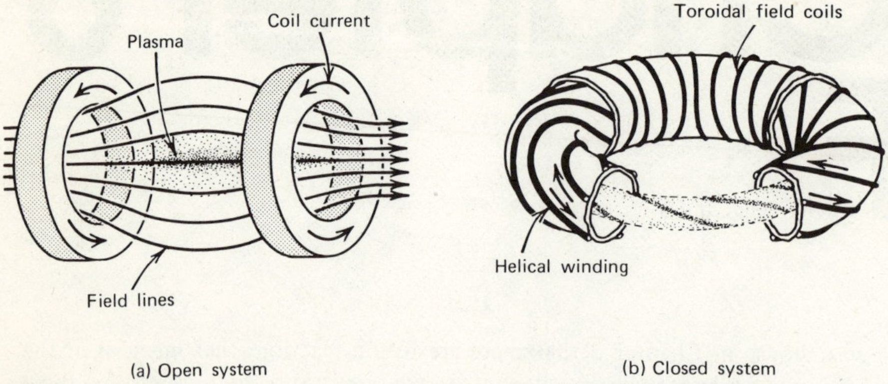

Figure 8.1 Open and closed systems for magnetic confinement of plasma.

higher flux density at the ends. This is the "mirror" system of Figure 8.1*a*. It can be shown that under proper conditions the charged particles will actually be reflected by the higher flux density, thus "closing" the system. There are sufficient exceptions to the proper conditions, however, for leakage to remain a very serious problem in open systems.

The closed system is one in which the lines of force close on themselves. A toroidal system (Figure 8.1*b*) is one obvious configuration. A refinement of the toroidal system, called the Tokamak, was developed in Russia. Experimental devices in this class have given the most promise of achieving fusion reactor conditions. Currently, most effort is being devoted to variations of the Tokamak principle. The closed systems are certainly not without their problems, however. One of them is how to extract reaction products from the plasma and—a parallel problem—how to inject new reactant materials into the plasma. The open systems offer a relatively simple answer to the extraction and injection problems. In the following sections we attempt to develop an elementary but quantitative picture of the requirements for a fusion-reacting plasma.

The Rate of the Fusion Reaction

In Chapter 4, we developed the expression for the rate of neutron-nuclei interactions as

$$R = \Sigma n\bar{v} = N\sigma n\bar{v} \tag{4.13}$$

when n and N are the number densities of neutrons and nuclei, respectively. In fusion, an analogous rate equation can be derived

$$R = n_1 n_2 \overline{\sigma v} \tag{8.1}$$

where n_1 and n_2 are the number densities of the two species of ions or nuclei reacting. If the two nuclei are of the same species (for example, deuterium) then one writes $n^2/2$ instead of $n_1 n_2$. The other two terms, σ and v, are the cross section and the relative speed, respectively. Since σ is a function of v, and since the nuclei exhibit a range of speeds, an effective average product, $\overline{\sigma v}$, is used. Kammash[1] gives equations for $\overline{\sigma v}$ for the DD and DT reactions as

$$\text{DD:}\quad \overline{\sigma v} = 2.3 \times 10^{-14} \frac{\exp\left(-18.8/(kT)^{1/3}\right)}{(kT)^{2/3}} \text{ cm}^3/\text{s} \tag{8.2}$$

$$\text{DT:}\quad \overline{\sigma v} = 3.7 \times 10^{-12} \frac{\exp\left(-20/(kT)^{1/3}\right)}{(kT)^{2/3}} \text{ cm}^3/\text{s} \tag{8.3}$$

In these equations, kT is in units of keV. We can calculate $\overline{\sigma v}$ values. These are given in Table 8.1. From Table 8.1, it is apparent that the cross sections increase with temperature. One might conclude that the higher the temperature the faster the reaction rate. This is not true, however, because, according to Eq. 8.1, the reaction rate also includes the ion densities. These densities obey the perfect gas law and, accordingly

$$n \propto \frac{1}{T} \tag{8.4}$$

For any given pressure, the density varies inversely as the absolute temperature. The reaction rate, therefore, can be expressed as the proportionality

$$R \propto \frac{\overline{\sigma v}}{T^2} \tag{8.5}$$

This quantity for the DD and DT reactions can be computed from Eq. 8.2 and 8.3, respectively.

A very important conclusion from Table 8.2 is that for a fixed pressure there is a temperature at which the reaction rate is a maximum. For both the DD and DT reactions, this maximum rate temperature is in the range of 10–20 keV (100–200 million °K). A third candidate for a fusion reaction, $^3\text{He}(d, p)^4\text{He}$, has a maximum rate temperature much closer to 100 keV, which is one reason it is not currently the target reaction in the fusion reactor development program.

Another important conclusion from both Tables 8.1 and 8.2 is that at any given plasma temperature the DT reaction rate is nearly 100 times the DD reaction rate. At a given temperature and pressure, the rate of the DD reaction will be about $\frac{1}{25}$ the DT reaction because, by Eq. 8.1, the rate is proportional to n_D^2; n_D in a pure deuterium plasma would be twice that of n_D or n_T in a 50–50 mixture. The important point is that the DT reaction will proceed under much

[1] See Reference 1 at end of chapter.

Table 8.1 DD and DT Reaction Cross Sections

kT (keV)	T (10^6 °K)	DD REACTION $\overline{\sigma v}$ (cm^3/s)	DT REACTION $\overline{\sigma v}$ (cm^3/s)
1	11.6	1.6×10^{-22}	7.6×10^{-21}
2	23.2	4.8×10^{-21}	3.0×10^{-19}
5	58.0	1.3×10^{-19}	1.0×10^{-17}
10	116	8.0×10^{-19}	7.4×10^{-17}
20	232	3.1×10^{-18}	3.2×10^{-16}
50	580	1.0×10^{-17}	1.2×10^{-15}
100	1160	1.9×10^{-17}	2.3×10^{-15}
200	2320	2.7×10^{-17}	3.5×10^{-15}
500	5800	3.4×10^{-17}	4.7×10^{-15}

Table 8.2 The Parameter $\overline{\sigma v}/T^2$ for the DD and DT Reactions

kT (keV)	T (10^6 °K)	DD REACTION $\overline{\sigma v}/T^2$ ($cm^3/s \cdot K^2$)	DT REACTION $\overline{\sigma v}/T^2$ ($cm^3/s \cdot K^2$)
1	11.6	1.6×10^{-22}	7.6×10^{-21}
2	23.2	1.2×10^{-21}	7.4×10^{-20}
5	58.0	5.3×10^{-21}	4.2×10^{-19}
10	116	8.0×10^{-21}	7.4×10^{-19}
20	232	7.7×10^{-21}	7.9×10^{-19}
50	580	4.1×10^{-21}	4.8×10^{-19}
100	1160	1.9×10^{-21}	2.3×10^{-19}
200	2320	6.8×10^{-22}	8.9×10^{-20}
500	5800	1.4×10^{-22}	1.9×10^{-20}

less severe conditions of temperature and density than the DD reaction. It is for this reason that the DT reaction is the focus of most attention in the fusion energy program, in spite of the fact that there would be advantages to the DD reaction (such as not needing to produce tritium).

The Pressure in a Magnetically Confined Plasma

We know from Eq. 8.1 that the fusion reaction rate is proportional to the ion density in the plasma. The achievable ion density is itself proportional to the pressure. It can be shown that the maximum pressure exerted by the ions of the plasma that can be resisted or confined by the magnetic field is

$$p_{\max} = \frac{B_0^2}{2\mu_0} = \frac{B_0^2}{8\pi \times 10^{-7}} \tag{8.6}$$

where B_0 is the magnetic flux density in a vacuum and μ_0 is the permeability of a vacuum. In Eq. 8.6, B_0 must be in units of tesla (1 tesla or 1 T equals 10,000 gauss or 10 kilogauss) and the pressure is in pascals (Pa or N/m^2). For example, in a magnetic field of 1 Tesla, the maximum ion pressure would be 3.98×10^5 Pa (3.93 atmospheres).

Since the plasma constituents (ions and electrons) obey the perfect gas laws, we are now in a position to calculate the ion density. In macroscopic units the perfect gas equation is

$$p = \rho RT \tag{8.7}$$

where R is the universal gas constant and ρ is density in moles per unit volume. We can divide R by Avogadro's number to obtain k, the Boltzmann constant, and multiply ρ by Avogadro's number to obtain n, the number density or ions per unit volume. The perfect gas relationship becomes

$$p = nkT \tag{8.8}$$

Remember that in the totally ionized plasma, the electrons are free and also exert a pressure according to the perfect gas law.

We have already seen that the DT and DD reaction rates will have a maximum at a kT value of about 15 keV. Let us now calculate the maximum ion density with a magnetic flux density of 1 T. Combining Eq. 8.6 and 8.8 gives

$$n = \frac{B_0^2}{8\pi kT \times 10^{-7}} \tag{8.9}$$

We must express kT in consistent SI units, joules in this case. A kT value of 15 keV is 2.403×10^{-15} J.

$$n = \frac{1}{8\pi \times 2.403 \times 10^{-15} \times 10^{-7}} = 1.656 \times 10^{20} \text{ ions/m}^3$$

It will perhaps give some perspective to compare this density to the density of hydrogen atoms in gas at room temperature and pressure (300°K and 100,000 Pa). Under these conditions, the density of hydrogen atoms would be 4.8×10^{25} atoms/m^3, a density 300,000 times as great as the ions in the plasma. The plasma is a very rarified medium.

Example 8.1

Calculate the reaction rate and the energy release (MJ/m^3) in a 50-50 DT plasma at a temperature of 20 keV, confined in a magnetic field of 5T. Assume that the pressure is 10% of the maximum.

Solution: From Eq. 8.9, the maximum ion density is

$$n = \frac{25}{8\pi \times 20 \times 1.602 \times 10^{-16} \times 10^{-7}} = 3.10 \times 10^{21} \text{ ions/m}^3$$

The actual density will be 10% of this value. The density of the deuterium and tritium ions will each be half the total, or 1.55×10^{20} ions/m^3. From Table 8.1, the value of $\overline{\sigma v}$ is 3.2×10^{-16} cm^3/s or 3.2×10^{-22} m^3/s. From Eq. 8.1

$$R = n_1 n_2 \overline{\sigma v} = (1.55)^2 \times 10^{40} \times 3.2 \times 10^{-22}$$

$$R = 7.68 \times 10^{18} \text{ reactions/m}^3 \cdot \text{s}$$

The Q-value of the DT reaction is 17.6 MeV. The energy release, therefore, is

$$E = 7.68 \times 10^{-18} \times 17.6 \times 1.602 \times 10^{-13} = 2.17 \times 10^7 \text{ J/m}^3 \cdot \text{s}$$

$$= 21.7 \text{ MW/m}^3$$

$$= 21.7 \text{ W/cm}^3$$

It must be remembered that the neutron gets 80% of the DT reaction Q-value (Eq. 2.8). The plasma density is so low that essentially all the neutrons leave the plasma without any interaction. The ^{4}He nucleus produced has 20% or 3.5 MeV. Being charged, it remains in the plasma and contributes to the plasma heating.

Example 8.2

Using the ion density computed in Example 8.1, calculate the average residence time for an ion if $n\tau = 2 \times 10^{14}$ s/cm^3. What total distance will the average deuterium ion travel in this time?

Solution: The ion density calculated in Example 8.1 is 3.10×10^{20} ions/m^3, or 3.10×10^{14} ions/cm^3.

$$\tau = \frac{n\tau}{n} = \frac{2 \times 10^{14}}{3.10 \times 10^{14}} = 0.65 \text{ s}$$

The temperature of the plasma is 20 keV. In a Maxwellian population, the average speed is

$$\bar{v} = \left(\frac{8kT}{\pi m}\right)^{1/2}$$

Put the energy and mass quantities in the proper units.

$$kT = 20 \times 1.602 \times 10^{-16} = 3.20 \times 10^{-15} \text{ J}$$

$$m = 2 \times 1.66 \times 10^{-27} = 3.32 \times 10^{-27} \text{ kg}$$

$$\bar{v} = \left(\frac{8 \times 3.20 \times 10^{-15}}{\pi \times 3.32 \times 10^{-27}}\right)^{1/2} = 1.57 \times 10^6 \text{ m/s}$$

At this speed, the distance traveled in 0.65 seconds is about one million meters or 1000 kilometers (the distance from New York to Chicago).

Ignition of the Fusion Plasma

Most campers at one time or another have had the frustrating experience of not being able to start a wood fire, even with a plentiful supply of matches. They experience an ignition problem. In the case of a wood fire, ignition depends on raising the temperature of some of the wood to the point at which the wood decomposes. The flammable gaseous decomposition products ignite, and they in turn raise the temperature of more wood; the process becomes self-sustaining. There are other processes that operate to prevent ignition. A strong wind may dissipate the flammable gases. If the wood is damp, the vaporization of free water may absorb too much of the energy.

An analogous problem exists in fusion. In order to become self-sustaining, fusion reactions must take place at a rate sufficiently high to balance energy-dissipating processes. In the case of the DT reaction, each fusion reaction supplies to the plasma only the energy of the ^4He ion, or 3.5 MeV, since the neutron easily escapes the plasma. An important energy-dissipating process is the emission of electromagnetic radiation by the electrons of the plasma. The principal source of this radiation is the bremsstrahlung phenomenon (Chapter 4), which occurs when an electron is accelerated in the field of an ion. In the case of a fusion plasma, this radiation is relatively low energy and, therefore, not very penetrating compared to, say, gamma radiation. However, the density of the plasma is so low that the photons escape the plasma region, and their energy is "lost" in the sense of being unavailable to help sustain the fusion reaction.

Kammash[2] derives an approximate formula for the rate of energy loss by bremsstrahlung

$$P_b = 5.35 \times 10^{-31}\, n_e (kT_e)^{1/2} \sum_i n_i Z_i^2 \quad \mathrm{W/cm^3} \tag{8.10}$$

where n_e and n_i are the number density of electrons and ions, respectively. T_e is the electron temperature, which, due to nonequilibrium conditions in the plasma, can differ from the ion temperature. In this equation, the electron kT_e value must be in units of keV.

The simplest fusion ignition condition is obtained by setting the fusion energy release rate in the plasma equal to the energy loss rate due to bremsstrahlung. First, we consider the DT reaction and specify a 50–50 DT mixture.

2 See Reference 1 at end of chapter.

$$n = n_D + n_T$$

$$n_D n_T = \frac{n^2}{4}$$

We now combine Eq. 8.1 and 8.3 to obtain the energy release rate to the plasma by the DT reaction.

$$P_f = 3.5 \times 3.7 \times 10^{-12} \times 1.602 \times 10^{-13} \frac{n^2}{4} \frac{\exp(-20/(kT)^{1/3})}{(kT)^{2/3}} \text{ W/cm}^3$$

$$P_f = 5.19 \times 10^{-25}\, n^2 \frac{\exp(-20/(kT)^{1/3})}{(kT)^{2/3}} \text{ W/cm}^3 \qquad (8.11)$$

Equation 8.10, for the 50-50 DT mixture becomes

$$P_b = 5.35 \times 10^{-31}\, n^2 (kT)^{1/2} \qquad (8.12)$$

We assume that the ion and electron temperatures are equal. The ignition condition, therefore, is

$$5.19 \times 10^{-25} \frac{\exp(-20/(kT)^{1/3})}{(kT)^{2/3}} = 5.35 \times 10^{-31} (kT)^{1/2} \qquad (8.13)$$

The solution gives a kT value of 4.63 keV, or an ignition temperature of 54 million °K.

This same evaluation of ignition temperature can be applied to the DD reaction in a pure deuterium plasma. As discussed in Chapter 3, the DD reaction goes two ways:

$$^2\text{H} + {}^2\text{H} \longrightarrow {}^1\text{H} + {}^3\text{H} \quad Q = 4.0 \text{ MeV}$$

$$^2\text{H} + {}^2\text{H} \longrightarrow n + {}^3\text{He} \quad Q = 3.2 \text{ MeV}$$

each occurring about 50% of the time. All the energy released in the first reaction is deposited in the plasma, because the particles are charged. In the second reaction, only the energy of the ^{3}He ion, 0.8 MeV, is deposited. The average energy deposited per reaction is, therefore, 2.4 MeV. The ignition temperature for the DD reaction works out to 530 million °K ($kT = 46$ keV). This value is much higher than for the DT reaction and, once again, emphasizes the reason most current attention is directed to DT reaction.

An additional point regarding bremsstrahlung losses should be made. Equation 8.10 shows that these losses are proportional to the square of the atomic number of the plasma constituents (Z_i^2). This means that if the plasma becomes contaminated with nonhydrogen constituents, the bremsstrahlung losses could rise sharply. The helium products of the fusion reactions are one source of such contamination. The possibility of material from the reaction vessel wall getting into the plasma also is of concern.

Energy and Material Balances in a Fusion Plasma

In the energy balance used to compute the ignition temperature, we assumed that the only energy lost to the plasma is bremsstrahlung. There is another important loss, namely, the energy carried off by the ions and electrons that "leak," or are diverted, from the confinement zone. The energy is not lost from the system, of course; rather, it is transferred and partially converted to electricity. At a temperature, T, the average energy of an ion or electron is $\frac{3}{2}kT$. Therefore, if the ion density is n, the thermal energy of the plasma per unit volume is $3nkT$. If the average residence time in the plasma is designated τ, the rate of energy loss due to leakage is $3nkT/\tau$.

We now write the energy balance

$$P_f = P_b + \frac{3nkT}{\tau} \tag{8.14}$$

where, for a 50–50 DT mixture ($n = 2n_D = 2n_T = n_D + n_T$)

$$P_f = 5.19 \times 10^{-25}\ n^2\ \frac{\exp\,[-20/(kT)^{1/3}]}{(kT)^{2/3}}\ \text{W/cm}^3 \tag{8.11}$$

$$P_b = 5.35 \times 10^{-31}\ n^2\ (kT)^{1/2}\ \text{W/cm}^3 \tag{8.12}$$

Again, we assume that ion and electron temperatures are the same. Substituting and rearranging gives

$$5.19 \times 10^{-25}\ \frac{\exp\,[-20/(kT)^{1/3}]}{(kT)^{2/3}} = 5.35 \times 10^{-31}\ (kT)^{1/2} + \frac{4.81 \times 10^{-16}\ kT}{n\tau} \tag{8.15}$$

This equation can be solved for $n\tau$ as a function of kT. Some solutions are:

kT (keV)	$n\tau$ (s/cm^3)
5	1.2×10^{16}
10	5.5×10^{14}
20	2.3×10^{14}
50	1.5×10^{14}
100	1.5×10^{14}
200	2×10^{14}

The value of $n\tau$ has a rather broad minimum around 50 keV. The term $n\tau$ is frequently quoted as a measure of confinement achieved in a fusion experimental device. It is called the Lawson criterion.

Since the plasma constituents react and also escape, it follows that the constituents must be supplied. The feed rate, S, is simply the sum of the reaction rate and the leakage rate.

$$S = 2R + \frac{n}{\tau} \tag{8.16}$$

The reaction rate is multiplied by 2, because two ions are lost per reaction. We can express a very important quantity, the burnup fraction per pass, f_b, in terms of these quantities.

$$f_b \equiv \frac{2R}{S} \tag{8.17}$$

It is the fraction of ions injected into the reacting plasma that reacts before escaping. Again we emphasize that escaping ions can be reclaimed and recycled to the reactor. Substituting for R and S in Eq. 8.16, we obtain for the 50-50 DT mixture

$$\frac{n^2\overline{\sigma v}}{2f_b} = \frac{n^2\overline{\sigma v}}{2} + \frac{n}{\tau}$$

or

$$1/f_b = 1 + \frac{2}{n\tau\overline{\sigma v}} \tag{8.18}$$

We have previously computed $n\tau$ and $\overline{\sigma v}$ as a function of temperature

kT (keV)	$n\tau$ (s/cm^3)	$\overline{\sigma v}$ (cm^3/s)	f_b
10	5.5×10^{14}	7.4×10^{-17}	0.02
20	2.3×10^{14}	3.2×10^{-16}	0.035
50	1.5×10^{14}	1.2×10^{-15}	0.081

The fractional burnup is a very important operating characteristic. If, for example, the value is 0.02, it means that on the average a given deuterium or tritium atom or ion makes 50 trips through the reactor. Each trip through, the material must be reclaimed, purified, and processed for reinjection. In the case of tritium, these operations must be performed with a very high recovery (low loss)—first, because tritium is radioactive, and second, because tritium itself must be produced in the fusion reactor, and there will be little to spare.

Most current fusion reactor conceptual designs have a target fractional burnup of about 0.05, meaning that the average reactants go through the reactor about 20 times. The reader might also note that if f_b becomes large, the reaction rate will decrease because of the depletion of the reactants, deuterium and tritium, and the buildup of the reaction product, helium. The reaction rate will decrease as indicated by Eq. 8.1; the radiation losses will also increase, following Eq. 8.10, because the helium product has a higher Z value.

Example 8.3
What is the thermal energy (J/cm^3) in a plasma in which the ions and electrons have an average energy of 12 keV and the ion (and electron) density is 5×10^{13} per cm^3? If the total plasma volume is 5000 m^3 and is enclosed by 2000 m^2 of steel wall 2cm thick, by how much would the wall temperature rise if all the thermal energy of the plasma was deposited in it?

Solution: The thermal energy is

$$2 \times 5 \times 10^{13} \times 12 \times 10^3 \times 1.602 \times 10^{-19} = 0.192 \text{ J/cm}^3$$

For comparison, the heat capacity of liquid water is 4.184 J/cm$^3 \cdot$ K. The thermal energy of the entire plasma is

$$5000 \times 10^6 \times 0.192 = 9.61 \times 10^8 \text{ J} = 0.961 \text{ GJ}$$

The steel wall surrounding the plasma has a density of 7.8 g/cm^3 and a mass of

$$2000 \times 10^4 \times 2 \times 7.8 = 3.12 \times 10^8 \text{ g}$$

Taking the heat capacity of the steel to be 0.4 J/g $\cdot$ K, the expected temperature rise, should all the thermal energy of the plasma be instantaneously deposited in the wall, would be

$$\frac{9.61 \times 10^8}{3.12 \times 10^8 \times 0.4} = 7.7 \text{ °K}$$

It is this relatively modest amount of thermal energy stored in the plasma that is perceived as a significant safety feature of fusion reactors.

Fusion Power Plant Design

Thus far we have tried to present some of the simpler portions of the theory on which the design of a fusion power reactor would be based. Over the past 25 years there have been experimental programs, in the United States and other countries, directed primarily toward an understanding of a fusion plasma confinement. Although these efforts have yet to meet the target of confinement ($n\tau$) necessary for fusion energy to be as large as input (energy breakeven), they have made substantial progress toward that goal. More recently a number of studies have been directed to the detailed, although conceptual, design of a fusion power plant to generate electricity as a facility in a utility grid. Although premature in the sense that the fundamental physics basis for such a plant has yet to be established, these studies serve the valuable purpose of uncovering many technological difficulties, quite apart from plasma physics, that must be resolved if economic fusion energy is to become a reality.

In the course of these design studies, several classes of problems have been uncovered. These include the magnet system, plasma ignition and heating systems, materials transfer systems, energy transfer and conversion systems, tritium production and handling systems, and system maintenance. We shall discuss each of these briefly. Although there have been several design studies, for the sake of simplicity we shall focus on a single design, the University of Wisconsin UWMAK-II.[3] This design is based on the Tokamak concept, which falls in the toroidal class of magnetic confinement system (Figure 8.1). In the Tokamak concept, the confinement produced by current in the external windings is enhanced by a current induced in the plasma itself (Figure 8.2*a*). This current is

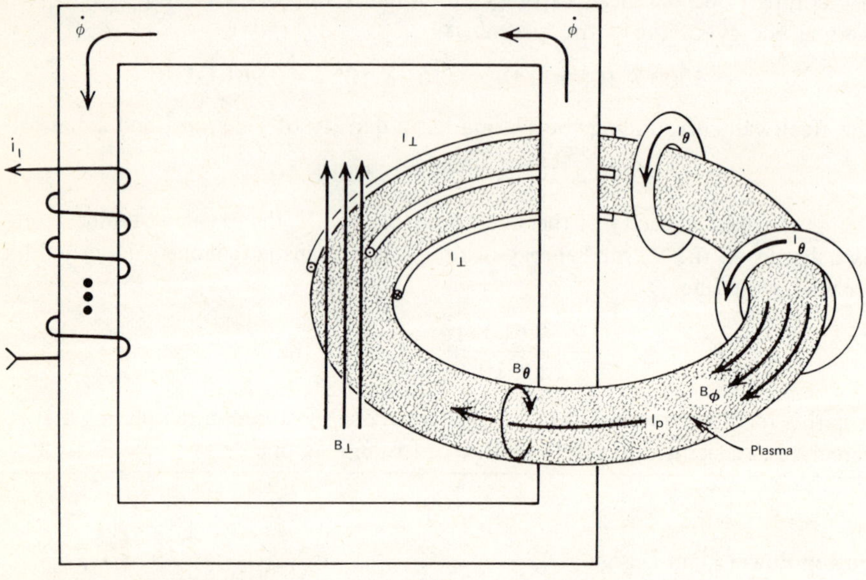

(*a*)

Figure 8.2 Evolution of a fusion reactor from physics concept to engineering concept. (*a*) Adapted from G. L. Kulcinski, et al., "UWMAK-I, The Wisconsin Tokamak Reactor Design," Report UWFDM-68, University of Wisconsin, March 1974. (*b*) and (*c*) From F. L. Ribe, "Fusion Reactor Systems," Proceedings, International Conference on Nuclear Solutions to World Energy Programs, American Nuclear Society, 1973. (*d*) Adapted from R. W. Conn, et al., "UWMAK-II, A Conceptual Tokamak Reactor Design," Report UWFDM-112, University of Wisconsin, December 1975.

[3] See Reference 2 at end of chapter.

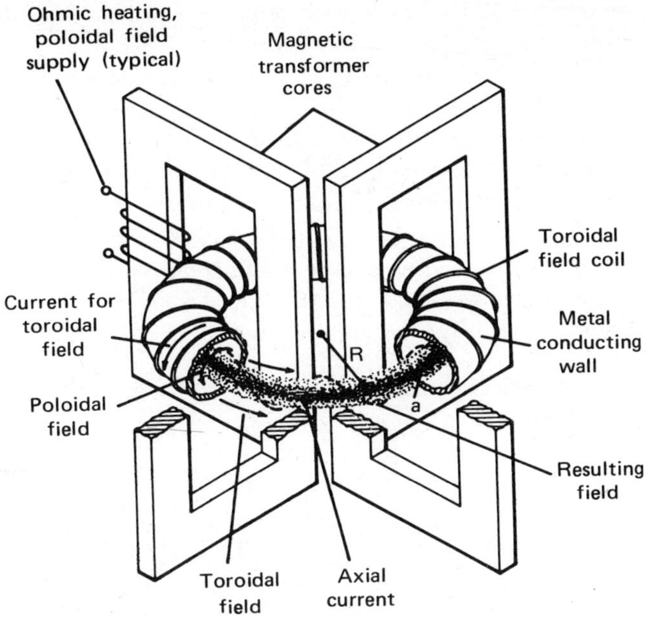

(b)

Figure 8.2 *(Continued)*

Table 8.3 Some Characteristics of the UWMAK-II Conceptual Fusion Reactor Design

Average electrical power	1716 MW(*e*)
Average thermal power	4712 MW(*t*)
Plasma dimensions	
Major radius	13 m
Plasma radius	5 m
First wall radius	5.5 m
Plasma volume	6415 m^3
Plasma current	14.9 MA
Ion temperature	13.2 keV
Electron temperature	12.0 keV
Deuterium plus tritium density	7.33×10^{13} cm^{-3}
Alpha (Helium) density	1.87×10^{12} cm^{-3}
Fractional burnup	0.0485
Particle confinement time (τ)	8.28 s

induced by an alternating current in an external coil, which plays the role of a transformer primary to the plasma secondary.

Figure 8.2 shows the "evolution" of a Tokamak concept to the detailed design of the UWMAK-II and other power plant designs based on the same concept. The increase in complexity is worthy of note. Some important characteristics of the UWMAK-II are given in Table 8.3. There are a few features of this design that present an interesting contrast with modern fossil fuel plants. The outer diameter of the toroid is about 52 meters. Two of them would not fit on an American football field. The reactor is also about 40 meters high. We see from Table 8.3 that into this huge reactor we feed some 22 kg/d of fuel (deuterium plus tritium). Actually, the net feed is only about $\frac{1}{20}$ of this value, but

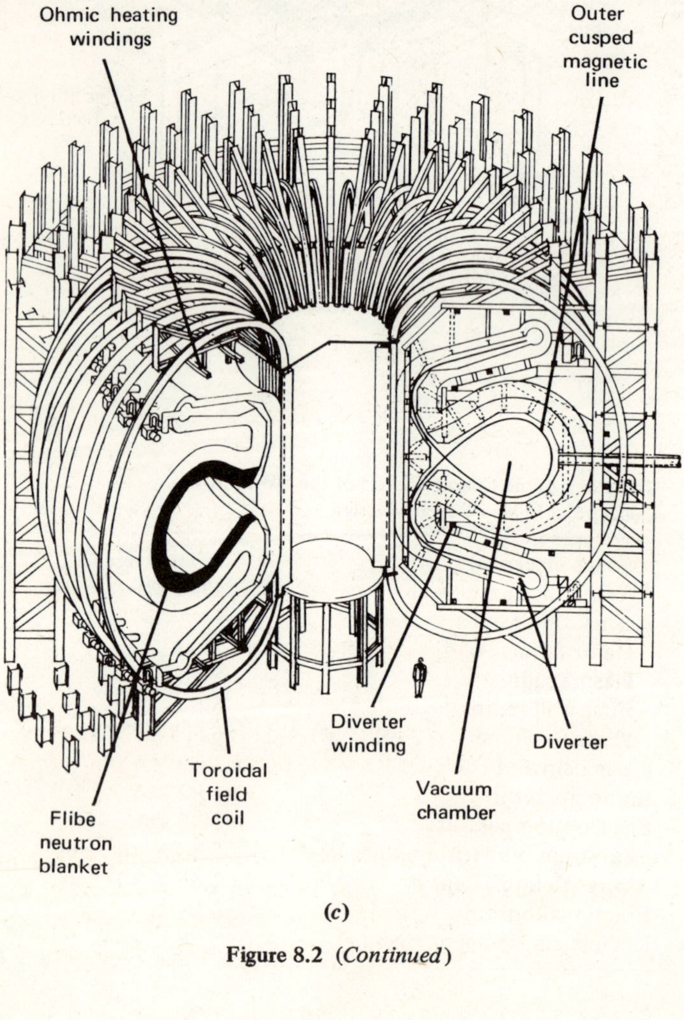

(c)

Figure 8.2 (*Continued*)

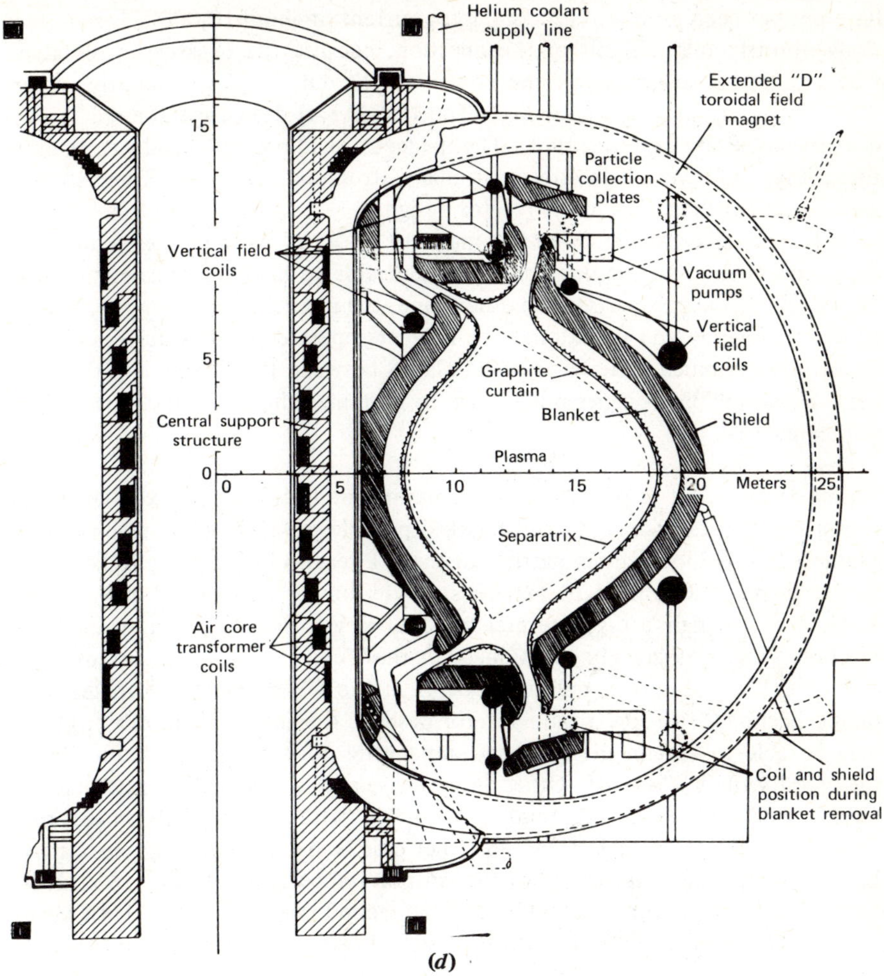

(*d*)

Figure 8.2 (*Continued*)

because the fractional burnup is only 4.85%, there is considerable recycle. A coal-fired plant of the same electrical output would require some 6 million kg/d of fuel.

Reactor Magnet System. The strength and the configuration of the magnetic field that confines the plasma is the single most important feature of any fusion reactor of this kind. The magnetic flux density, of course, varies with position. The UWMAK-II design specifies a field of 3.67 tesla (36.7 kilogauss) at the axis of the toroid. The plasma volume is 6415 m^3. Fields of this strength and size have not yet been produced. A conductor current of about 10,000 A is required. Conventional coils are out of the question, because the resistive losses alone would consume too much of the power plant's output. Consequently, superconducting magnets are a necessity. The UWMAK-II design specifies NbTi conductors cooled by liquid helium. The magnetic field pressure produces reactive forces that cause very high stresses in many structural members. These stresses are a major consideration in the reactor design.

The fact that superconducting magnets must be used creates a difficult design problem. On the one hand, it is desirable to have the magnets as close as possible to the plasma. On the other hand, the plasma is a high-density source of thermal energy and radiation. The magnets operate at a few degrees Kelvin. Consequently, much material must be placed between the plasma and the magnet. However, the designer must strive to minimize the thickness of the intervening material.

Plasma Heating. It is planned to heat plasmas in a variety of ways. Once the fusion reaction is underway, any charged particle product of the reaction (for example, the 3.5-MeV alpha particle in the DT reaction) will give up its energy to the plasma. The reactant materials, deuterium or tritium or both, may be accelerated to high energy (several hundred keV) prior to injection into the reaction zone. As already mentioned, in the Tokamak design a current is induced in the plasma (15 million amperes in the UWMAK-II), which leads to ohmic heating. In some fusion reactor designs, the fusion reaction's products may be able to sustain the reaction temperature once it is under way. This is the case for the UWMAK-II. Other designs have such a high leakage rate that the plasma will have to be continuously supplied with energy from an external source such as particle injection. In either case, some such energy source must be available to raise the plasma to the ignition temperature. In the UWMAK-II, ohmic heating is used and, in addition, a 10-second injection of a 200-MW beam of 500-keV neutral deutrium atoms is used to bring the plasma up to reaction temperature. Once this heating phase is over, supplementary heating is not supposed to be necessary; the reactor is to operate at steady state for 90 minutes. The reactor is then shut down for about 300 seconds to purge the plasma of impurities, and the cycle is repeated. During the steady-state operation, it is expected that the deuterium and tritium can be injected into the system, perhaps in the form of frozen pellets, without the need to accelerate them beforehand.

Example 8.4

A fusion reactor is to generate 4000 MW(t) of fusion energy. An average power density of 2 W/cm^3 is attainable. If the plasma region is to be toroidal in shape, with a ratio of major radius, R, to minor radius, r, (the aspect ratio) of 3.0, what are the dimensions?

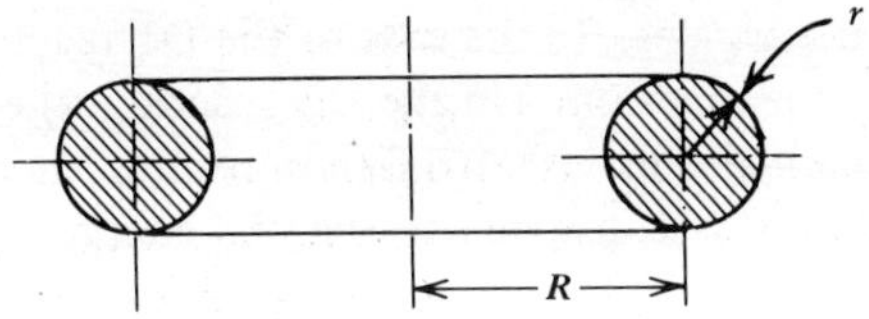

Solution: The volume is

$$V = \frac{4000}{2} = 2000 \text{ m}^3$$

The volume of the toroid is

$$V = 2\pi^2 r^2 R$$

and, if $R = 3r$

$$V = 6\pi^2 r^3$$

$$r = \left(\frac{2000}{6\pi^2}\right)^{1/3} = 3.23 \text{ m}$$

$$R = 9.70 \text{ m}$$

Example 8.5
During the UWMAK-II heating phase, 200 MW of 500-keV deuterons are injected for 10 seconds. How many deuterons are injected, and how does this number compare to the steady-state number of deuterons in the plasma?

Solution: A 500-keV deuteron possesses 8.01×10^{-14} J of kinetic energy. In a 200-MW beam, there are

$$\frac{200 \times 10^6}{8.01 \times 10^{-14}} = 2.50 \times 10^{21} \text{ deuterons/s}$$

In 10 seconds there are some 2.50×10^{22} deuterons added to the plasma. The steady-state ion density is 7.33×10^{13} ions/cm^3, of which half are deuterons. The number of deuterons in the 6415 m^3 of plasma at steady state is

$$\frac{7.33 \times 10^{13}}{2} \times 10^6 \times 6415 = 2.35 \times 10^{23}$$

The deuterons added during the heating phase are about 10% of the number at steady state.

Blanket Design. One of the more important features of a fusion reactor based on the DT reaction is the fact that so much of the fusion energy is deposited (that is, converted to thermal energy) outside the reacting zone or plasma. It is carried away primarily by neutrons and, to a lesser extent, by electromagnetic radiation and charged particles. In the case of the DT reaction, neutrons possess 80% of the energy of the reaction. (In the DD reaction, they possess only 11%.) This situation is in marked contrast to a fission reactor, in which the preponderance of the fission energy is converted to thermal energy within micrometers of each fission reaction site.

Since the energy of the neutrons from the DT reaction must be recovered, and at the highest practicable temperature (for the sake of efficient conversion to electricity), it is necessary to surround the plasma with a neutron-catching region designed with a suitable heat-transfer system. This region is called the blanket. Its design is complicated, because the blanket serves two additional functions: production of tritium necessary to conduct the DT reaction, and protection of the magnet system from radiation and thermal energy.

The blanket design of UWMAK-II is sketched in Figure 8.3; additional information is supplied in Table 8.4. The plasma region itself is 5 meters in radius (in the actual design, Figure 8.3*d*, the cross section is not perfectly circular). The first solid material is a woven graphite curtain, which is proposed by the designers to reduce the chances of high-*Z* material being knocked off the first structural wall and contaminating the plasma, with the result that bremsstrahlung radiation losses become excessive. The first structural wall is a metal wall, which must sustain the vacuum around the plasma. This is a very important part of the blanket and is referred to as the "first wall." It is subject to intensive neutron irradiation (see Example 8.6), which promises to limit severely its useful life.

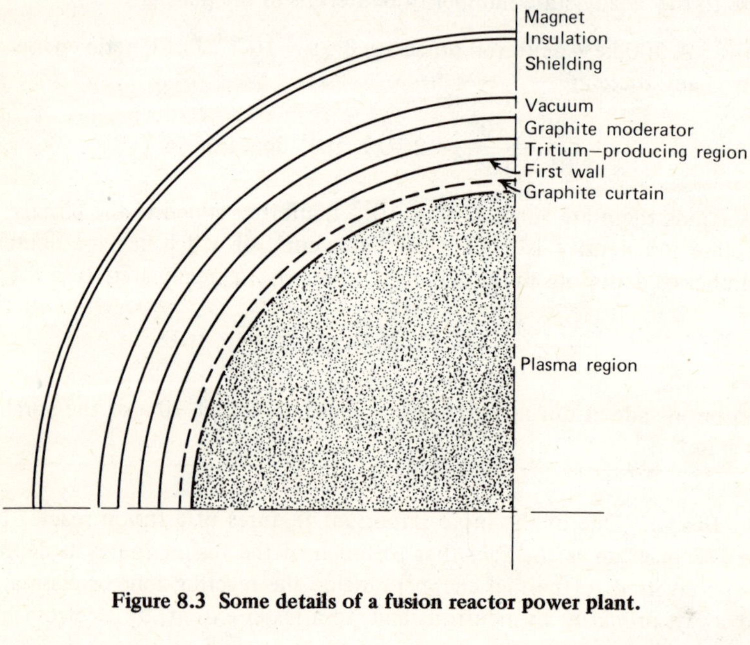

Figure 8.3 Some details of a fusion reactor power plant.

Table 8.4 Description of UWMAK-II Blanket and Shield Regions

NAME	PRINCIPAL CONSTITUENTS	OUTER RADIUS (cm)	REGION THICKNESS (cm)	PRINCIPAL FUNCTION
Plasma	^2H, ^3H	500	–	DT reactions
Curtain	Graphite	525	–	Reduce plasma contamination
Vacuum	–	550	50	
First wall	Stainless	552	2	Contain vacuum
Breeder	Beryllium $^6Li_2Al_2O_3$	583	31	Increase neutrons, produce tritium
Moderator	Graphite	621	38	Slow neutrons
Vacuum	Stainless	639	18	Thermal insulation Coolant plumbing
Shield	Lead B_4C	739	100	Protect magnet from radiation
Insulation	Mylar	754	15	Reduce heat transfer to magnet
Magnet	TiNb	–	–	Produce field for plasma confinement

It also must function at high temperatures (400 to 1000°C). Because of the intensive neutron irradiation, the potential for radioactivation of the first wall exists, depending, of course, on the material chosen. In addition to stainless steel, proposed by UWMAK-II designers, other materials such as molybdenum, niobium, and vanadium have been suggested. These have high temperature properties superior to stainless steel, but each has its disadvantages. Niobium, for example, would become very radioactive, forming 10.2-day ^{92}Zr, 35-day ^{95}Mo, and 2×10^4-year ^{94}Mo. The selection of the "best" first wall material is an unresolved issue.

Immediately after the first wall comes a series of regions whose function is to produce tritium. A portion of these regions is made up of lithium in the form of solid $LiAl_2O_3$. Tritium is formed primarily by the slow-neutron reaction

$$^6\text{Li} + n \longrightarrow {}^3\text{H} + {}^4\text{He} \quad Q = 4.8 \text{ MeV}$$

and, to a much lesser extent, by the endothermic (negative Q-value) reaction (see Example 8.7)

$$^7\text{Li} + n \longrightarrow {}^3\text{H} + {}^4\text{He} + n$$

The energy produced in the ^{6}Li(n, α) reaction is very substantial. In the UWMAK-II system, for example, the average energy produced per DT fusion reaction is 21.56 MeV, compared to a Q-value of 17.6 MeV.

An essential function of the blanket is to slow down the neutrons, since it is in this manner that much of the fusion energy must be recovered. Some of this slowing down in the UWMAK-II blanket design is accomplished by graphite, but some slowing down also occurs in the beryllium "neutron multiplier" (Table 8.4). The function of the beryllium is to increase the number of neutrons by the (*n*, 2*n*) reaction and thus increase tritium production. In the UWMAK-II design, the lithium in the $LiAl_2O_3$ is enriched in the isotope ^6Li in order to reduce the amount needed and, thus, the tritium inventory in the blanket. The tritium produced is obtained by pumping a vacuum on the $LiAl_2O_3$.

The remainder of the blanket, as depicted in Figure 8.3, serves primarily to protect the magnet. What is not shown, however, is the very important cooling system that interlaces most of the blanket, extracting thermal energy. In the UWMAK-II, the selected cooling medium is helium. Other designs have selected liquid lithium, which could serve not only as coolant but also as a neutron-moderating and tritium-producing medium. One problem with liquid lithium is that, being an electric conductor, whenever it crosses magnetic flux lines, pumping work must be done. Furthermore, the extraction of tritium from the lithium coolant promises to be a difficult task. Consequently, in the UWMAK-II case, helium is the preferred coolant.

Example 8.6

Compute the minimum neutron energy required to cause the $^7Li(n, n'\ \alpha)$ reaction.

Solution: Compute the *Q*-value.

$$^7_3Li + n \longrightarrow n' + {}^4_2He + {}^3_1H$$

From Eq. 2.4 and Table A.3

$$Q = 14.907 - 2.425 - 14.950 = -2.468 \text{ MeV}$$

The reactant neutron must have at least this energy.

Power Cycle of the Fusion Reactor. The purpose of the fusion power plant under discussion is to supply electrical energy to a utility network. The fusion energy released in the reacting plasma must be converted to electrical energy. In the case of the DT reaction, this conversion must involve a thermal-electric conversion process, because the neutron's energy can only be captured as thermal energy. In the case of the DD reaction or the D–^{3}He reaction, where most or all of the energy appears as the kinetic energy of charged particles, it would, in principle, be possible to convert the energy directly to electricity, with a much higher conversion efficiency. Schemes for such direct conversion of fusion energy have been discussed but are of less immediate interest, because of the primary focus on the DT reaction.

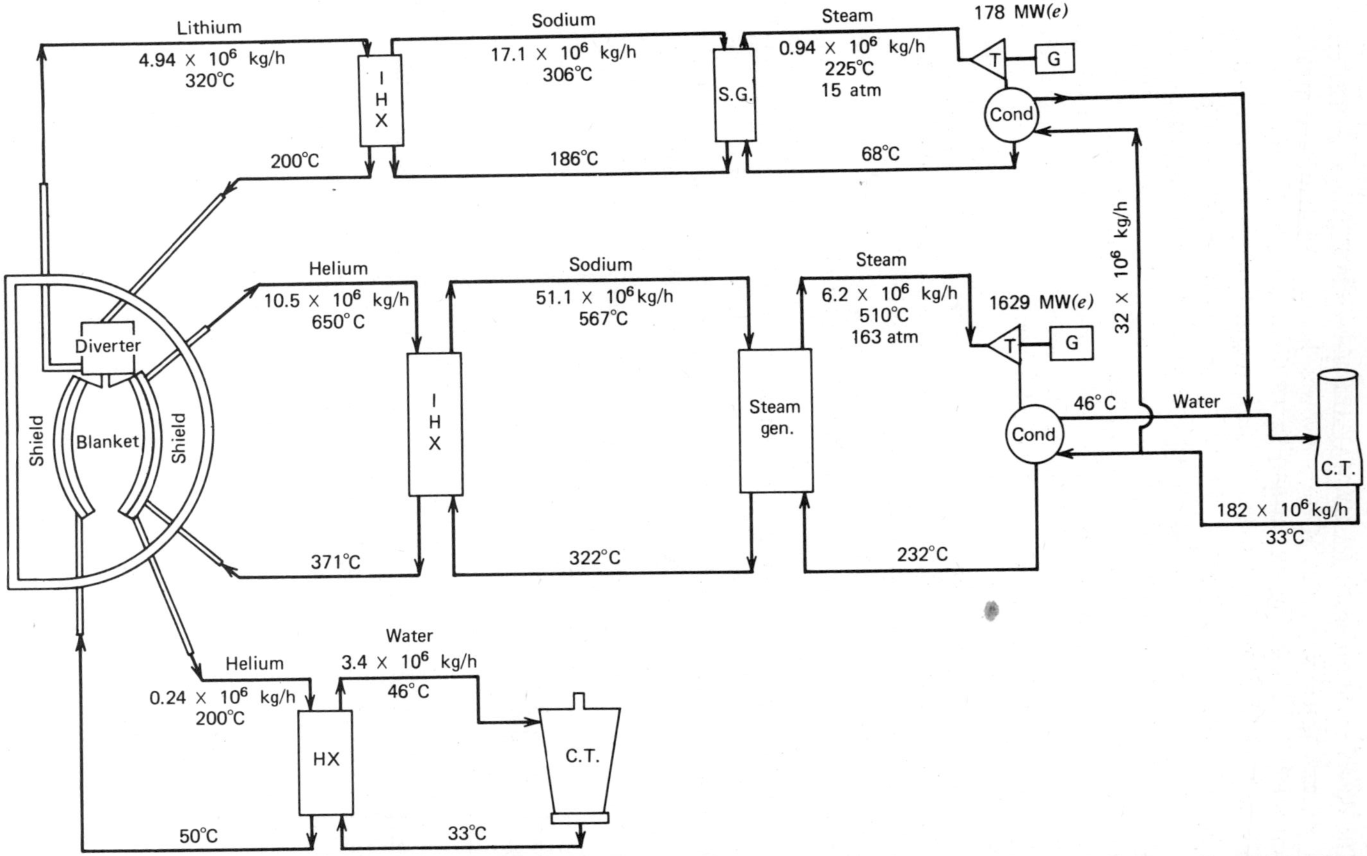

Figure 8.4 The power cycle of UWMAK-II, (*Note:* HX = heat exchanger; C. T. = cooling tower; S. G. = steam generator; T = turbine; G = generator.) Adapted from R. W. Conn, et al., "UWMAK-II, A Conceptual Tokamak Reactor Design," Report UWFDM-112, University of Wisconsin, December 1975.

The power cycle proposed for UWMAK-II is shown in Figure 8.4. The primary system, generating 1629 MW(e), involves helium, sodium, and steam. Helium emerges from the blanket at 650°C. Its thermal energy is transferred to an intermediate loop of liquid sodium, whose thermal energy is transferred in turn to water, generating steam in a more or less conventional steam-electric plant. A separate circuit is shown in which thermal energy is extracted from the diverter by liquid lithium. The diverter is a magnetic device that guides and collects charged particles diffusing out of the plasma. The rate of thermal energy deposition in the diverter is high, requiring liquid lithium for its removal.

The thermal efficiency of the UWMAK-II power plant is 36.4%; that is, 36.4% of the nuclear energy released in the DT, ^{6}Li(n, α), and other nuclear reactions is converted to electricity and, presumably, is available for sale to the utility customer. In this connection, the fusion power plant has a potentially very difficult problem, namely, net efficiency. The problem is shown schematically in Figure 8.5. In any power plant, fission or fossil, for example, there are in-plant requirements for electricity that subtract from the net output. The difference in the fusion power plant is that there probably will be higher in-plant requirements. A cryogenic refrigeration system will be required for the supercooled magnets. Tritium must be extracted from lithium and purified. Electricity is required for heating the plasma, most probably to drive particle accelerators. If the fractional burnup is low and the deuterium and tritium reactants must be recycled many times, the energy requirements could be sub-

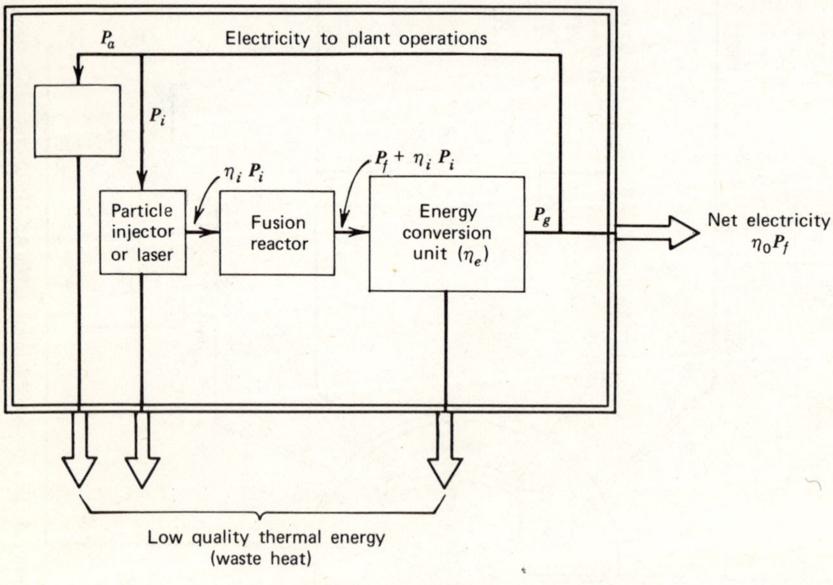

Figure 8.5 Schematic of a fusion reactor power plant.

stantial. In the case of the UWMAK-II design, particle acceleration is proposed only for a 10-second heating period. During a 90-minute self-sustained burn period, it is proposed that makeup deuterium and tritium be added to the reactor in the form of small, frozen pellets. If all of the deuterium and tritium introduced to the reaction zone should have to be accelerated, the energy requirements would be prohibitive. It can be argued that the fuel costs for fusion energy may be so low that it is not necessary to convert the energy to electricity with high efficiency. A weakness of this argument is that a low efficiency means a large quantity of waste heat (Figure 8.5) and can create a severe thermal pollution problem. For the time being, the emphasis in the development program must be on plasma confinement, but a very difficult problem of plant net efficiency lies in the future.

Example 8.7

Consider the net efficiency of a fusion power plant represented by Figure 8.5. The plant auxiliaries require electric power, P_a, equal to 10% of the gross electric power, P_g. The particle accelerator has an efficiency, η_i, of 0.9. The energy required to sustain the reaction, $\eta_i P_i$, is 10% of P_f. The energy conversion unit has a thermal efficiency, η_e, of 0.4. Calculate the overall efficiency of the plant and also the thermal energy rejected per unit of net electrical energy.

Solution: The overall efficiency of the plant is

$$\eta_0 \equiv \frac{P_{\text{net}}}{P_f}$$

where P_f is the energy released in the fusion reactions and also in any side nuclear reactions such as the $^6\text{Li}(n, \alpha)$ reaction.

$$\eta_0 = \frac{P_g - P_a - P_i}{P_f} = \frac{\eta_e(P_f + \eta_i P_i) - P_a - P_i}{P_f}$$

$$\eta_0 = \eta_e\left(1 + \frac{\eta_i P_i}{P_f}\right) - 0.1\left(\eta_e\left(1 + \frac{\eta_i P_i}{P_f}\right)\right) - \frac{P_i}{P_f}$$

Substituting

$$\eta_0 = 0.4(1 + 0.1) - 0.1(0.4 \times 1.1) - \frac{0.1}{0.9}$$

$$\eta_0 = 0.285 \ (28.5\%)$$

If the net electrical energy production rate is $\eta_0 P_f$, then the thermal energy rejection rate must be $(1 - \eta_0)P_f$. The thermal energy rejected per net electric energy is, therefore,

$$\frac{1 - \eta_0}{\eta_0} = 2.51$$

The key parameter in these calculations, and the one that will remain most uncertain until an actual plant is built and operated, is the external energy required to get the fusion reaction started and keep it going (that is, P_i/P_f). Unless this value can be kept low, the plant efficiency will be low.

Example 8.8
Idealize a reacting plasma as an infinite cylinder 5 m in diameter, surrounded by a wall of inside diameter 6 m. The DT reaction is occurring in the plasma at a fusion power density of 0.8 MW(t)/m^3-plasma. Compute the current of 14-MeV neutrons (neutrons/cm^2 · s) into the inner wall.

Solution: The Q-value of the DT reaction is 17.6 MeV

$$^2\text{H} + {}^3\text{H} \longrightarrow {}^4\text{He} + n \qquad Q = 17.6 \text{ MeV}$$

(Chapter 3). The volumetric reaction rate is, therefore

$$R = \frac{0.8 \times 10^6 \times 6.242 \times 10^{12}}{17.6} = 2.837 \times 10^{17} \text{ reactions/m}^3 \cdot \text{s}$$

$$\frac{\text{J}}{\text{m}^3 \cdot \text{s}} \times \frac{\text{MeV}}{\text{J}} \times \frac{\text{reaction}}{\text{MeV}}$$

This means that 2.837×10^{17} 14-MeV neutrons are created each second in a cubic meter of plasma. The plasma itself is of such a low density that the neutrons proceed to the wall essentially without scattering. Since the neutron current into the wall will be everywhere the same, we need only compute the area of wall that encloses the plasma.

$$\frac{S}{V} = \frac{\pi d}{\pi r^2} = \frac{6}{\frac{25}{4}} = 0.96 \text{ m}^2/\text{m}^3$$

The current of neutrons into the wall is

$$\text{J} = \frac{2.837 \times 10^{17}}{0.96} = 2.955 \times 10^{17} \text{ neutrons/m}^2 \cdot \text{s}$$

$$= 2.955 \times 10^{13} \text{ neutrons/cm}^2 \cdot \text{s}$$

There will be other neutrons entering the wall, of course, because of backscattering. They are not included in this calculation.

System Maintenance. One of the most challenging problems presented by the fusion reactor of the UWMAK-II design is that of maintenance. It is anticipated, for example, that the first wall will have to be replaced about every two years as a result of the radiation damage sustained (Example 8.8). The blanket, which has just been described, is 2 m thick and surrounds the first wall, which is roughly 11 m in diameter. The total blanket weighs thousands of tons. The proposal for replacement of the first wall is to construct the blankets in segments or modules that can be individually removed from the reactor toroid. Although such a design does appear to make maintenance possible, it remains a formidable problem.

LASER FUSION (INERTIAL CONFINEMENT)

In recent years, an approach to a fusion reactor quite different from the magnetic confinement concept has emerged. One way of expressing the difference is that the new approach seeks to achieve an acceptable value of $n\tau$ by creating a high density, n, rather than a high residence time, τ. The concept is to add energy to small quantities of reactants (such as deuterium and tritium) by means of pulsed laser beams. The density and temperature of the reactants are increased enormously. Fusion ignition and burn conditions are achieved so quickly that expansion does not have time to take place before some reactions occur. The process is variously referred to as "laser fusion" or "inertial confinement." The reacting time is of the order of 10^{-9} seconds. Eq. 8.1 tells us that the reaction rate is proportional to the square of the density, however, so that 10^{-9} seconds is long enough if the density and temperature are high enough.

The microscopic details of the laser heating of the deuterium-tritium pellets (spheres of frozen DT of about 1 to 2 mm diameter are visualized) are not thoroughly understood. Experimental information is only beginning to be obtained. Emmett[4] gives one theoretical picture. The laser beams, impinging on the DT sphere, penetrate for a short distance, in which the photon energy is transferred to electrons. These electrons collide with ions, and thermal energy is produced in a thin annular shell at the outside of the sphere. This thermal energy travels inward by conduction, raising the temperature and pressure. There is also a rapid ablation of plasma from the pellet surface which, by reaction, causes compression of the inner sphere. This compression is very high, resulting in a density in the inner sphere of perhaps 1000 times solid density. Such compression is very necessary in order that, when ignition of the DT reaction occurs, the resulting 3.5-MeV alpha particle will give up its energy within the sphere and not escape, thus raising the temperature and the reaction rate. Ignition and "burning" of the plasma continue until the density of reactants is reduced, either by expansion or by consumption in the fusion reaction. We have had a kind of miniature fusion explosion. It should be mentioned that means other than laser beams are being considered for energy addition to the reactant spheres. These include electron beams[5] and ion (for example, deuterium ion) beams.[6]

[4] See Reference 3 at end of chapter.
[5] See Reference 4 at end of chapter.
[6] See Reference 5 at end of chapter.

Laser Fusion Power Plant Design

As in the case of magnetic confinement, there have also been design studies of full-scale fusion-electric power plants based on laser-induced fusion. We will not discuss these in any length here, because they do not differ widely from studies such as UWMAK-II. The laser fusion reactor (Fig. 8.6) must provide for very rapid insertion of deuterium-tritium pellets (100 per second is a common figure). The pellets are ignited by impinging laser beam pulses, which must be exquisitely timed. As many as 50 laser beams impinging simultaneously have been proposed in order to obtain symmetrical heating. Surrounding the reacting channel is a blanket that serves much the same functions as in UWMAK-II: thermal energy recovery and tritium production.

Of course, there are no magnets required in the laser fusion reactor. Instead, a very high-performance laser system is required. A suitable laser should possess four qualities: wavelength in the range of 1 to 2 μm, a high efficiency of conver-

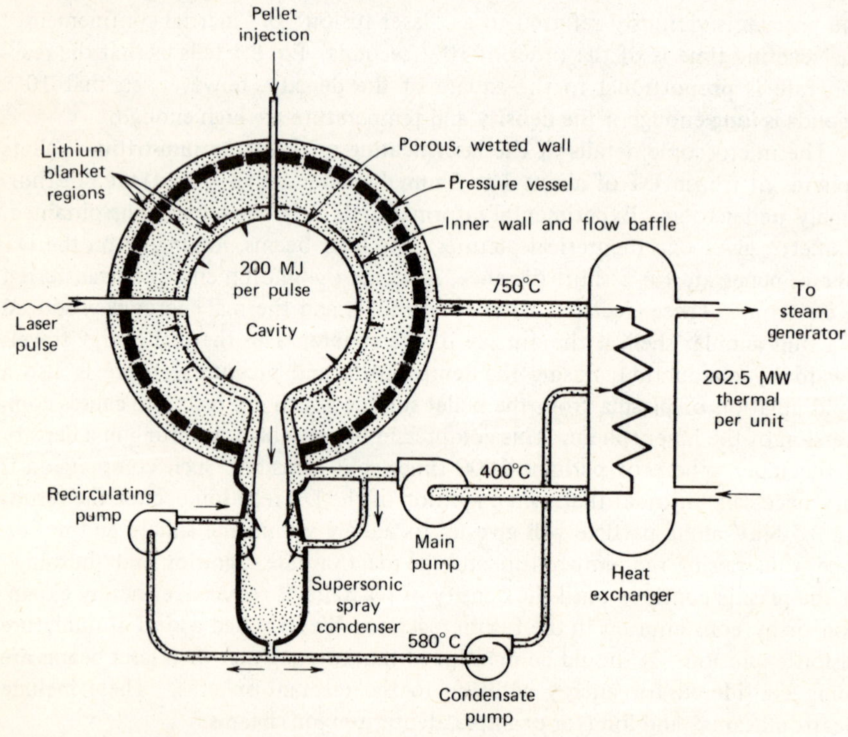

Figure 8.6 Concept of a laser fusion reactor. From F. L. Ribe, "Fusion Reactor Systems," Proceedings, International Conference on Nuclear Solutions to World Energy Programs, American Nuclear Society, 1973.

sion of electrical into photon energy (perhaps 20%), high energy of light output per pulse (10^5 J), and a high repetition rate (10 to 100 Hz). The two lasers being used in current large-scale experiments are the neodymium glass laser and the CO_2 laser. Each of these fails to meet some of the above specifications, but it is believed that a gas laser of suitable characteristics can be developed. In the meantime, very valuable physics information can be obtained in the Nd-glass and CO_2 experiments.

The laser fusion power system faces a net electrical efficiency problem similar to that of the magnetically confined systems (Figure 8.5 and Example 8.7). Brueckner[7] proposes that lasers that convert 20% of input electric energy to laser energy should be possible (that is, in Figure 8.5, $\eta_i = 0.2$). If, following Example 8.7, we assign an η_e value of 0.4 and a P_a/P_g of 0.05, we find that in order to obtain an overall plant efficiency of 25% ($\eta_0 = 0.25$), the value of $\eta_i P_i/P_f$ must be held to 0.028. In other words, the energy of the fusion reactions triggered by the laser beams must be 36 times greater than the energy of the beams themselves. Just as in the magnetic confinement case, the fusion power plant based on laser fusion may have a severe thermal efficiency problem.

Example 8.9

If 8% fuel burnup of deuterium and tritium is achieved in a sphere of 2mm diameter and 175 kg/m^3 density, at what rate would these spheres have to be reacted in a 5000 MW(t) laser fusion reactor? The pellets have an equal number of deuterium and tritium atoms.

Solution: The pellets weigh

$$\frac{4\pi}{3} r^3 \rho = \frac{\pi}{4} (.001)^3 \times 175 = 7.33 \times 10^{-7} \text{ kg}$$

$$= 7.33 \times 10^{-4} \text{ g}$$

Since the average atomic weight is 2.5, the number of atoms is

$$\frac{7.33 \times 10^{-4}}{2.5} \times 6.023 \times 10^{23} = 1.766 \times 10^{20}$$

If 8% react, we have

$$\frac{1.766 \times 10^{20}}{2} \times 0.08 = 7.06 \times 10^{18} \text{ reactions}$$

Each DT reaction releases 17.6 MeV. (We ignore any contribution from such side reactions as ^{6}Li(n, α).) The energy release per pellet is, therefore,

$$7.06 \times 10^{18} \times 17.6 \times 1.602 \times 10^{-13} = 1.99 \times 10^7 \text{ W/pellet}$$

[7]See Reference 6 at end of chapter.

In order to generate 5000 MW(t), we would need

$$\frac{5000 \times 10^6}{1.99 \times 10^7} = 251 \text{ pellets/s}$$

REFERENCES

1. T. Kammash, *Fusion Reactor Physics*, Ann Arbor Science, Ann Arbor, 1975.
2. B. Badger et al., "UWMAK-II, A Conceptual Tokamak Power Reactor Design," UWFDM-112, Nuclear Engineering Department, The University of Wisconsin, 1975.
3. J. L. Emmett, J. Nuckolls, L. Wood, "Fusion Power by Laser Implosion," *Scientific American* Vol. 230, No. 6 (June 1974), 24–37.
4. G. Yonas et al., "Electron Beam Focusing and Application to Pulsed Fusion," *Nuclear Fusion* Vol. 14, No. 5 (1974), 731–740.
5. J. W. Shearer, "Ion Beam Compression of Thermonuclear Pellets," *Nuclear Fusion* Vol. 15, No. 5 (1975), 952–955.
6. K. A. Brueckner, "The Potential of Laser Compression as the Basis of a Fusion Reactor," *Nuclear Energy Maturity*, Proceedings of the European Nuclear Conference, Pergamon Press, Oxford, 1975.

PROBLEMS

8.1 In a fusion (or thermonuclear) plasma, a 50–50 mixture of deuterium and tritium ions should have a density of 10^{21} ions/m^3 and a temperature equivalence of 10 keV. What is the density (kg/m^3) of such a mixture? What is the pressure?

8.2 In a fusion reactor, the reaction $^3\text{H}(d, n)$ is proceeding at a rate to give a power density of 100 W/cm^3. Focus on a single cubic centimeter of this reactor. What is the neutron dose equivalent (mrem/h) at a distance of 10 meters? Assume no shielding of the neutrons over this distance.

8.3 A 5000 MW(t) fusion plant is operating on the DT reaction. Tritium to replace that which is consumed is produced by reactions with lithium in the blanket. How many curies of tritium are produced each day? Suppose 1% of this tritium escapes into the environment. Is this a serious problem? See Table 5.8 for MPC values of tritium.

8.4 Consider a 1-cm^3 volume of a plasma initially at a pressure of 100 bar and a temperature of 10 keV, consisting of a 50–50 (atomic) mixture of ^{2}H and ^{3}H.

(a) What is the initial fusion power (W/cm^3)? Consider the DT reaction only.

(b) The neutron escapes, and 25% of the remainder of the reaction energy leaves as electromagnetic radiation. What is the initial value of the rate of temperature change (keV/s)?

Appendix

Table A.1a Mass-Energy Conversion Factors

UNIT	SYMBOL	kg	J	eV	u	erg	cal	Btu	MWd	lb
1 Kilogram	kg	×	8.988E 16	5.610E 35	6.022E 26	8.988E 23	2.148E 16	8.519E 13	1.040E 06	2.205E 00
1 Joule	J	1.113E−17	×	6.242E 18	6.701E 09	1.000E 07	2.390E−01	9.478E−04	1.157E−11	2.453E−17
1 Electron volt	eV	1.783E−36	1.602E−19	×	1.074E−09	1.602E−12	3.829E−20	1.518E−22	1.854E−30	3.929E−36
1 Atomic mass unit	u	1.660E−27	1.492E−10	9.315E 08	×	1.492E−03	3.567E−11	1.414E−13	1.727E−21	3.660E−27
1 Erg	erg	1.113E−24	1.000E−07	6.242E 11	6.701E 02	×	2.390E−08	9.478E−11	1.157E−18	2.453E−24
1 Gram calorie	cal	4.655E−17	4.184E 00	2.612E 19	2.804E 10	4.184E 07	×	3.694E−03	4.843E−11	1.026E−16
1 British Thermal Unit	Btu	1.174E−14	1.055E 03	6.586E 21	7.070E 12	1.055E 10	2.522E 02	×	1.221E−08	2.588E−14
1 Megawatt day	MWd	9.616E−07	8.640E 10	5.393E 29	5.790E 20	8.640E 17	2.065E 10	8.189E 07	×	2.120E−06
1 Pound	lb	4.536E−01	4.077E 16	2.545E 35	2.732E 26	4.077E 23	9.743E 15	3.864E 13	4.717E 05	×

Table A.1b Conversion Factors

DIMENSION	TO CONVERT FROM	TO	MULTIPLY BY
Length	inches	meters (m)	2.5400 E-02[a]
	feet		3.0480 E-01
	miles (U.S.)		1.6093 E-03
Area	ft^2	m^2	9.2903 E-02
	$mile^2$		2.5900 E+06
	cm^2		1.0000 E-04
	barn		1.0000 E-28
Volume	ft^3	m^3	2.8317 E-02
	gallons (U.S.)		3.7854 E-03
	cm^3		1.0000 E-06
	liters		1.0000 E-03
Mass	pounds mass	kg	4.5359 E-01
	short ton (2000 lbm)		9.0718 E+02
	metric ton		1.0000 E+03
Time	years	s	3.1558 E+07
	days		8.6400 E+04
	hours		3.6000 E+03
	minutes		6.0000 E+01
Pressure	atmospheres	pascals (Pa)	1.0132 E+05
	bar		1.0000 E+05
	lbf/in^2 (psi)		6.8948 E+03
	mm Hg (torr)		1.3332 E+02
Density	lbm/ft^3	kg/m^3	1.6018 E+01
	g/cm^3		1.0000 E+03
Power	Btu/h	W	2.9307 E-01
	horsepower		7.4567 E+02
	MeV/s		1.6022 E-13
Heat capacity	$cal/g \cdot {}^\circ C$	$J/kg \cdot {}^\circ K$	4.1868 E+03
	$Btu/lbm \cdot {}^\circ F$		4.1868 E+03
Heat flux	$Btu/ft^2 \cdot h$	W/m^2	3.1544 E+00
	W/cm^2		1.0000 E+04
Power Density	$Btu/ft^3 \cdot h$	W/m^3	1.0350 E+01

[a] $2.5400\ E{-}02 = 2.54 \times 10^{-2}$

Table A.2 Some Fundamental Physical Constants

CONSTANT	SYMBOL	VALUE	UNITS
Speed of light in a vacuum	c	2.9979×10^{8}	m/s
Elementary charge	e	1.6021×10^{-19}	coulomb
Planck constant	h	6.6256×10^{-34}	J · s
Avogadro's number	A_0	6.0225×10^{23}	mole^{-1}
Mass unit	u	1.6604×10^{-27}	kg
Electron rest mass	m_e	5.4860×10^{-4}	u
Neutron rest mass	m_n	1.008665	u
Proton rest mass	m_p	1.007277	u
Hydrogen atom rest mass	m_H	1.007825	u
Ideal gas constant	R	8.3143	J/mole · K

Table A.3 Properties of Selected Elements and Nuclides

ELEMENT					ISOTOPE					
NAME	Z	ATOMIC WEIGHT	DENSITY (g/cm^3)	SYMBOL	A	NATURAL ABUNDANCE (a/o)	MASS (m-A) (MeV)	HALF-LIFE	PRINCIPAL DECAY MODE	DECAY Q-VALUE (MeV)
Neutron	0			n	1		8.0714	12 min	β^-	0.7825
Hydrogen	1	1.0079	gas	H						
					1	99.9852	7.2890			
					2	0.0148	13.1359			
					3		14.9500	12.3 y	β^-	0.0186
Helium	2	4.0026	gas	He						
					3	1.3×10^{-4}	14.9313			
					4	100	2.4248			
Lithium	3	6.941	0.53	Li						
					6	7.42	14.088			
					7	92.58	14.907			
					8		20.946	0.84 s	β^-	16.00
Beryllium	4	9.0122	1.85	Be						
					9	100	11.351			
					10		12.607	2.5×10^6 y	β^-	0.555
Boron	5	10.81	2.3	B						
					10	19.8	12.052			
					11	80.2	8.6677			
					12		13.370	0.02 s	β^-	13.37
Carbon	6	12.011	1.60	C						
					12	98.89	≡0			
					13	1.11	3.125			
					14		3.0198	5730 y	β^-	0.1561

Table A.3 (*Continued*)

ELEMENT					ISOTOPE					
NAME	Z	ATOMIC WEIGHT	DENSITY (g/cm^3)	SYMBOL	A	NATURAL ABUNDANCE (a/o)	MASS (m-A) (MeV)	HALF-LIFE	PRINCIPAL DECAY MODE	DECAY Q-VALUE (MeV)
Nitrogen	7	14.007	gas	N						
					14	99.635	2.8637			
					15	0.365	0.100			
					16		5.685	7.2 s	β^-	10.42
Oxygen	8	15.999	gas	O						
					16	99.759	- 4.7366			
					17	0.037	- 0.808			
					18	0.204	- 0.7824			
					19		3.333	29 s	β^-	4.819
Fluorine	9	18.998	gas	F						
					19	100	- 1.486			
					20		- 0.012	11.4 s	β^-	7.030
Sodium	11	22.990	0.97	Na						
					22		- 5.182	2.6 y	β^+	2.843
					23	100	- 9.528			
					24		- 8.418	15.0 h	β^-	5.515
Magnesium	12	24.305	1.74	Mg						
					24	78.8	-13.933			
					25	10.2	-13.191			
					26	11.1	-16.214			
Aluminum	13	26.982	2.7	Al						
					27	100	-17.196			
					28		-16.855	2.31 min	β^-	4.635
Silicon	14	28.086	2.33	Si						
Argon	18	39.948	gas	Ar						
					40	99.6	-35.038			
					41		-33.061	1.83 h	β^-	2.491

Potassium	19	39.098	0.86	K	39	93.22	−33.803			
					40		−33.533	1.26×10^9 y	β^-	1.314
					41	6.77	−35.552			
Titanium	22	47.90	4.51	Ti						
Vanadium	23	50.941	6.1	V	51	99.75	−52.199			
					52		−51.44	2.0 min	β^-	3.5
Chromium	24	51.996	7.19	Cr						
Manganese	25	54.938	7.43	Mn	55	100	−57.705			
					56		−56.904	2.58 h	β^-	3.702
Iron	26	55.847	7.87	Fe						
Cobalt	27	58.933	8.8	Co	59	100	−62.233			
					60		−61.651	5.26 y	β^-	2.819
Nickel	28	58.71	8.9	Ni						
Copper	29	63.546	8.96	Cu						
Zinc	30	65.38	7.13	Zn						
Krypton	36	83.80	gas	Kr	84	57.0	−82.433			
					85		−81.48	10.76 y	β^-	0.67
					86	17.3	−83.259			
Strontium	38	87.62	2.6	Sr	88	82.6	−87.89			
					89		−86.22	52 d	β^-	1.463
					90		−85.95	28.1 y	β^-	0.546
Yttrium	39	88.906	5.51	Y	89	100	−87.678			
					90		−86.50	64 h	β^-	2.27
Zirconium	40	91.22	6.5	Zr						
Niobium	41	92.906	8.57	Nb						
Molybdenum	42	95.94	10.2	Mo						
Technetium	43			Tc	99		−87.33	2.1×10^5 y	β^-	0.292

Table A.3 (*Continued*)

ELEMENT					ISOTOPE					
NAME	Z	ATOMIC WEIGHT	DENSITY (g/cm^3)	SYMBOL	A	NATURAL ABUNDANCE (a/o)	MASS (m-A) (MeV)	HALF-LIFE	PRINCIPAL DECAY MODE	DECAY Q-VALUE (MeV)
Silver	47	107.87	10.5	Ag						
Cadmium	48	112.40	8.65	Cd						
Indium	49	114.82	7.31	In						
					115	95.7	-89.21			
					116		-88.20	54 min	β^-	3.33
Iodine	53	126.90	4.93	I						
					127	100	-88.984			
					129		-88.50	1.7×10^7 y	β^-	0.189
					131		-87.441	8.05 d	β^-	0.97
					135		-84.	6.7 h	β^-	2.8
Xenon	54	131.30	gas	Xe						
					135		-86.6	9.2 h	β^-	1.16
Cesium	55	132.91	1.9	Cs						
					134		-86.79	2.05 y	β^-	2.06
					137		-86.9	30 y	β^-	1.176
Hafnium	72	178.49	13.4	Hf						
Tantalum	73	180.95	16.6	Ta						
Tungsten	74	183.85	19.2	W						
Mercury	80	200.59	13.6	Hg						
Lead	82	207.2	11.3	Pb						
					210		-14.73	21 y	β^-	0.061
Bismuth	83	208.98	9.7	Bi						
					210		-14.79	5 d	β^-	1.160
Polonium	84		9.24	Po						
					210		-15.95	138.4 d	α	5.408
Radon	86		gas	Rn						
					222		16.39	3.82 d	α	5.587

Radium	88		5.0	Ra	226		23.69	1602 y	α	4.869
Thorium	90	232.04	11.7	Th	228		26.77	1.91 y	α	5.521
					230		30.87	8×10^4 y	α	4.767
					232	100	35.47	1.41×10^{10} y	α	4.08
					233		38.76	22.2 min	β^-	1.246
Protactinium	91			Pa	233		37.51	27 d	β^-	0.571
Uranium	92	238.03	19.1	U	232		34.60	72 y	α	5.414
					233		36.94	1.62×10^5 y	α	4.909
					234	0.0055	38.16	2.47×10^5 y	α	4.856
					235	0.71	40.93	7.1×10^8 y	α	4.681
					236		42.46	2.4×10^7 y	α	4.573
					237		45.41	6.75 d	β^-	0.517
					238	99.28	47.33	4.51×10^9 y	α	4.268
					239		50.60	23.5 min	β^-	1.28
Neptunium	93			Np	237		44.89	2.1×10^6 y	α	4.268
Plutonium	94		19.6	Pu	236		42.90	2.85 y	α	5.868
					238		46.18	86 y	α	5.592
					239		48.60	2.44×10^4 y	α	5.243
					240		50.14	6580 y	α	5.255
					241		52.98	13.2 y	β^-	0.021
					242		54.74	3.8×10^5 y	α	4.98
					243		57.77	5.0 h	β^-	0.59
Americium	95			Am	241		52.96	458 y	α	5.640
					243		57.18	7950 y	α	5.439
Curium	96			Cm	242		54.82	163 d	α	6.217
					244		58.47	17.6 y	α	5.902

Table A.3 (*Continued*)

NAME	Z	SYMBOL	A	NEUTRON CROSS SECTIONS (BARNS)					
						NEUTRON ENERGY			
				SCATTERING	KIND OF REACTION	0.0253 eV	RESONANCE INTEGRAL	2 MeV	14 MeV
Neutron	0	n	1	–		–	–	–	
Hydrogen	1	H	a	20.4	(n, γ)	0.332	–		32 μb
			1						
			2	3.4	(n, γ)	5.3×10^{-4}			29 μb
			3	1.3					
Helium	2	He		0.76	abs.	<0.05			
			3		abs.	5327			
			4	0.76					
Lithium	3	Li							
			6	0.72	(n, α)	940			
			7		(n, γ)	0.037			
Beryllium	4	Be							
			9	6.14	(n, γ)	0.0093	0.004		
Boron	5	B		3.6	abs.	759	341		
			10	2.2	(n, α)	3837	1722	0.4	
			11	3.9	(n, γ)	0.0055		0.008	
Carbon	6	C		4.75	(n, γ)	0.0034	0.0015		
Nitrogen	7	N		10.6	abs.	1.85	0.9		
			14		(n, p)	1.81			0.013
					(n, γ)	0.075			
Oxygen	8	O		3.76	(n, γ)	0.00027			
			16		(n, γ)	0.00018	0.00027		
Fluorine	9	F							
			19	4.0	(n, γ)	0.0095			

Sodium	11	Na						
			23	3.2	(n, γ)	0.53	0.31	
Magnesium	12	Mg		3.42	(n, γ)	0.063	0.038	
			24	3.8	(n, γ)	0.052		
Aluminum	13	Al						
			27	1.49	(n, γ)	0.23	0.17	0.0002
Silicon	14	Si		2.2	(n, γ)	0.16		
Argon	18	Ar		0.64	(n, γ)	0.68	0.42	
Potassium	19	K		1.5	(n, γ)	2.1	1.0	
Titanium	22	Ti		4.0	(n, γ)	6.1		0.003
Vanadium	23	V		4.9	(n, γ)	5.0	2.7	
Chromium	24	Cr		3.8	(n, γ)	3.1	1.7	0.004
Manganese	25	Mn						
			55	2.1	(n, γ)	13.3	14.0	0.0024
Iron	26	Fe		10.9	(n, γ)	2.55	1.4	0.001
Cobalt	27	Co						
			59	6.7	(n, γ)	37.2	75.5	0.0027
Nickel	28	Ni		17.3	(n, γ)	4.43	2.2	0.007
Copper	29	Cu		7.9	(n, γ)	3.79	3.2	0.01
Zinc	30	Zn		4.2	(n, γ)	1.1	2.3	0.015
Zirconium	40	Zr		6.4	(n, γ)	0.185	1.10	0.001
Niobium	41	Nb			(n, γ)	1.15	8.5	0.01
Molybdenum	42	Mo		5.8	(n, γ)	2.65	22	0.01
Technetium	43	Te						
			99		(n, γ)	19	340	
Silver	47	Ag			(n, γ)	63.6	747	0.1
Cadmium	48	Cd		5.6	(n, γ)	2450		0.04
Indium	49	In			(n, γ)	193.5	3200	0.2
Hafnium	72	Hf		8.2	(n, γ)	102	2000	

Table A.3 (*Continued*)

				NEUTRON CROSS SECTIONS (BARNS)					
						NEUTRON ENERGY			
NAME	Z	SYMBOL	A	SCATTERING	KIND OF REACTION	0.0253 eV	RESONANCE INTEGRAL	2 MeV	14 MeV
Tantalum	73	Ta		6.2	(n, γ)	21	710	0.055	
Tungsten	74	W			(n, γ)	18.5	352	0.1	
Mercury	80	Hg			(n, γ)	375	73		
Lead	82	Pb		11.4	(n, γ)	0.17	0.16	0.003	
Bismuth	83	Bi			(n, γ)	0.033	0.19		
Thorium	90	Th		12.7	(n, γ)	7.4	85	0.065	0.005
					(n, f)	0	0	0.1	0.36
Protactinium	91	Pa							
			233		(n, γ)	41			
Uranium	92	U							
			232	14.7	(n, γ)	73.1	280		
					(n, f)	75.2	320		
			233	8.2	(n, γ)	47.7	140	0.02	
					(n, f)	531.1	764	2	2.2
			234		(n, γ)	100.2	630		
			235	13.8	(n, γ)	98.6	144	0.1	
					(n, f)	582.2	275	1.3	2
			236		(n, γ)	5.2	365	0.09	
			237		(n, γ)	411	290		
			238	8.9	(n, γ)	2.70	275	0.054	
					(n, f)	0	0	0.55	1.1
Neptunium	93	Np							

			237		(n, γ)	169	660		
					(n, f)	0.02	0	1.5	2.5
Plutonium	94	Pu							
			238		(n, γ)	547			
					(n, f)	16.5		3	
			239	7.7	(n, γ)	268.8	200	0.04	0.1
					(n, f)	742.5	301	2	2.7
			240		(n, γ)	289.5	8013		
					(n, f)	0.03		1.6	
			241	11	(n, γ)	368	162		
					(n, f)	1009	570	1.8	2.5
			242		(n, γ)	18.5			
Americium	95	Am							
			241		(n, γ)	832	1477		
					(n, f)	3.2	21		
			243		(n, γ)	160	1930		

[a]When no value of A is given, the properties are those of the natural element.

Index

CONVERSION FACTORS

Dimension	To convert from	to	multiply by
Length	inches	meters (m)	2.5400 E−02
	feet		3.0480 E−01
	miles (U.S.)		1.6093 E−03
Area	ft^2	m^2	9.2903 E−02
	$mile^2$		2.5900 E+06
	cm^2		1.0000 E−04
	barn		1.0000 E−28
Volume	ft^3	m^3	2.8317 E−02
	gallons (U.S.)		3.7854 E−03
	cm^3		1.0000 E−06
	litres		1.0000 E−03
Mass	pounds mass	kg	4.5359 E−01
	short ton (2000 lbm)		9.0718 E+02
	metric ton		1.0000 E+03
Time	years	s	3.1558 E+07
	days		8.6400 E+04
	hours		3.6000 E+03
	minutes		6.0000 E+01
Pressure	atmospheres	pascals	1.0132 E+05
	bar	(Pa)	1.0000 E+05
	lbf/in^2 (psi)		6.8948 E+03
	mm Hg (torr)		1.3332 E+02
Density	lbm/ft^3	kg/m^3	1.6018 E+01
	g/cm^2		1.0000 E+03
Power	Btu/hr	W	2.9307 E−01
	horsepower		7.4567 E+02
	MeV/s		1.6022 E−13
Heat capacity	cal/g · °C	J/kg · K	4.1868 E+03
	Btu/lbm · °F		4.1868 E+03
Heat flux	$Btu/ft^2 \cdot h$	W/m^2	3.1544 E+00
	W/cm^2		1.0000 E+04
Power density	$Btu/ft^3 \cdot h$	W/m^3	1.0350 E+01